AF696616

Contractor's Pricing Guide 2003

Means Residential **Repair & Remodeling** Costs

Senior Editor
Robert W. Mewis, CCC

Contributing Editors
Barbara Balboni
Robert A. Bastoni
John H. Chiang, PE
J. Robert Lang
Robert C. McNichols
Melville J. Mossman, PE
John J. Moylan
Jeannene D. Murphy
Peter T. Nightingale
Stephen C. Plotner
Michael J. Regan
Marshall J. Stetson
Phillip R. Waier, PE

Senior Engineering Operations Manager
John H. Ferguson, PE

Vice President, Product Management
Roger J. Grant

Senior Vice President
Charles Spahr

Sales Director
John M. Shea

Production Manager
Michael Kokernak

Production Coordinator
Marion E. Schofield

Technical Support
Thomas J. Dion
Jonathan Forgit
Mary Lou Geary
Gary L. Hoitt
Paula Reale-Camelio
Karen Moran
Robin Richardson
Kathryn S. Rodriguez
Sheryl A. Rose
Elizabeth Testa

Book & Cover Design
Norman R. Forgit

RSMeans

Means Residential Repair & Remodeling Costs

Contractor's Pricing Guide 2003

- Quick Costs to Help You Estimate
- Easy-to-Use Format . . . Organized the Way You Build
- Hundreds of Work Items with Material, Labor & Equipment

$38.95 per copy. (In United States).
Price subject to change without prior notice.

RSMeans

Copyright 2002

Construction Publishers & Consultants

63 Smiths Lane

Kingston, MA 02364-0800

(781) 422-5000

R.S. Means Company, Inc. ("R.S. Means"), its authors, editors and engineers, apply diligence and judgment in locating and using reliable sources for the information published. **However, R.S. Means makes no express or implied warranty or guarantee in connection with the content of the information contained herein, including the accuracy, correctness, value, sufficiency, or completeness of the data, methods and other information contained herein. R.S. Means makes no express or implied warranty of merchantability or fitness for a particular purpose.** R.S. Means shall have no liability to any customer or third party for any loss, expense, or damage, including consequential, incidental, special or punitive damages, including lost profits or lost revenue, caused directly or indirectly by any error or omission, or arising out of, or in connection with, the information contained herein.

No part of this publication may be reproduced, stored in a retrieval system, or transmitted in any form or by any means without prior written permission of R.S. Means Company, Inc.

Printed in the United States of America

10 9 8 7 6 5 4 3 2 1

ISSN Pending

ISBN 0-87629-687-8

Reed Construction Data

Foreword

R.S. Means Co., Inc. is a subsidiary of Reed Construction Data, a leading provider of construction information, products, and services in North America and globally. Reed Construction Data's project information products include more than 100 regional editions, national construction data, sales leads, and over 70 local plan rooms in major business centers. Reed Construction Data's PlansDirect provides surveys, plans and specifications. The First Source suite of products consists of *First Source for Products*, SPEC-DATA™, MANU-SPEC™, CADBlocks, Manufacturer Catalogs and First Source Exchange (www.firstsourceexchange.com) for the selection of nationally available building products. Reed Construction Data also publishes ProFile, a database of more than 20,000 U.S. architectural firms. R.S. Means provides construction cost data, training, and consulting services in print, CD-ROM and online. Reed Construction Data is headquartered in Atlanta and has 1,400 employees worldwide. Reed Construction Data is owned by Reed Business Information (www.cahners.com), a leading provider of critical information and marketing solutions to business professionals in the media, manufacturing, electronics, construction and retail industries. Its market-leading properties include more than 135 business-to-business publications, over 125 Webzines and Web portals, as well as online services, custom publishing, directories, research and direct-marketing lists. Reed Business Information is a member of the Reed Elsevier plc group (NYSE: RUK and ENL)—a world-leading publisher and information provider operating in the science and medical, legal, education and business-to-business industry sectors.

Our Mission

Since 1942, R.S. Means Company, Inc. has been actively engaged in construction cost publishing and consulting throughout North America.

Today, over 50 years after the company began, our primary objective remains the same: to provide you, the construction and facilities professional, with the most current and comprehensive construction cost data possible.

Whether you are a contractor, an owner, an architect, an engineer, a facilities manager, or anyone else who needs a fast and reliable construction cost estimate, you'll find this publication to be a highly useful and necessary tool.

Today, with the constant flow of new construction methods and materials, it's difficult to find the time to look at and evaluate all the different construction cost possibilities. In addition, because labor and material costs keep changing, last year's cost information is not a reliable basis for today's estimate or budget.

That's why so many construction professionals turn to R.S. Means. We keep track of the costs for you, along with a wide range of other key information, from city cost indexes . . . to productivity rates . . . to crew composition . . . to contractor's overhead and profit rates.

R.S. Means performs these functions by collecting data from all facets of the industry, and organizing it in a format that is instantly accessible to you. From the preliminary budget to the detailed unit price estimate, you'll find the data in this book useful for all phases of construction cost determination.

The Staff, the Organization, and Our Services

When you purchase one of R.S. Means' publications, you are in effect hiring the services of a full-time staff of construction and engineering professionals.

Our thoroughly experienced and highly qualified staff works daily at collecting, analyzing, and disseminating comprehensive cost information for your needs. These staff members have years of practical construction experience and engineering training prior to joining the firm. As a result, you can count on them not only for the cost figures, but also for additional background reference information that will help you create a realistic estimate.

The Means organization is always prepared to help you solve construction problems through its five major divisions: Construction and Cost Data Publishing, Electronic Products and Services, Consulting Services, Insurance Services, and Educational Services.

Besides a full array of construction cost estimating books, Means also publishes a number of other reference works for the construction industry. Subjects include construction estimating and project and business management; special topics such as HVAC, roofing, plumbing, and hazardous waste remediation; and a library of facility management references.

In addition, you can access all of our construction cost data through your computer with Means CostWorks 2003 CD-ROM, an electronic tool that offers over 50,000 lines of Means detailed construction cost data, along with assembly and whole building cost data. You can also access Means cost information from our Web site at www.rsmeans.com

What's more, you can increase your knowledge and improve your construction estimating and management performance with a Means Construction Seminar or In-House Training Program. These two-day seminar programs offer unparalleled opportunities for everyone in your organization to get updated on a wide variety of construction-related issues.

Means also is a worldwide provider of construction cost management and analysis services for commercial and government owners and of claims and valuation services for insurers.

In short, R.S. Means can provide you with the tools and expertise for constructing accurate and dependable construction estimates and budgets in a variety of ways.

Robert Snow Means Established a Tradition of Quality That Continues Today

Robert Snow Means spent years building his company, making certain he always delivered a quality product.

Today, at R.S. Means, we do more than talk about the quality of our data and the usefulness of our books. We stand behind all of our data, from historical cost indexes... to construction materials and techniques... to current costs.

If you have any questions about our products or services, please call us toll-free at 1-800-334-3509. Our customer service representatives will be happy to assist you or visit our Web site at www.rsmeans.com

Table of Contents

Foreword	v
How the Book Is Built: An Overview	ix
How To Use the Book: The Details	xi
Unit Price Section	1
Reference Section	
Location Factors	293
Abbreviations	299
Index	300
Other Means Publications and Reference Works	
Installing Contractor's Overhead & Profit	Inside Back Cover

How the Book is Built: An Overview

A Powerful Construction Tool

You have in your hands one of the most powerful construction tools available today. A successful project is built on the foundation of an accurate and dependable estimate. This book will enable you to construct just such an estimate.

For the casual user the book is designed to be:

- quickly and easily understood so you can get right to your estimate
- filled with valuable information so you can understand the necessary factors that go into the cost estimate

For the regular user, the book is designed to be:

- a handy desk reference that can be quickly referred to for key costs
- a comprehensive, fully reliable source of current construction costs, so you'll be prepared to estimate any project
- a source book for project cost, product selections, and alternate materials and methods

To meet all of these requirements we have organized the book into the following clearly defined sections.

Unit Price Section

The cost data has been divided into 23 sections representing the order of construction. Within each section is a listing of components and variations to those components applicable to that section. Costs are shown for the various actions that must be done to that component.

Reference Section

This section includes information on Location Factors, and a listing of Abbreviations.

Location Factors: Costs vary depending upon regional economy. You can adjust the "national average" costs in this book to over 930 major cities throughout the U.S. and Canada by using the data in this section.

Abbreviations: A listing of the abbreviations used throughout this book, along with the terms they represent, is included.

Index

A comprehensive listing of all terms and subjects in this book to help you find what you need quickly when you are not sure where it falls in the order of construction.

The Scope of This Book

This book is designed to be as comprehensive and as easy to use as possible. To that end we have made certain assumptions and limited its scope in three key ways:

1. We have established material prices based on a "national average."
2. We have computed labor costs based on a 7 major region average of open shop wage rates.
3. We have targeted the data for projects of a certain size range.

Project Size

This book is intended for use by those involved primarily in Residential Repair and Remodeling construction costing between $10,000-$100,000.

With reasonable exercise of judgment the figures can be used for any building work. For other types of projects, such as new home construction or commercial buildings, consult the appropriate Means publication for more information.

How to Use the Book: The Details

What's Behind the Numbers? The Development of Cost Data

The staff at R.S. Means continuously monitors developments in the construction industry in order to ensure reliable, thorough and up-to-date cost information.

While *overall* construction costs may vary relative to general economic conditions, price fluctuations within the industry are dependent upon many factors. Individual price variations may, in fact, be opposite to overall economic trends. Therefore, costs are continually monitored and complete updates are published yearly. Also, new items are frequently added in response to changes in materials and methods.

Costs–$ (U.S.)

All costs represent U.S. national averages and are given in U.S. dollars. The Means Location Factors can be used to adjust costs to a particular location. The Location Factors for Canada can be used to adjust U.S. national averages to local costs in Canadian dollars.

Material Costs

The R.S. Means staff contacts manufacturers, dealers, distributors, and contractors all across the U.S. and Canada to determine national average material costs. If you have access to current material costs for your specific location, you may wish to make adjustments to reflect differences from the national average.

Included within material costs are fasteners for a normal installation. R.S. Means engineers use manufacturers' recommendations, written specifications and/or standard construction practice for size and spacing of fasteners. Adjustments to material costs may be required for your specific application or location. Material costs do not include sales tax.

Labor Costs

Labor costs are based on the average of open shop wages from across the U.S. for the current year. Rates along with overhead and profit markups are listed on the inside back cover of this book.

- If wage rates in your area vary from those used in this book, or if rate increases are expected within a given year, labor costs should be adjusted accordingly.

Labor costs reflect productivity based on actual working conditions. These figures include time spent during a normal workday on tasks other than actual installation, such as material receiving and handling, mobilization at site, site movement, breaks, and cleanup.

Productivity data is developed over an extended period so as not to be influenced by abnormal variations and reflects a typical average.

Equipment Costs

Equipment costs include not only rental costs, but also operating costs such as fuel, oil, and routine maintenance. Equipment and rental rates are obtained from industry sources throughout North America—contractors, suppliers, dealers, manufacturers, and distributors.

Factors Affecting Costs

Costs can vary depending upon a number of variables. Here's how we have handled the main factors affecting costs.

Quality—The prices for materials and the workmanship upon which productivity is based represent sound construction work. They are also in line with U.S. government specifications.

Overtime—We have made no allowance for overtime. If you anticipate premium time or work beyond normal working hours, be sure to make an appropriate adjustment to your labor costs.

Productivity—The labor costs for each line item are based on working an eight-hour day in daylight hours in moderate temperatures. For work that extends beyond normal work hours or is performed under adverse conditions, productivity may decrease.

Size of Project—The size, scope of work, and type of construction project will have a significant impact on cost. Economies of scale can reduce costs for large projects. Unit costs can often run higher for small projects. Costs in this book are intended for the size and type of project as previously described in "How the Book Is Built: An Overview." Costs for projects of a significantly different size or type should be adjusted accordingly.

Location—Material prices in this book are for metropolitan areas. However, in dense urban areas, traffic and site storage limitations may increase costs. Beyond a 20-mile radius of large cities, extra trucking or transportation charges may also increase the material costs slightly. On the other hand, lower wage rates may be in effect. Be sure to consider both these factors when preparing an estimate, particularly if the job site is located in a central city or remote rural location.

In addition, highly specialized subcontract items may require travel and per diem expenses for mechanics.

Other factors–

- season of year
- contractor management
- weather conditions
- building code requirements
- availability of:
    - adequate energy
    - skilled labor
    - building materials
- owner's special requirements/ restrictions
- safety requirements
- environmental considerations

General Conditions—The extreme right-hand column of each chart gives the "Totals." These figures contain the installing contractor's O&P (in other words, the contractor doing the work). Therefore, it is necessary for a general contractor to add a percentage of all subcontracted items. For a detailed breakdown of O&P see the inside back cover of this book.

Unpredictable Factors—General business conditions influence "in-place" costs of all items. Substitute materials and construction methods may have to be employed. These may affect the installed cost and/or life cycle costs. Such factors may be difficult to evaluate and cannot necessarily be predicted on the basis of the job's location in a particular section of the country. Thus, where these factors apply, you may find significant, but unavoidable cost variations for which you will have to apply a measure of judgment to your estimate.

Final Checklist

Estimating can be a straightforward process provided you remember the basics. Here's a checklist of some of the items you should remember to do before completing your estimate.

Did you remember to . . .

- factor in the Location Factor for your locale
- take into consideration which items have been marked up and by how much
- mark up the entire estimate sufficiently for your purposes
- include all components of your project in the final estimate
- double check your figures to be sure of your accuracy
- call R.S. Means if you have any questions about your estimate or the data you've found in our publications

Remember, R.S. Means stands behind its publications. If you have any questions about your estimate . . . about the costs you've used from our books . . . or even about the technical aspects of the job that may affect your estimate, feel free to call the R.S. Means editors at 1-800-334-3509.

Unit Price Section

Table of Contents

Page

Job Costs

Construction Fees 1
Construction Permits 1
Trade Labor 2
Temporary Utilities 3
Equipment Rental 3
Move / Reset Contents 6
Cover and Protect 6
Temporary Bracing 6
Jack and Relevel 7
Dumpster Rental 7
Debris Hauling 7
Dump Fees 7
Scaffolding 7
Construction Clean-up 8

Job Preparation

Building Demolition 9
General Clean-up 10
Flood Clean-up 10
Asbestos Removal 12
Room Demolition 13
Selective Demolition 14

Excavation

Excavation 17
Backfill 17

Foundation

Piles 19
Foundation Post 22
Concrete Footing 24
Concrete Slab 24

Rough Frame / Structure

Beam / Girder 27
Column 29
Floor Framing System 29
Subflooring 33
Wall Framing System 34
Rough-in Opening 38
Glue-Laminated Beams 42
Metal Stud Framing 44
Metal Joist 45
Exterior Sheathing 45
Plywood Sheathing 46
Stairs 47
Ceiling Framing 47
Roof Framing 49
Residential Trusses 54
Commercial Trusses 55
Roof Sheathing 56
Wood Deck 57
Porch 58

Page

Exterior Trim

Exterior Trim 61
Fascia 62
Soffit 64

Roofing

Metal Flashing 67
Roof Shingles 69
Rolled Roofing 69
Composition Shingles 69
Wood Shingles 71
Slate Tile Roofing 72
Flat Clay Tile Roofing 73
Mission Tile Roofing 73
Spanish Tile Roofing 75
Concrete Tile Roofing 76
Aluminum Sheet 77
Fiberglass Sheet 79
Galvanized Steel 79
Standing Seam 80
Elastomeric Roofing 81
Modified Bitumen Roofing 83
Built-up Roofing 83

Exterior Doors and Windows

Exterior Type Door 85
Residential Metal Door 88
Metal Door 88
Solid Core Door 89
Components 92
Sliding Patio Door 95
Screen Door 97
Storm Door 98
Garage Door 98
Commercial Metal Door 100
Aluminum Window 102
Vinyl Window 110
Wood Window 111
Window Screens 116
Window Glass 116
Storefront System 118
Curtain Wall System 120
Skylight 120
Industrial Steel Window 122

Masonry

Masonry Wall 123
Chimney 127
Fireplace 127

Siding

Exterior Siding 131

Site Work

Grading 137
Asphalt Paving 137
Masonry Pavers 137
Fencing 138

Page

Rough Mechanical

Plumbing 145
Water Heater 147
Ductwork 148
Heating 151
Furnace Unit 152
Boiler 154
Heat Pump 155
Heating 156
Heating Control 157
Air Conditioning 157
Sprinkler System 161

Rough Electrical

Electrical Per S.F. 163
Main Service 163

Insulation

Floor Insulation 165
Wall Insulation 166
Ceiling Insulation 168
Exterior Insulation 170
Roof Insulation Board 170

Walls / Ceilings

Wall Plaster 173
Ceiling Plaster 174
Gypsum Wallboard 175
Ceramic Wall Tile 178
Wall Tile 179
Wall Paneling 180
Wallcovering 182
Acoustical Ceiling 185

Finish Carpentry

Hollow Core Door 189
Solid Core Door 193
Bifold Door 194
Louver Door 197
Bypass Sliding Door 198
Base Molding 199
Molding 202
Window Trim Set 203
Window Sill 204
Shelving 204
Closet Shelving 205
Stair Assemblies 206
Stairs 207
Stair Components 207

Cabinets and Countertops

Wall Cabinets 209
Tall Cabinets 216
Base Cabinets 220
Countertop 227
Vanity Cabinets 229
Vanity Cabinet Tops 232

Unit Price Section

Table of Contents

Page

Painting

Paint Preparation, Walls ... 233
Paint Preparation, Ceilings ... 234
Paint / Texture, Walls ... 234
Paint / Texture, Ceilings ... 235

Flooring

Wood Plank Flooring ... 237
Wood Parquet Tile Floor ... 239
Ceramic Tile Flooring ... 240
Hard Tile Flooring ... 241
Sheet Vinyl Per S.F. ... 242
Sheet Vinyl ... 243
Vinyl Tile ... 243
Carpeting Per S.F. ... 243
Carpeting ... 245
Carpet Pad Per S.F. ... 246
Carpet Pad ... 246
Stair Components ... 247

Finish Mechanical

Commode ... 249
Sink (assembly) ... 250
Sink Only ... 251
Bathtub ... 253
Shower ... 255
Ductwork ... 256

Page

Finish Electrical

Electrical Per S.F. ... 259
Circuits ... 259
Light Switch ... 260
Cover for Outlet / Switch ... 262
Residential Light Fixture ... 262
Commercial Light Fixture ... 264
Fan ... 266
Low Voltage Systems ... 267

Improvements / Appliances / Treatments

Window Treatment ... 271
Stairs ... 273
Elevator ... 273
Rangetop ... 274
Rangehood ... 274
Oven ... 275
Dryer ... 276
Washer ... 277
Refrigerator ... 277
Freezer ... 278
Dishwasher ... 279
Garbage Disposal ... 279
Trash Compactor ... 279
Kitchenette Unit ... 279
Bath Accessories ... 280
Fire Prevention / Protection ... 281

Page

Gutters ... 282
Downspouts ... 282
Shutters ... 283
Awning ... 285
Vent ... 286
Cupolas ... 287
Aluminum Carport ... 288
Storage Shed ... 289
Gazebo ... 289
Pool ... 290

Close Out

Construction Clean-up ... 291

How to Use the Unit Price Pages

The prices in this publication include all overhead and profit mark-ups for the installing contractor. If a general contractor is involved, an additional 10% should be added to the cost shown in the "Total" column.

The costs for material, labor and any equipment are shown here. Material costs include delivery, and a 10% mark-up. No sales taxes are included. Labor costs include all mark-ups as shown on the inside back cover of this book. Equipment costs also include a 10% mark-up.

The cost in the total column should be multiplied by the quantity of each item to determine the total cost for that item. The sum of all item's total costs becomes the project's total cost.

The book is arranged into 23 categories that follow the order of construction for residential projects. The category is shown as white letters on a black background at the top of each page.

A brief specification of the cost item is shown here.

Graphics representing some of the cost items are found at the beginning of each category.

The cost data is arranged in a tree structure format. Major cost sections are left justified and are shown as bold headings within a grey background. Cost items and sub-cost items are found under the bold headings. The complete description should read as "bold heading, cost item, sub-cost item." For example, the identified item's description is "Beam/Girder, Solid Wood Beam, 2″ x 8″."

Each action performed to the cost item is conveniently listed to the right of the description. See the explanation below for the scope of work included as part of the action.

The unit of measure upon which all the costs are based is shown here. An abbreviations list is included within the reference section found at the back of this book.

Rough Frame / Structure

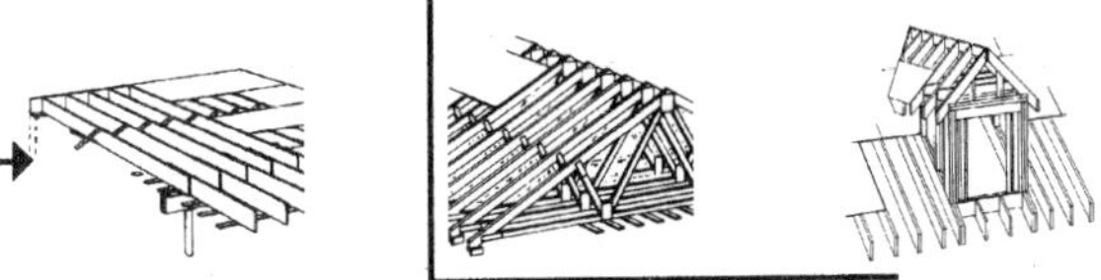

Rough Framing

Beam / Girder	Unit	Material	Labor	Equip.	Total	Specification
Solid Wood Beam						
2″ x 8″						
Demolish	L.F.		.59		.59	Cost includes material and labor to install 2″ x 8″ beam.
Install	L.F.	.88	.93		1.81	
Demolish and Install	L.F.	.88	1.52		2.40	
Clean	L.F.	.09	.19		.28	
Paint	L.F.	.12	1.43		1.55	
Minimum Charge	Job		150		150	
2″ x 10″						
Demolish	L.F.		.73		.73	Cost includes material and labor to install 2″ x 10″ beam.
Install	L.F.	1.27	1		2.27	
Demolish and Install	L.F.	1.27	1.73		3	
Clean	L.F.	.11	.23		.34	
Paint	L.F.	.15	.88		1.03	
Minimum Charge	Job		150		150	
2″ x 12″						
Demolish	L.F.		.88		.88	Cost includes material and labor to install 2″ x 12″ beam.
Install	L.F.	1.72	1.09		2.81	
Demolish and Install	L.F.	1.72	1.97		3.69	
Clean	L.F.	.12	.27		.39	
Paint	L.F.	.18	2.11		2.29	
Minimum Charge	Job		150		150	
4″ x 8″						
Demolish	L.F.		3.01		3.01	Cost includes material and labor to install 4″ x 8″ beam.
Install	L.F.	3.83	1.35	.70	5.88	
Demolish and Install	L.F.	3.83	4.36	.70	8.89	
Clean	L.F.	.11	.23		.34	
Paint	L.F.	.15	.88		1.03	
Minimum Charge	Job		150		150	
4″ x 10″						
Demolish	L.F.		3.76		3.76	Cost includes material and labor to install 4″ x 10″ beam.
Install	L.F.	4.79	1.42	.74	6.95	
Demolish and Install	L.F.	4.79	5.18	.74	10.71	
Clean	L.F.	.12	.27		.39	
Paint	L.F.	.18	1.03		1.21	
Minimum Charge	Job		150		150	
4″ x 12″						
Demolish	L.F.		4.51		4.51	Cost includes material and labor to install 4″ x 12″ beam.
Install	L.F.	5.70	1.50	.78	7.98	
Demolish and Install	L.F.	5.70	6.01	.78	12.49	
Clean	L.F.	.14	.31		.45	
Paint	L.F.	.18	2.11		2.29	
Minimum Charge	Job		150		150	

Action	Scope
Demolish	Includes removal of the item and hauling the debris to a truck, dumpster, or storage area.
Install	Includes the installation of the item with normal placement and spacing of fasteners.
Demolish and Install	Includes the demolish action and the install action.
Reinstall	Includes the reinstallation of a removed item.
Clean	Includes cleaning of the item.
Paint	Includes up to two coats of finish on the item (unless otherwise indicated).
Minimum Charge	Includes the minimum labor or the minimum labor and equipment cost. Use this charge for small quantities of material.

Job Costs

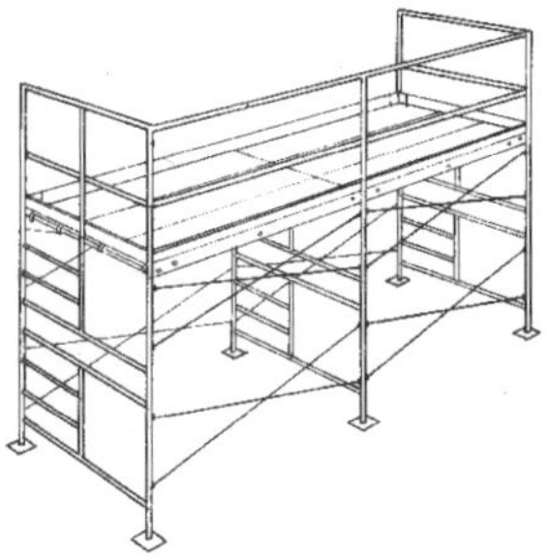

Scaffolding

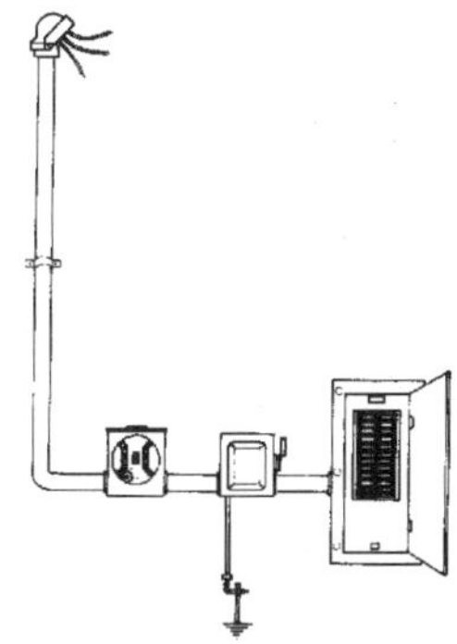

Temporary Electrical Service

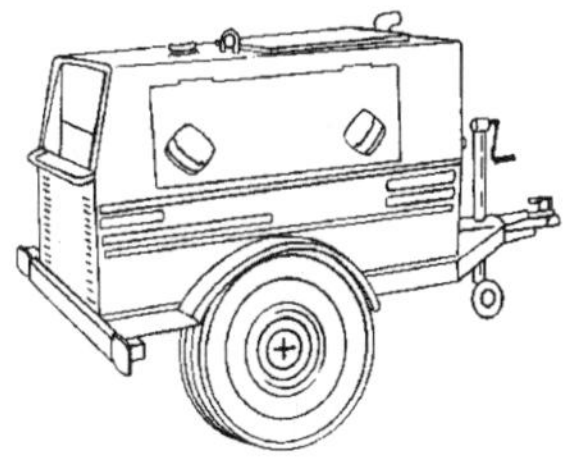

Air Compressor

Construction Fees		Unit	Material	Labor	Equip.	Total	Specification
Asbestos Test							
	Minimum Charge	Job		176		176	Minimum charge for testing.
Architectural							
	Minimum Charge	Day	480			480	Minimum labor charge for architect.
Engineering							
	Minimum Charge	Day	250			250	Minimum labor charge for field engineer.

Construction Permits		Unit	Material	Labor	Equip.	Total	Specification
Job (less than 5K)							
	Minimum Charge	Job	110			110	Permit fees, job less than 5K.
Job (5K to 10K)							
	Minimum Charge	Job	220			220	Permit fees, job more than 5K, but less than 10K.
Job (10K to 25K)							
	Minimum Charge	Job	550			550	Permit fees, job more than 10K, but less than 25K.
Job (25K to 50K)							
	Minimum Charge	Job	1100			1100	Permit fees, job more than 25K, but less than 50K.
Job (50K to 100K)							
	Minimum Charge	Job	2200			2200	Permit fees, job more than 50K, but less than 100K.
Job (100K to 250K)							
	Minimum Charge	Job	5500			5500	Permit fees, job more than 100K, but less than 250K.
Job (250K to 500K)							
	Minimum Charge	Job	11000			11000	Permit fees, job more than 250K, but less than 500K.
Job (500K to 1M)							
	Minimum Charge	Job	22000			22000	Permit fees, job more than 500K, but less than 1 million.

Trade Labor		Unit	Material	Labor	Equip.	Total	Specification
Carpenter							
Daily (8 hours)							
	Install	Day		300		300	Daily labor rate for one carpenter.
	Minimum Charge	Job		150		150	
Weekly (40 hours)							
	Install	Week		1500		1500	Weekly labor rate for one carpenter.
	Minimum Charge	Job		150		150	
Drywaller							
Daily (8 hours)							
	Install	Day		300		300	Daily labor rate for one carpenter.
	Minimum Charge	Job		150		150	
Weekly (40 hours)							
	Install	Week		1500		1500	Weekly labor rate for one carpenter.
	Minimum Charge	Job		150		150	
Roofer							
Daily (8 hours)							
	Install	Day		282		282	Daily labor rate for one skilled roofer.
	Minimum Charge	Job		141		141	
Weekly (40 hours)							
	Install	Week		1400		1400	Weekly labor rate for one skilled roofer.
	Minimum Charge	Job		141		141	
Painter							
Daily (8 hours)							
	Install	Day		264		264	Daily labor rate for one painter.
	Minimum Charge	Job		132		132	
Weekly (40 hours)							
	Install	Week		1325		1325	Weekly labor rate for one painter.
	Minimum Charge	Job		132		132	
Mason							
Daily (8 hours)							
	Install	Day		305		305	Daily labor rate for one skilled mason.
	Minimum Charge	Job		152		152	
Weekly (40 hours)							
	Install	Week		1525		1525	Weekly labor rate for one skilled mason.
	Minimum Charge	Job		152		152	
Electrician							
Daily (8 hours)							
	Install	Day		330		330	Daily labor rate for an electrician.
	Minimum Charge	Job		165		165	
Weekly (40 hours)							
	Install	Week		1650		1650	Weekly labor rate for an electrician.
	Minimum Charge	Job		165		165	
Plumber							
Daily (8 hours)							
	Install	Day		325		325	Daily labor rate for a plumber.
	Minimum Charge	Job		163		163	
Weekly (40 hours)							
	Install	Week		1625		1625	Weekly labor rate for a plumber.
	Minimum Charge	Job		163		163	
Common Laborer							
Daily (8 hours)							
	Install	Day		218		218	Daily labor rate for a common laborer.
	Minimum Charge	Job		109		109	

Job Costs

Trade Labor		Unit	Material	Labor	Equip.	Total	Specification
Weekly (40 hours)							
	Install	Week		1100		1100	Weekly labor rate for a common laborer.
	Minimum Charge	Job		109		109	

Temporary Utilities		Unit	Material	Labor	Equip.	Total	Specification
Power Pole							
Job Site Electricity							
	Install	Month	33			33	Includes average cost of electricity per month.
	Minimum Charge	Job		165		165	
Water Service							
Job Site Water							
	Install	Month	62.50			62.50	Includes average costs for water for one month.
	Minimum Charge	Month	62.50			62.50	
Telephone							
	Install	Month	224			224	Includes average cost for local and long distance telephone service for one month.
	Minimum Charge	Month	224			224	
Chemical Toilet							
	Install	Month			242	242	Monthly rental of portable toilet.
	Minimum Charge	Day			12.10	12.10	
Office Trailer							
Place and Remove							
	Install	Job	213			213	Includes cost to a contractor to deliver, place and level an office trailer and remove at end of job.
	Minimum Charge	Job		122		122	
Storage Container							
	Install	Job	144			144	Includes cost to a contractor to deliver and place a storage container and remove at end of job.
	Minimum Charge	Job		122		122	
Fencing							
	Install	L.F.	1.80			1.80	Includes material costs for temporary fencing.
	Minimum Charge	L.F.	275			275	
Railing							
	Install	L.F.	7.05	8.70	.49	16.24	Includes material and labor to install primed steel pipe.
	Minimum Charge	L.F.	275			275	

Equipment Rental		Unit	Material	Labor	Equip.	Total	Specification
Flatbed Truck							
Daily Rental							
	Install	Day			132	132	Daily rental of flatbed truck.
	Minimum Charge	Job			132	132	
Weekly Rental							
	Install	Week			660	660	Weekly rental of flatbed truck.
	Minimum Charge	Job			132	132	
Monthly Rental							
	Install	Month			2650	2650	Monthly rental of flatbed truck.
	Minimum Charge	Job			132	132	

Equipment Rental		Unit	Material	Labor	Equip.	Total	Specification
Dump Truck							
Daily Rental							
	Install	Day			335	335	Daily rental of dump truck.
	Minimum Charge	Day			335	335	
Weekly Rental							
	Install	Week			1675	1675	Weekly rental of dump truck.
	Minimum Charge	Day			335	335	
Monthly Rental							
	Install	Month			6725	6725	Monthly rental of dump truck.
	Minimum Charge	Day			335	335	
Forklift							
Daily Rental							
	Install	Day			252	252	Daily rental of forklift.
	Minimum Charge	Day			252	252	
Weekly Rental							
	Install	Week			1250	1250	Weekly rental of forklift.
	Minimum Charge	Day			252	252	
Monthly Rental							
	Install	Month			5025	5025	Monthly rental of forklift.
	Minimum Charge	Day			252	252	
Bobcat							
Daily Rental							
	Install	Day			168	168	Daily rental of wheeled, skid steer loader.
	Minimum Charge	Day			168	168	
Weekly Rental							
	Install	Week			840	840	Weekly rental of wheeled, skid steer loader.
	Minimum Charge	Day			168	168	
Monthly Rental							
	Install	Month			3375	3375	Monthly rental of wheeled, skid steer loader.
	Minimum Charge	Day			168	168	
Air Compressor							
Daily Rental							
	Install	Day			62	62	Daily rental of air compressor.
	Minimum Charge	Day			62	62	
Weekly Rental							
	Install	Week			310	310	Weekly rental of air compressor.
	Minimum Charge	Day			62	62	
Monthly Rental							
	Install	Month			1250	1250	Monthly rental of air compressor.
	Minimum Charge	Day			62	62	
Jackhammer							
Daily Rental							
	Install	Day			16.10	16.10	Daily rental of air tool, jackhammer.
	Minimum Charge	Day			16.10	16.10	
Weekly Rental							
	Install	Week			80.50	80.50	Weekly rental of air tool, jackhammer.
	Minimum Charge	Day			16.10	16.10	
Monthly Rental							
	Install	Month			320	320	Monthly rental of air tool, jackhammer.
	Minimum Charge	Day			16.10	16.10	
Concrete Mixer							
Daily Rental							
	Install	Day			97	97	Daily rental of concrete mixer.
	Minimum Charge	Day			97	97	

Equipment Rental		Unit	Material	Labor	Equip.	Total	Specification
Weekly Rental							
	Install	Week			485	485	Weekly rental of concrete mixer.
	Minimum Charge	Day			97	97	
Monthly Rental							
	Install	Month			1925	1925	Monthly rental of concrete mixer.
	Minimum Charge	Day			97	97	
Concrete Bucket							
Daily Rental							
	Install	Day			17.35	17.35	Daily rental of concrete bucket.
	Minimum Charge	Day			17.35	17.35	
Weekly Rental							
	Install	Week			87	87	Weekly rental of concrete bucket.
	Minimum Charge	Day			17.35	17.35	
Monthly Rental							
	Install	Month			345	345	Monthly rental of concrete bucket.
	Minimum Charge	Day			17.35	17.35	
Concrete Pump							
Daily Rental							
	Install	Day			810	810	Daily rental of concrete pump.
	Minimum Charge	Day			810	810	
Weekly Rental							
	Install	Week			4050	4050	Weekly rental of concrete pump.
	Minimum Charge	Day			810	810	
Monthly Rental							
	Install	Month			16200	16200	Monthly rental of concrete pump.
	Minimum Charge	Day			810	810	
Generator							
Daily Rental							
	Install	Day			30	30	Daily rental of generator.
	Minimum Charge	Day			30	30	
Weekly Rental							
	Install	Week			150	150	Weekly rental of generator.
	Minimum Charge	Day			30	30	
Monthly Rental							
	Install	Month			600	600	Monthly rental of generator.
	Minimum Charge	Day			30	30	
Welder							
Daily Rental							
	Install	Day			70.50	70.50	Daily rental of welder.
	Minimum Charge	Day			70.50	70.50	
Weekly Rental							
	Install	Week			355	355	Weekly rental of welder.
	Minimum Charge	Day			70.50	70.50	
Monthly Rental							
	Install	Month			1425	1425	Monthly rental of welder.
	Minimum Charge	Day			70.50	70.50	
Sandblaster							
Daily Rental							
	Install	Day			24.50	24.50	Daily rental of sandblaster.
	Minimum Charge	Day			24.50	24.50	
Weekly Rental							
	Install	Week			123	123	Weekly rental of sandblaster.
	Minimum Charge	Day			24.50	24.50	

Equipment Rental		Unit	Material	Labor	Equip.	Total	Specification
Monthly Rental							
	Install	Month			495	495	Monthly rental of sandblaster.
	Minimum Charge	Day			24.50	24.50	
Space Heater							
Daily Rental							
	Install	Day			19.60	19.60	Daily rental of space heater.
	Minimum Charge	Day			19.60	19.60	
Weekly Rental							
	Install	Day			98	98	Weekly rental of space heater.
	Minimum Charge	Day			19.60	19.60	
Monthly Rental							
	Install	Day			390	390	Monthly rental of space heater.
	Minimum Charge	Day			19.60	19.60	

Move / Reset Contents		Unit	Material	Labor	Equip.	Total	Specification
Small Room							
	Install	Room		26.50		26.50	Includes cost to remove and reset the contents of a small room.
	Minimum Charge	Job		109		109	
Average Room							
	Install	Room		32		32	Includes cost to remove and reset the contents of an average room.
	Minimum Charge	Job		109		109	
Large Room							
	Install	Room		44		44	Includes cost to remove and reset the contents of a large room.
	Minimum Charge	Job		109		109	
Extra Large Room							
	Install	Room		70		70	Includes cost to remove and reset the contents of a extra large room.
	Minimum Charge	Job		109		109	
Store Contents							
Weekly							
	Install	Week	152			152	Storage fee, per week.
Monthly							
	Install	Month	455			455	Storage fee, per month.

Cover and Protect		Unit	Material	Labor	Equip.	Total	Specification
Cover / Protect Walls							
	Install	SF Wall		.08		.08	Includes labor costs to cover and protect walls.
	Minimum Charge	Job		109		109	
Cover / Protect Floors							
	Install	SF Flr.		.08		.08	Includes labor costs to cover and protect floors.
	Minimum Charge	Job		109		109	

Temporary Bracing		Unit	Material	Labor	Equip.	Total	Specification
Shim Support Piers							
	Install	Ea.	3.18	87		90.18	Includes labor and material to install shims support for piers.
	Minimum Charge	Job		261		261	

Job Costs

Jack and Relevel		Unit	Material	Labor	Equip.	Total	Specification
Single Story							
	Install	Ea.	1.20	1.05		2.25	Includes labor and equipment to jack and relevel single-story building.
	Minimum Charge	Job		261		261	
Two-story							
	Install	Ea.	1.20	1.74		2.94	Includes labor and equipment to jack and relevel two-story building.
	Minimum Charge	Job		261		261	

Dumpster Rental		Unit	Material	Labor	Equip.	Total	Specification
Small Load							
	Install	Week			450	450	Dumpster, weekly rental, 2 dumps, 5 yard container.
	Minimum Charge	Week			450	450	
Medium Load							
	Install	Week			560	560	Dumpster, weekly rental, 2 dumps, 10 yard container.
	Minimum Charge	Week			560	560	
Large Load							
	Install	Week			950	950	Dumpster, weekly rental, 2 dumps, 30 yard container.
	Minimum Charge	Week			950	950	

Debris Hauling		Unit	Material	Labor	Equip.	Total	Specification
Per Ton							
	Install	Ton		68.50	36.50	105	Includes hauling of debris by trailer or dump truck.
	Minimum Charge	Ton		515	275	790	
Per Cubic Yard							
	Install	C.Y.		23	12.20	35.20	Includes hauling of debris by trailer or dump truck.
	Minimum Charge	Job		515	275	790	
Per Pick-up Truck Load							
	Install	Ea.		121		121	Includes hauling of debris by pick-up truck.
	Minimum Charge	Job		515	275	790	

Dump Fees		Unit	Material	Labor	Equip.	Total	Specification
Per Cubic Yard							
	Install	C.Y.	6.95			6.95	Dump charges, building materials.
	Minimum Charge	C.Y.	6.95			6.95	
Per Ton							
	Install	Ton	55.50			55.50	Dump charges, building materials.
	Minimum Charge	Ton	55.50			55.50	

Scaffolding		Unit	Material	Labor	Equip.	Total	Specification
4′ to 6′ High							
	Install	Month	51.50			51.50	Includes monthly rental of scaffolding, steel tubular, 30″ wide, 7′ long, 5′ high.
	Minimum Charge	Month	51.50			51.50	
7′ to 11′ High							
	Install	Month	61.50			61.50	Includes monthly rental of scaffolding, steel tubular, 30″ wide, 7′ long, 10′ high.
	Minimum Charge	Month	61.50			61.50	

Scaffolding	Unit	Material	Labor	Equip.	Total	Specification
12′ to 16′ High						
Install	Month	71			71	Includes monthly rental of scaffolding, steel tubular, 30″ wide, 7′ long, 15′ high.
Minimum Charge	Month	71			71	
17′ to 21′ High						
Install	Month	95			95	Includes monthly rental of scaffolding, steel tubular, 30″ wide, 7′ long, 20′ high.
Minimum Charge	Month	95			95	
22′ to 26′ High						
Install	Month	105			105	Includes monthly rental of scaffolding, steel tubular, 30″ wide, 7′ long, 25′ high.
Minimum Charge	Month	105			105	
27′ to 30′ High						
Install	Month	114			114	Includes monthly rental of scaffolding, steel tubular, 30″ wide, 7′ long, 30′ high.
Minimum Charge	Month	114			114	

Construction Clean-up	Unit	Material	Labor	Equip.	Total	Specification
Initial						
Clean	S.F.		.27		.27	Includes labor for general clean-up.
Minimum Charge	Job		109		109	
Progressive						
Clean	S.F.		.14		.14	Includes labor for progressive site/job clean-up on a weekly basis.
Minimum Charge	Job		109		109	

Building Demolition	Unit	Material	Labor	Equip.	Total	Specification
Wood-framed Building						
1 Story Demolition						
Demolish	S.F.		1.03	1.30	2.33	Includes demolition of 1 floor only light wood-framed building, including hauling of debris. Dump fees not included.
Minimum Charge	Job		1625	2050	3675	
2nd Story Demolition						
Demolish	S.F.		1.05	1.32	2.37	Includes demolition of 2nd floor only light wood-framed building, including hauling of debris. Dump fees not included.
Minimum Charge	Job		1625	2050	3675	
3rd Story Demolition						
Demolish	S.F.		1.36	1.72	3.08	Includes demolition of 3rd floor only light wood-framed building, including hauling of debris. Dump fees not included.
Minimum Charge	Job		1625	2050	3675	
Masonry Building						
Demolish	S.F.		1.47	1.85	3.32	Includes demolition of one story masonry building, including hauling of debris. Dump fees not included.
Minimum Charge	Job		1625	2050	3675	
Concrete Building						
Unreinforced						
Demolish	S.F.		2.99	.92	3.91	Includes demolition of a one story non-reinforced concrete building, including hauling of debris. Dump fees not included.
Minimum Charge	Job		1625	2050	3675	
Reinforced						
Demolish	S.F.		3.43	1.06	4.49	Includes demolition of a one story reinforced concrete building, including hauling of debris. Dump fees not included.
Minimum Charge	Job		1625	2050	3675	
Steel Building						
Demolish	S.F.		1.95	2.46	4.41	Includes demolition of a metal building, including hauling of debris. Dump fees not included. No salvage value assumed for scrap metal.
Minimum Charge	Job		1625	2050	3675	

Job Preparation

General Clean-up		Unit	Material	Labor	Equip.	Total	Specification
Remove Debris							
	Clean	S.F.		.67	.13	.80	Includes removal of debris caused by the loss including contents items and loose debris.
	Minimum Charge	Job		335	66	401	
Muck-out							
	Clean	S.F.		.42		.42	Includes labor to remove 2" of muck and mud after a flood loss.
	Minimum Charge	Job		109		109	
Deodorize / Disinfect							
	Clean	SF Flr.		.15		.15	Includes deodorizing building with application of mildicide and disinfectant to floors and walls - will vary depending on the extent of deodorization required.
	Minimum Charge	Job		109		109	
Clean Walls							
Heavy							
	Clean	S.F.		.11		.11	Includes labor and material for heavy cleaning (multiple applications) with detergent and solvent.
	Minimum Charge	Job		109		109	
Clean Ceiling							
Light							
	Clean	S.F.		.18		.18	Includes labor and material to clean light to moderate smoke and soot.
	Minimum Charge	Job		109		109	
Heavy							
	Clean	S.F.		.26		.26	Includes labor and material to clean heavy smoke and soot.
	Minimum Charge	Job		109		109	

Flood Clean-up		Unit	Material	Labor	Equip.	Total	Specification
Emergency Service Call							
After Hours / Weekend							
	Minimum Charge	Job		145		145	Includes minimum charges for deflooding clean-up after hours or on weekends.
Grey Water / Sewage							
	Minimum Charge	Job		109		109	Includes minimum charges for deflooding clean-up.
Water Extraction							
Clean (non-grey)							
	Clean	SF Flr.	.02	.44		.46	Includes labor to remove clean uncontaminated water from the loss.
	Minimum Charge	Job		109		109	
Grey (non-solids)							
	Clean	SF Flr.	.02	.87		.89	Includes labor to remove non-solid grey water from the loss.
	Minimum Charge	Job		109		109	
Sewage / Solids							
	Clean	SF Flr.	.03	1.09		1.12	Includes labor to remove water from the loss that may contain sewage or solids.
	Minimum Charge	Job		109		109	

Flood Clean-up		Unit	Material	Labor	Equip.	Total	Specification
Mildicide Walls							
Topographical							
	Clean	SF Wall		.20		.20	Clean-up of deflooding including topographical mildicide application.
	Minimum Charge	Job		109		109	
Injection Treatments							
	Clean	SF Flr.	25	3.35		28.35	Clean-up of deflooding including mildicide injection.
	Minimum Charge	Job		109		109	
Carpet Cleaning							
Uncontaminated							
	Clean	SF Flr.		.14		.14	Includes labor to clean carpet.
	Minimum Charge	Job		109		109	
Contaminated							
	Clean	SF Flr.		.22		.22	Includes labor to clean contaminated carpet.
	Minimum Charge	Job		109		109	
Stairway							
	Clean	Ea.	.10	2.56		2.66	Includes labor to clean contaminated carpet on a stairway.
	Minimum Charge	Job		109		109	
Carpet Treatment							
Disinfect / Deodorize							
	Clean	SF Flr.		.14		.14	Includes disinfecting and deodorizing of carpet.
	Minimum Charge	Job		109		109	
Disinfect For Sewage							
	Clean	SF Flr.		.35		.35	Includes labor to disinfect and deodorize carpet exposed to sewage.
	Minimum Charge	Job		109		109	
Mildicide							
	Clean	SF Flr.	.01	.14		.15	Includes labor and materials to apply mildicide to carpet.
	Minimum Charge	Job		109		109	
Thermal Fog Area							
	Clean	SF Flr.	.03	.34		.37	Includes labor and materials to perform thermal fogging during deflooding clean-up.
	Minimum Charge	Job		109		109	
Wet Fog / ULV Area							
	Clean	SF Flr.	.02	.19		.21	Includes labor and materials to perform wet fogging / ULV during deflooding clean-up.
	Minimum Charge	Job		109		109	
Steam Clean Fixtures							
	Clean	Ea.	.01	22		22.01	Includes labor and materials to steam clean fixtures during deflooding clean-up.
	Minimum Charge	Job		109		109	
Pressure Wash Sewage							
	Clean	SF Flr.		1.98		1.98	Includes labor to pressure wash sewage contamination.
	Minimum Charge	Job		109		109	
Airmover							
	Minimum Charge	Day			19.60	19.60	Daily rental of air mover.

Job Preparation

Flood Clean-up		Unit	Material	Labor	Equip.	Total	Specification
Dehumidifier							
Medium Size							
	Minimum Charge	Day			33.50	33.50	Daily rental of dehumidifier, medium.
Large Size							
	Minimum Charge	Day			66.50	66.50	Daily rental of dehumidifier, large.
Air Purifier							
HEPA Filter-Bacteria							
	Minimum Charge	Ea.	195			195	HEPA filter for air purifier.

Asbestos Removal		Unit	Material	Labor	Equip.	Total	Specification
Asbestos Analysis							
	Minimum Charge	Job		176		176	Minimum charge for testing.
Asbestos Pipe Insulation							
1/2″ to 3/4″ Diameter							
	Demolish	L.F.	.34	4.56		4.90	Minimum charge to remove asbestos. Includes full Tyvek suits for workers changed four times per eight-hour shift with respirators, air monitoring and supervision by qualified professionals.
	Minimum Charge	Job		2825		2825	
1″ to 3″ Diameter							
	Demolish	L.F.	1.06	14.15		15.21	Minimum charge to remove asbestos. Includes full Tyvek suits for workers changed four times per eight-hour shift with respirators, air monitoring and supervision by qualified professionals.
	Minimum Charge	Job		2825		2825	
Asbestos-based Siding							
	Demolish	S.F.	2.34	31.50		33.84	Minimum charge to remove asbestos. Includes full Tyvek suits for workers changed four times per eight-hour shift with respirators, air monitoring and supervision by qualified professionals.
	Minimum Charge	Job		2825		2825	
Asbestos-based Plaster							
	Demolish	S.F.	1.80	24		25.80	Minimum charge to remove asbestos. Includes full Tyvek suits for workers changed four times per eight-hour shift with respirators, air monitoring and supervision by qualified professionals.
	Minimum Charge	Job		2825		2825	
Encapsulate Asbestos Ceiling							
	Install	S.F.	.29	.13		.42	Includes encapsulation with penetrating sealant sprayed onto ceiling and walls with airless sprayer.
	Minimum Charge	Job		2825		2825	
Scrape Acoustical Ceiling							
	Demolish	S.F.	.07	.80		.87	Minimum charge to remove asbestos. Includes full Tyvek suits for workers changed four times per eight-hour shift with respirators, air monitoring and supervision by qualified professionals.
	Minimum Charge	Job		2825		2825	

Asbestos Removal		Unit	Material	Labor	Equip.	Total	Specification
Seal Ceiling Asbestos							
	Install	S.F.	1.10	2.66		3.76	Includes attaching plastic cover to ceiling area and sealing where asbestos exists, taping all seams, caulk or foam insulation in cracks and negative air system.
	Minimum Charge	Job		2825		2825	
Seal Wall Asbestos							
	Install	S.F.	1.10	2.37		3.47	Includes attaching plastic cover to wall area and sealing where asbestos exists, taping all seams, caulk or foam insulation in cracks and negative air system.
	Minimum Charge	Job		2825		2825	
Seal Floor Asbestos							
	Install	S.F.	1.76	4.73		6.49	Includes attaching plastic cover to floor area and sealing where asbestos exists, taping all seams, caulk or foam insulation in cracks and negative air system.
	Minimum Charge	Job		2825		2825	
Negative Air Vent System							
	Install	Day			43.50	43.50	Daily rental of negative air vent system.
	Minimum Charge	Week			218	218	
HEPA Vacuum Cleaner							
	Install	Day			24.50	24.50	Daily rental of HEPA vacuum, 16 gallon.
	Minimum Charge	Week			123	123	
Airless Sprayer							
	Install	Day			36	36	Daily rental of airless sprayer.
	Minimum Charge	Week			180	180	
Decontamination Unit							
	Install	Day			133	133	Daily rental of decontamination unit.
	Minimum Charge	Week			665	665	
Light Stand							
	Install	Day			12.90	12.90	Daily rental of floodlight w/tripod.
	Minimum Charge	Week			64.50	64.50	

Room Demolition		Unit	Material	Labor	Equip.	Total	Specification
Strip Typical Room							
	Demolish	S.F.		2.18		2.18	Includes 4 labor hours for general demolition.
	Minimum Charge	Job		121		121	
Strip Typical Bathroom							
	Demolish	S.F.		2.91		2.91	Includes 4 labor hours for general demolition.
	Minimum Charge	Job		121		121	
Strip Typical Kitchen							
	Demolish	S.F.		2.79		2.79	Includes 4 labor hours for general demolition.
	Minimum Charge	Job		121		121	
Strip Typical Laundry							
	Demolish	S.F.		2.49		2.49	Includes 4 labor hours for general demolition.
	Minimum Charge	Job		121		121	

Selective Demolition	Unit	Material	Labor	Equip.	Total	Specification
Concrete Slab						
4″ Unreinforced						
Demolish	S.F.		1.46	.24	1.70	Includes labor and equipment to remove unreinforced concrete slab and haul debris to truck or dumpster.
Minimum Charge	Job		620	102	722	
6″ Unreinforced						
Demolish	S.F.		2.07	.34	2.41	Includes labor and equipment to remove unreinforced concrete slab and haul debris to truck or dumpster.
Minimum Charge	Job		620	102	722	
8″ Unreinforced						
Demolish	S.F.		3.40	.56	3.96	Includes labor and equipment to remove unreinforced concrete slab and haul debris to truck or dumpster.
Minimum Charge	Job		620	102	722	
4″ Reinforced						
Demolish	S.F.		1.84	.30	2.14	Includes labor and equipment to remove reinforced concrete slab and haul debris to truck or dumpster.
Minimum Charge	Job		620	102	722	
6″ Reinforced						
Demolish	S.F.		2.76	.46	3.22	Includes labor and equipment to remove reinforced concrete slab and haul debris to truck or dumpster.
Minimum Charge	Job		620	102	722	
8″ Reinforced						
Demolish	S.F.		4.51	.75	5.26	Includes labor and equipment to remove reinforced concrete slab and haul debris to truck or dumpster.
Minimum Charge	Job		620	102	722	
Brick Wall						
4″ Thick						
Demolish	S.F.		2.03	.34	2.37	Includes labor and equipment to remove brick wall and haul debris to a central location for disposal.
Minimum Charge	Job		620	102	722	
8″ Thick						
Demolish	S.F.		2.79	.46	3.25	Includes labor and equipment to remove brick wall and haul debris to a central location for disposal.
Minimum Charge	Job		620	102	722	
12″ Thick						
Demolish	S.F.		3.72	.61	4.33	Includes labor and equipment to remove brick wall and haul debris to a central location for disposal.
Minimum Charge	Job		620	102	722	
Concrete Masonry Wall						
4″ Thick						
Demolish	S.F.		2.26	.37	2.63	Includes minimum labor and equipment to remove 4″ block wall and haul debris to a central location for disposal.
Minimum Charge	Job		620	102	722	
6″ Thick						
Demolish	S.F.		2.79	.46	3.25	Includes minimum labor and equipment to remove 6″ block wall and haul debris to a central location for disposal.
Minimum Charge	Job		620	102	722	

Selective Demolition		Unit	Material	Labor	Equip.	Total	Specification
8" Thick							
	Demolish	S.F.		3.44	.57	4.01	Includes minimum labor and equipment to remove 8" block wall and haul debris to a central location for disposal.
	Minimum Charge	Job		620	102	722	
12" Thick							
	Demolish	S.F.		4.97	.82	5.79	Includes minimum labor and equipment to remove 12" block wall and haul debris to a central location for disposal.
	Minimum Charge	Job		620	102	722	
Tree Removal							
Small, 4" - 8"							
	Demolish	Ea.		78.50	63.50	142	Minimum charge, tree removal.
	Minimum Charge	Job		109		109	
Medium, 9" - 12"							
	Demolish	Ea.		118	95	213	Minimum charge, tree removal.
	Minimum Charge	Job		109		109	
Large, 13" - 18"							
	Demolish	Ea.		142	114	256	Minimum charge, tree removal.
	Minimum Charge	Job		360	131	491	
Very Large, 19" - 24"							
	Demolish	Ea.		177	142	319	Minimum charge, tree removal.
	Minimum Charge	Job		360	131	491	
Stump Removal							
	Demolish	Ea.		49	112	161	Includes minimum charges for a site demolition crew and backhoe / loader.
	Minimum Charge	Job		200	67	267	

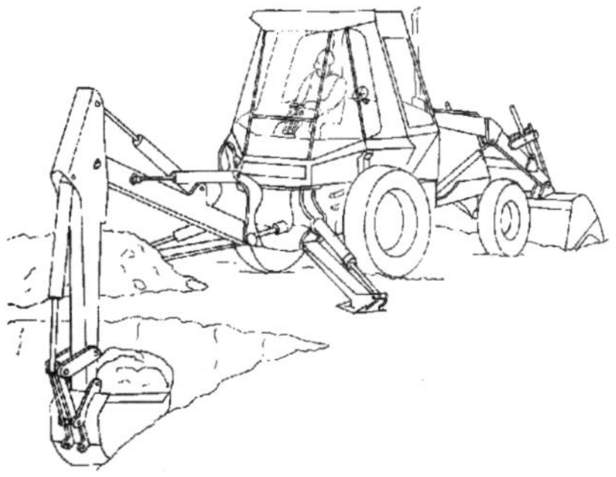

Backhoe/Loader - Wheel Type

Tractor Loader - Wheel Type

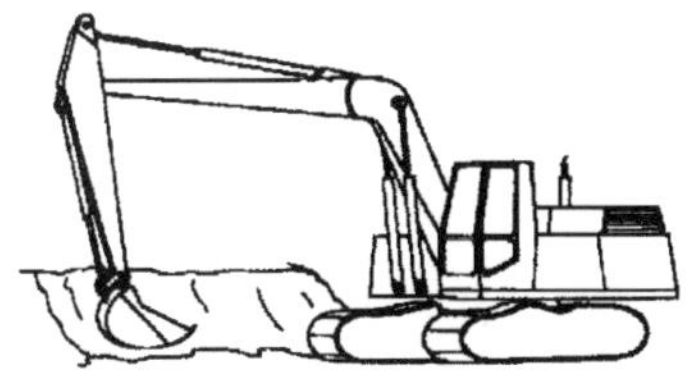

Backhoe - Crawler Type

Excavation	Unit	Material	Labor	Equip.	Total	Specification
Footings						
Hand Trenching						
Install	L.F.		1.36		1.36	Includes hand excavation for footing or trench up to 10″ deep and 12″ wide.
Minimum Charge	Job		54.50		54.50	
Machine Trenching						
Install	L.F.		.52	.24	.76	Includes machine excavation for footing or trench up to 10″ deep and 12″ wide.
Minimum Charge	Job		258	120	378	
Trenching w / Backfill						
Hand						
Install	L.F.		2.18		2.18	Includes hand excavation, backfill and compaction.
Minimum Charge	Job		54.50		54.50	
Machine						
Install	L.F.		2.58	8.10	10.68	Includes machine excavation, backfill and compaction.
Minimum Charge	Job		109	43.50	152.50	

Backfill	Unit	Material	Labor	Equip.	Total	Specification
Hand (no compaction)						
Install	C.Y.		12.80		12.80	Includes backfill by hand in average soil without compaction.
Minimum Charge	Job		109		109	
Hand (w / compaction)						
Install	C.Y.		25.50		25.50	Includes backfill by hand in average soil with compaction.
Minimum Charge	Job		109		109	
Machine (no compaction)						
Install	C.Y.		2.58	1.20	3.78	Includes backfill, by 55 HP wheel loader, of trenches from loose material piled adjacent without compaction.
Minimum Charge	Job		258	120	378	
Machine (w / compaction)						
Install	C.Y.		6.90	3.21	10.11	Includes backfill, by 55 HP wheel loader, of trenches from loose material piled adjacent with compaction by vibrating tampers.
Minimum Charge	Job		258	120	378	

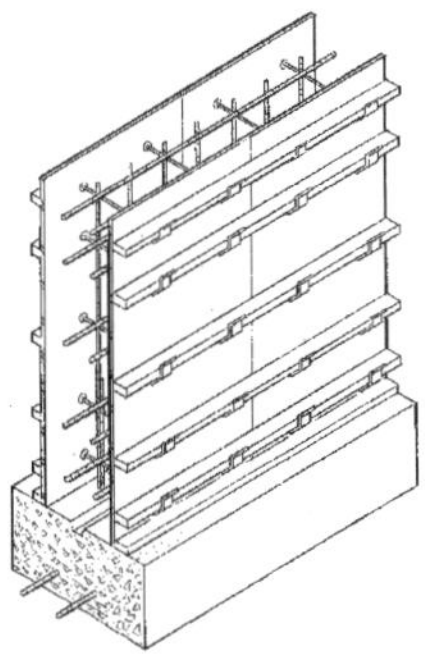

Concrete Footing and Formwork

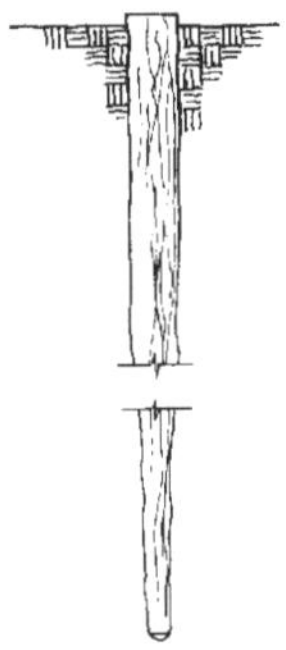

Wood Pile

Piles	Unit	Material	Labor	Equip.	Total	Specification
Treated Wood, 12" Butt						
To 30'						
Demolish	L.F.		3.48	1.16	4.64	Includes material, labor and equipment to install up to 30' long treated wood piling.
Install	V.L.F.	10	3.41	2.98	16.39	
Demolish and Install	V.L.F.	10	6.89	4.14	21.03	
Paint	L.F.	.37	1.51		1.88	
Minimum Charge	Job		2125	1850	3975	
30' to 40'						
Demolish	L.F.		3.48	1.16	4.64	Includes material, labor and equipment to install 30' to 40' long treated wood piling.
Install	V.L.F.	10.20	3.04	2.66	15.90	
Demolish and Install	V.L.F.	10.20	6.52	3.82	20.54	
Paint	L.F.	.37	1.51		1.88	
Minimum Charge	Job		2125	1850	3975	
40' to 50'						
Demolish	L.F.		3.48	1.16	4.64	Includes material, labor and equipment to install 40' to 50' long treated wood piling.
Install	V.L.F.	10.10	2.96	2.59	15.65	
Demolish and Install	V.L.F.	10.10	6.44	3.75	20.29	
Paint	L.F.	.37	1.51		1.88	
Minimum Charge	Job		2125	1850	3975	
50' to 60'						
Demolish	L.F.		3.48	1.16	4.64	Includes material, labor and equipment to install 50' to 60' long treated wood piling.
Install	V.L.F.	11.05	2.66	2.33	16.04	
Demolish and Install	V.L.F.	11.05	6.14	3.49	20.68	
Paint	L.F.	.37	1.51		1.88	
Minimum Charge	Job		2125	1850	3975	
60' to 80'						
Demolish	L.F.		3.48	1.16	4.64	Includes material, labor and equipment to install 60' to 80' long treated wood piling.
Install	V.L.F.	18.90	2.95	2.21	24.06	
Demolish and Install	V.L.F.	18.90	6.43	3.37	28.70	
Paint	L.F.	.37	1.51		1.88	
Minimum Charge	Job		2125	1850	3975	
Untreated Wood, 12" Butt						
To 30'						
Demolish	L.F.		3.48	1.16	4.64	Includes material, labor and equipment to install up to 30' long untreated wood piling.
Install	V.L.F.	6.45	3.41	2.98	12.84	
Demolish and Install	V.L.F.	6.45	6.89	4.14	17.48	
Paint	L.F.	.37	1.51		1.88	
Minimum Charge	Job		2125	1850	3975	

Piles		Unit	Material	Labor	Equip.	Total	Specification
30′ to 40′							
	Demolish	L.F.		3.48	1.16	4.64	Includes material, labor and equipment to install 30′ to 40′ long untreated wood piling.
	Install	V.L.F.	6.45	3.04	2.66	12.15	
	Demolish and Install	V.L.F.	6.45	6.52	3.82	16.79	
	Paint	L.F.	.37	1.51		1.88	
	Minimum Charge	Job		2125	1850	3975	
40′ to 50′							
	Demolish	L.F.		3.48	1.16	4.64	Includes material, labor and equipment to install 40′ to 50′ long untreated wood piling.
	Install	V.L.F.	6.45	2.96	2.59	12	
	Demolish and Install	V.L.F.	6.45	6.44	3.75	16.64	
	Paint	L.F.	.37	1.51		1.88	
	Minimum Charge	Job		2125	1850	3975	
50′ to 60′							
	Demolish	L.F.		3.48	1.16	4.64	Includes material, labor and equipment to install 50′ to 60′ long untreated wood piling.
	Install	V.L.F.	6.50	2.66	2.33	11.49	
	Demolish and Install	V.L.F.	6.50	6.14	3.49	16.13	
	Paint	L.F.	.37	1.51		1.88	
	Minimum Charge	Job		2125	1850	3975	
60′ to 80′							
	Demolish	L.F.		3.48	1.16	4.64	Includes material, labor and equipment to install 60′ to 80′ long untreated wood piling.
	Install	V.L.F.	8.15	2.53	2.22	12.90	
	Demolish and Install	V.L.F.	8.15	6.01	3.38	17.54	
	Paint	L.F.	.37	1.51		1.88	
	Minimum Charge	Job		2125	1850	3975	
Precast Concrete							
10″ Square							
	Demolish	L.F.		3.56	1.19	4.75	Includes material, labor and equipment to install a precast, prestressed, 40′ long, 10″ square concrete pile.
	Install	V.L.F.	7.90	3.04	2.66	13.60	
	Demolish and Install	V.L.F.	7.90	6.60	3.85	18.35	
	Paint	L.F.	.37	1.51		1.88	
	Minimum Charge	Job		2125	1850	3975	
12″ Square							
	Demolish	L.F.		3.81	1.27	5.08	Includes material, labor and equipment to install a precast, prestressed, 40′ long, 12″ square concrete pile.
	Install	V.L.F.	9.90	3.13	2.74	15.77	
	Demolish and Install	V.L.F.	9.90	6.94	4.01	20.85	
	Paint	L.F.	.37	1.51		1.88	
	Minimum Charge	Job		2125	1850	3975	
14″ Square							
	Demolish	L.F.		3.90	1.30	5.20	Includes material, labor and equipment to install a precast, prestressed, 40′ long, 14″ square concrete pile.
	Install	V.L.F.	11.70	3.55	3.10	18.35	
	Demolish and Install	V.L.F.	11.70	7.45	4.40	23.55	
	Paint	L.F.	.37	1.51		1.88	
	Minimum Charge	Job		2125	1850	3975	
16″ Square							
	Demolish	L.F.		4.10	1.37	5.47	Includes material, labor and equipment to install a precast, prestressed, 40′ long, 16″ square concrete pile.
	Install	V.L.F.	18.55	3.80	3.32	25.67	
	Demolish and Install	V.L.F.	18.55	7.90	4.69	31.14	
	Paint	L.F.	.37	1.51		1.88	
	Minimum Charge	Job		2125	1850	3975	
18″ Square							
	Demolish	L.F.		4.44	1.49	5.93	Includes material, labor and equipment to install a precast, prestressed, 40′ long, 18″ square concrete pile.
	Install	V.L.F.	22.50	4.76	3.57	30.83	
	Demolish and Install	V.L.F.	22.50	9.20	5.06	36.76	
	Paint	L.F.	.37	1.51		1.88	
	Minimum Charge	Job		2125	1850	3975	

Piles

	Unit	Material	Labor	Equip.	Total	Specification
Cross Bracing						
Treated 2" x 6"						
Demolish	L.F.		.71		.71	Cost includes material and labor to install 2" x 6" lumber for cross-bracing of foundation piling or posts.
Install	L.F.	.84	4.22		5.06	
Demolish and Install	L.F.	.84	4.93		5.77	
Remove and Install	L.F.		3.37		3.37	
Clean	S.F.		.27		.27	
Paint	L.F.	.17	1.02		1.19	
Minimum Charge	Job		525		525	
Treated 2" x 8"						
Demolish	L.F.		.71		.71	Cost includes material and labor to install 2" x 8" lumber for cross-bracing of foundation piling or posts.
Install	L.F.	1.05	4.36		5.41	
Demolish and Install	L.F.	1.05	5.07		6.12	
Remove and Install	L.F.		3.49		3.49	
Clean	S.F.		.27		.27	
Paint	L.F.	.21	1.06		1.27	
Minimum Charge	Job		525		525	
Treated 2" x 10"						
Demolish	L.F.		.73		.73	Cost includes material and labor to install 2" x 10" lumber for cross-bracing of foundation piling or posts.
Install	L.F.	1.55	4.51		6.06	
Demolish and Install	L.F.	1.55	5.24		6.79	
Remove and Install	L.F.		3.61		3.61	
Clean	S.F.		.27		.27	
Paint	L.F.	.25	1.08		1.33	
Minimum Charge	Job		525		525	
Treated 2" x 12"						
Demolish	L.F.		.88		.88	Cost includes material and labor to install 2" x 12" lumber for cross-bracing of foundation piling or posts.
Install	L.F.	1.99	4.67		6.66	
Demolish and Install	L.F.	1.99	5.55		7.54	
Remove and Install	L.F.		3.73		3.73	
Clean	S.F.		.27		.27	
Paint	L.F.	.30	1.10		1.40	
Minimum Charge	Job		525		525	
Metal Pipe						
6" Concrete Filled						
Demolish	L.F.		4.94	1.65	6.59	Includes material, labor and equipment to install concrete filled steel pipe pile.
Install	V.L.F.	11.65	4.43	3.88	19.96	
Demolish and Install	V.L.F.	11.65	9.37	5.53	26.55	
Paint	L.F.	.29	1.51		1.80	
Minimum Charge	Job		2125	1850	3975	
6" Unfilled						
Demolish	L.F.		4.94	1.65	6.59	Includes material, labor and equipment to install unfilled steel pipe pile.
Install	V.L.F.	9.90	4.05	3.55	17.50	
Demolish and Install	V.L.F.	9.90	8.99	5.20	24.09	
Paint	L.F.	.29	1.51		1.80	
Minimum Charge	Job		2125	1850	3975	
8" Concrete Filled						
Demolish	L.F.		4.94	1.65	6.59	Includes material, labor and equipment to install concrete filled steel pipe pile.
Install	V.L.F.	11.30	4.63	4.05	19.98	
Demolish and Install	V.L.F.	11.30	9.57	5.70	26.57	
Paint	L.F.	.29	1.51		1.80	
Minimum Charge	Job		2125	1850	3975	
8" Unfilled						
Demolish	L.F.		4.94	1.65	6.59	Includes material, labor and equipment to install unfilled steel pipe pile.
Install	V.L.F.	10.50	4.26	3.72	18.48	
Demolish and Install	V.L.F.	10.50	9.20	5.37	25.07	
Paint	L.F.	.29	1.51		1.80	
Minimum Charge	Job		2125	1850	3975	

Piles

Piles	Unit	Material	Labor	Equip.	Total	Specification
10" Concrete Filled						
Demolish	L.F.		5.10	1.70	6.80	Includes material, labor and equipment to install concrete filled steel pipe pile.
Install	V.L.F.	14.65	4.73	4.14	23.52	
Demolish and Install	V.L.F.	14.65	9.83	5.84	30.32	
Paint	L.F.	.29	1.51		1.80	
Minimum Charge	Job		2125	1850	3975	
10" Unfilled						
Demolish	L.F.		5.10	1.70	6.80	Includes material, labor and equipment to install unfilled steel pipe pile.
Install	V.L.F.	13.10	4.26	3.72	21.08	
Demolish and Install	V.L.F.	13.10	9.36	5.42	27.88	
Paint	L.F.	.29	1.51		1.80	
Minimum Charge	Job		2125	1850	3975	
12" Concrete Filled						
Demolish	L.F.		5.10	1.70	6.80	Includes material, labor and equipment to install concrete filled steel pipe pile.
Install	V.L.F.	17.15	5.15	4.49	26.79	
Demolish and Install	V.L.F.	17.15	10.25	6.19	33.59	
Paint	L.F.	.29	1.51		1.80	
Minimum Charge	Job		2125	1850	3975	
12" Unfilled						
Demolish	L.F.		5.10	1.70	6.80	Includes material, labor and equipment to install unfilled steel pipe pile.
Install	V.L.F.	16.25	4.48	3.92	24.65	
Demolish and Install	V.L.F.	16.25	9.58	5.62	31.45	
Paint	L.F.	.29	1.51		1.80	
Minimum Charge	Job		2125	1850	3975	
Steel H Section						
H Section 8 x 8 x 36#						
Demolish	L.F.		3.45	1.15	4.60	Includes material, labor and equipment to install steel H section pile.
Install	V.L.F.	9.25	3.33	2.91	15.49	
Demolish and Install	V.L.F.	9.25	6.78	4.06	20.09	
Paint	L.F.	.29	1.51		1.80	
Minimum Charge	Job		2125	1850	3975	
H Section 10 x 10 x 57#						
Demolish	L.F.		3.77	1.26	5.03	Includes material, labor and equipment to install steel H section pile.
Install	V.L.F.	14.65	3.49	3.05	21.19	
Demolish and Install	V.L.F.	14.65	7.26	4.31	26.22	
Paint	L.F.	.29	1.51		1.80	
Minimum Charge	Job		2125	1850	3975	
H Section 12 x 12 x 74#						
Demolish	L.F.		3.90	1.30	5.20	Includes material, labor and equipment to install steel H section pile.
Install	V.L.F.	19.25	4.20	3.15	26.60	
Demolish and Install	V.L.F.	19.25	8.10	4.45	31.80	
Paint	L.F.	.29	1.51		1.80	
Minimum Charge	Job		2125	1850	3975	
H Section 14 x 14 x 89#						
Demolish	L.F.		4.10	1.37	5.47	Includes material, labor and equipment to install steel H section pile.
Install	V.L.F.	23	4.59	3.44	31.03	
Demolish and Install	V.L.F.	23	8.69	4.81	36.50	
Paint	L.F.	.29	1.51		1.80	
Minimum Charge	Job		2125	1850	3975	

Foundation Post

Foundation Post	Unit	Material	Labor	Equip.	Total	Specification
Wood Foundation Post						
4" x 4"						
Demolish	L.F.		.56	.19	.75	Cost includes material and labor to install 4" x 4" treated wood post by hand in normal soil conditions.
Install	L.F.	.90	3.63		4.53	
Demolish and Install	L.F.	.90	4.19	.19	5.28	
Remove and Install	L.F.		2.91		2.91	
Minimum Charge	Job		525		525	

Foundation Post	Unit	Material	Labor	Equip.	Total	Specification
4" x 6"						
Demolish	L.F.		.56	.19	.75	Cost includes material and labor to install 4" x 6" treated wood post by hand in normal soil conditions.
Install	L.F.	1.36	3.63		4.99	
Demolish and Install	L.F.	1.36	4.19	.19	5.74	
Remove and Install	L.F.		2.91		2.91	
Minimum Charge	Job		525		525	
6" x 6"						
Demolish	L.F.		.56	.19	.75	Cost includes material and labor to install 6" x 6" treated wood post by hand in normal soil conditions.
Install	L.F.	2.02	3.63		5.65	
Demolish and Install	L.F.	2.02	4.19	.19	6.40	
Remove and Install	L.F.		2.91		2.91	
Minimum Charge	Job		525		525	
Water Jet						
Demolish	Ea.		64.50		64.50	Includes material, labor and equipment to water jet a 6" x 6" wood post under an existing building. Cost may vary depending upon location and access.
Install	Ea.	53.50	500		553.50	
Demolish and Install	Ea.	53.50	564.50		618	
Remove and Install	Ea.		498.29		498.29	
Minimum Charge	Job		435		435	
Shim						
Install	Ea.	2.97	52.50		55.47	Includes labor and material to trim or readjust existing foundation posts under an existing building. Access of 2 - 4 foot assumed.
Minimum Charge	Job		525		525	
Cross-bracing						
Treated 2" x 6"						
Demolish	L.F.		.71		.71	Cost includes material and labor to install 2" x 6" lumber for cross-bracing of foundation piling or posts.
Install	L.F.	.84	4.22		5.06	
Demolish and Install	L.F.	.84	4.93		5.77	
Remove and Install	L.F.		3.37		3.37	
Clean	S.F.		.27		.27	
Paint	L.F.	.17	1.02		1.19	
Minimum Charge	Job		525		525	
Treated 2" x 8"						
Demolish	L.F.		.71		.71	Cost includes material and labor to install 2" x 8" lumber for cross-bracing of foundation piling or posts.
Install	L.F.	1.05	4.36		5.41	
Demolish and Install	L.F.	1.05	5.07		6.12	
Remove and Install	L.F.		3.49		3.49	
Clean	S.F.		.27		.27	
Paint	L.F.	.21	1.06		1.27	
Minimum Charge	Job		525		525	
Treated 2" x 10"						
Demolish	L.F.		.73		.73	Cost includes material and labor to install 2" x 10" lumber for cross-bracing of foundation piling or posts.
Install	L.F.	1.55	4.51		6.06	
Demolish and Install	L.F.	1.55	5.24		6.79	
Remove and Install	L.F.		3.61		3.61	
Clean	S.F.		.27		.27	
Paint	L.F.	.25	1.08		1.33	
Minimum Charge	Job		525		525	
Treated 2" x 12"						
Demolish	L.F.		.88		.88	Cost includes material and labor to install 2" x 12" lumber for cross-bracing of foundation piling or posts.
Install	L.F.	1.99	4.67		6.66	
Demolish and Install	L.F.	1.99	5.55		7.54	
Remove and Install	L.F.		3.73		3.73	
Clean	S.F.		.27		.27	
Paint	L.F.	.30	1.10		1.40	
Minimum Charge	Job		525		525	

Concrete Footing

	Unit	Material	Labor	Equip.	Total	Specification
Continuous						
12" w x 6" d						
Demolish	L.F.		1.06	.93	1.99	Includes material and labor to install continuous reinforced concrete footing poured by chute cast against earth including reinforcing, forming and finishing.
Install	L.F.	3.07	2.48	.02	5.57	
Demolish and Install	L.F.	3.07	3.54	.95	7.56	
Minimum Charge	Job		600		600	
12" w x 12" d						
Demolish	L.F.		2.05	1.78	3.83	Includes material and labor to install continuous reinforced concrete footing poured by chute cast against earth including reinforcing, forming and finishing.
Install	L.F.	7.55	4.97	.04	12.56	
Demolish and Install	L.F.	7.55	7.02	1.82	16.39	
Minimum Charge	Job		600		600	
18" w x 10" d						
Demolish	L.F.		2.66	2.31	4.97	Includes material and labor to install continuous reinforced concrete footing poured by chute cast against earth including reinforcing, forming and finishing.
Install	L.F.	6.95	6.10	.06	13.11	
Demolish and Install	L.F.	6.95	8.76	2.37	18.08	
Minimum Charge	Job		600		600	
24" w x 24" d						
Demolish	L.F.		4.43	3.85	8.28	Includes material and labor to install continuous reinforced concrete footing poured by chute cast against earth including reinforcing, forming and finishing.
Install	L.F.	20.50	19.85	.18	40.53	
Demolish and Install	L.F.	20.50	24.28	4.03	48.81	
Minimum Charge	Job		600		600	
Stem Wall						
Single Story						
Demolish	L.F.		9.85	8.55	18.40	Includes material and labor to install stem wall 24" above and 18" below grade for a single story structure. Includes 12" wide and 8" deep footing.
Install	L.F.	9.55	52	45.50	107.05	
Demolish and Install	L.F.	9.55	61.85	54.05	125.45	
Minimum Charge	Job		600		600	
Two Story						
Demolish	L.F.		9.85	8.55	18.40	Includes material and labor to install stem wall 24" above and 18" below grade typical for a two story structure, includes 18" wide and 10" deep footing.
Install	L.F.	11.90	75	65	151.90	
Demolish and Install	L.F.	11.90	84.85	73.55	170.30	
Minimum Charge	Job		600		600	

Concrete Slab

	Unit	Material	Labor	Equip.	Total	Specification
Reinforced						
4"						
Demolish	S.F.		1.84	.30	2.14	Cost includes material and labor to install 4" reinforced concrete slab-on-grade poured by chute including forms, vapor barrier, wire mesh, 3000 PSI concrete, float finish, and curing.
Install	S.F.	1.60	.39		1.99	
Demolish and Install	S.F.	1.60	2.23	.30	4.13	
Clean	S.F.	.03	.24		.27	
Paint	S.F.	.12	.57		.69	
Minimum Charge	Job		137		137	

Concrete Slab	Unit	Material	Labor	Equip.	Total	Specification
6″						
Demolish	S.F.		2.76	.46	3.22	Cost includes material and labor to install 6″ reinforced concrete slab-on-grade poured by chute including forms, vapor barrier, wire mesh, 3000 PSI concrete, float finish, and curing.
Install	S.F.	2.20	.59	.01	2.80	
Demolish and Install	S.F.	2.20	3.35	.47	6.02	
Clean	S.F.	.03	.24		.27	
Paint	S.F.	.12	.57		.69	
Minimum Charge	Job		137		137	
8″						
Demolish	S.F.		4.51	.75	5.26	Cost includes material and labor to install 8″ reinforced concrete slab-on-grade poured by chute including forms, vapor barrier, wire mesh, 3000 PSI concrete, float finish, and curing.
Install	S.F.	2.84	.77	.01	3.62	
Demolish and Install	S.F.	2.84	5.28	.76	8.88	
Clean	S.F.	.03	.24		.27	
Paint	S.F.	.12	.57		.69	
Minimum Charge	Job		137		137	
Unreinforced						
4″						
Demolish	S.F.		1.46	.24	1.70	Cost includes material and labor to install 4″ concrete slab-on-grade poured by chute, including forms, vapor barrier, 3000 PSI concrete, float finish, and curing.
Install	S.F.	1.05	.68	.01	1.74	
Demolish and Install	S.F.	1.05	2.14	.25	3.44	
Clean	S.F.	.03	.24		.27	
Paint	S.F.	.12	.57		.69	
Minimum Charge	Job		137		137	
6″						
Demolish	S.F.		2.07	.34	2.41	Cost includes material and labor to install 6″ concrete slab-on-grade poured by chute, including forms, vapor barrier, 3000 PSI concrete, float finish, and curing.
Install	S.F.	1.53	.69	.01	2.23	
Demolish and Install	S.F.	1.53	2.76	.35	4.64	
Clean	S.F.	.03	.24		.27	
Paint	S.F.	.12	.57		.69	
Minimum Charge	Job		137		137	
8″						
Demolish	S.F.		3.40	.56	3.96	Cost includes material and labor to install 8″ concrete slab-on-grade poured by chute, including forms, vapor barrier, 3000 PSI concrete, float finish, and curing.
Install	S.F.	2.10	.73	.01	2.84	
Demolish and Install	S.F.	2.10	4.13	.57	6.80	
Clean	S.F.	.03	.24		.27	
Paint	S.F.	.12	.57		.69	
Minimum Charge	Job		137		137	
Lightweight						
Demolish	S.F.		.65	.19	.84	Includes material and labor to install lightweight concrete, 3″ - 4″ thick.
Install	S.F.	1.39	2.04		3.43	
Demolish and Install	S.F.	1.39	2.69	.19	4.27	
Clean	S.F.	.03	.24		.27	
Paint	S.F.	.12	.57		.69	
Minimum Charge	Job		137		137	
Gunite						
Demolish	S.F.		.80	.27	1.07	Includes material and labor to install gunite on flat plane per inch of thickness using 1″ as base price. No forms or reinforcing are included.
Install	S.F.	1.06	.87	.41	2.34	
Demolish and Install	S.F.	1.06	1.67	.68	3.41	
Clean	S.F.	.03	.24		.27	
Paint	S.F.	.12	.57		.69	
Minimum Charge	Job		765		765	
Epoxy Inject						
Install	L.F.	1.97	2.74		4.71	Includes labor and material to inject epoxy into cracks.
Minimum Charge	Job		137		137	

Concrete Slab		Unit	Material	Labor	Equip.	Total	Specification
Pressure Grout							
	Install	S.F.	2.44	2.03		4.47	Includes labor, material and equipment to install pressure grout under an existing building.
	Minimum Charge	Job		137		137	
Scrape and Paint							
	Paint	S.F.	.17	.66		.83	Includes labor and materials to scrape and paint a concrete floor.
	Minimum Charge	Job		132		132	
Vapor Barrier							
	Install	S.F.	.02	.08		.10	Cost includes material and labor to install 4 mil polyethylene vapor barrier 10′ wide sheets with 6″ overlaps.
	Minimum Charge	Job		150		150	

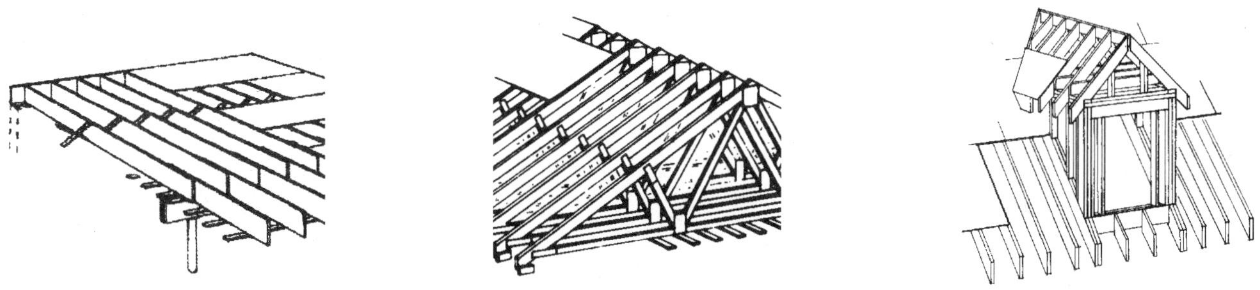

Rough Framing

Beam / Girder	Unit	Material	Labor	Equip.	Total	Specification
Solid Wood Beam						
2″ x 8″						
Demolish	L.F.		.59		.59	Cost includes material and labor to install 2″ x 8″ beam.
Install	L.F.	.88	.93		1.81	
Demolish and Install	L.F.	.88	1.52		2.40	
Clean	L.F.	.09	.19		.28	
Paint	L.F.	.12	1.43		1.55	
Minimum Charge	Job		150		150	
2″ x 10″						
Demolish	L.F.		.73		.73	Cost includes material and labor to install 2″ x 10″ beam.
Install	L.F.	1.27	1		2.27	
Demolish and Install	L.F.	1.27	1.73		3	
Clean	L.F.	.11	.23		.34	
Paint	L.F.	.15	.88		1.03	
Minimum Charge	Job		150		150	
2″ x 12″						
Demolish	L.F.		.88		.88	Cost includes material and labor to install 2″ x 12″ beam.
Install	L.F.	1.72	1.09		2.81	
Demolish and Install	L.F.	1.72	1.97		3.69	
Clean	L.F.	.12	.27		.39	
Paint	L.F.	.18	2.11		2.29	
Minimum Charge	Job		150		150	
4″ x 8″						
Demolish	L.F.		3.01		3.01	Cost includes material and labor to install 4″ x 8″ beam.
Install	L.F.	3.83	1.35	.70	5.88	
Demolish and Install	L.F.	3.83	4.36	.70	8.89	
Clean	L.F.	.11	.23		.34	
Paint	L.F.	.15	.88		1.03	
Minimum Charge	Job		150		150	
4″ x 10″						
Demolish	L.F.		3.76		3.76	Cost includes material and labor to install 4″ x 10″ beam.
Install	L.F.	4.79	1.42	.74	6.95	
Demolish and Install	L.F.	4.79	5.18	.74	10.71	
Clean	L.F.	.12	.27		.39	
Paint	L.F.	.18	1.03		1.21	
Minimum Charge	Job		150		150	
4″ x 12″						
Demolish	L.F.		4.51		4.51	Cost includes material and labor to install 4″ x 12″ beam.
Install	L.F.	5.70	1.50	.78	7.98	
Demolish and Install	L.F.	5.70	6.01	.78	12.49	
Clean	L.F.	.14	.31		.45	
Paint	L.F.	.18	2.11		2.29	
Minimum Charge	Job		150		150	

Beam / Girder		Unit	Material	Labor	Equip.	Total	Specification
6″ x 8″							
	Demolish	L.F.		4.51		4.51	Cost includes material and labor to install 6″ x 8″ beam.
	Install	L.F.	8.25	2.58	1.33	12.16	
	Demolish and Install	L.F.	8.25	7.09	1.33	16.67	
	Clean	L.F.	.12	.27		.39	
	Paint	L.F.	.18	1.03		1.21	
	Minimum Charge	Job		150		150	
6″ x 10″							
	Demolish	L.F.		5.65		5.65	Cost includes material and labor to install 6″ x 10″ triple beam.
	Install	L.F.	3.78	1.20		4.98	
	Demolish and Install	L.F.	3.78	6.85		10.63	
	Clean	L.F.	.14	.31		.45	
	Paint	L.F.	.20	1.17		1.37	
	Minimum Charge	Job		150		150	
6″ x 12″							
	Demolish	L.F.		6.70		6.70	Cost includes material and labor to install 6″ x 12″ triple beam.
	Install	L.F.	5.15	1.27		6.42	
	Demolish and Install	L.F.	5.15	7.97		13.12	
	Clean	L.F.	.17	.35		.52	
	Paint	L.F.	.22	1.32		1.54	
	Minimum Charge	Job		150		150	
Steel I-type							
W8 x 31							
	Demolish	L.F.		6.35	3.20	9.55	Includes material and labor to install W shaped steel beam / girder.
	Install	L.F.	23.50	2.05	1.23	26.78	
	Demolish and Install	L.F.	23.50	8.40	4.43	36.33	
	Minimum Charge	Job		375		375	
W8 x 48							
	Demolish	L.F.		6.35	3.20	9.55	Includes material and labor to install W shaped steel beam / girder.
	Install	L.F.	37	2.14	1.28	40.42	
	Demolish and Install	L.F.	37	8.49	4.48	49.97	
	Minimum Charge	Job		375		375	
W8 x 67							
	Demolish	L.F.		6.35	3.20	9.55	Includes material and labor to install W shaped steel beam / girder.
	Install	L.F.	51.50	2.24	1.34	55.08	
	Demolish and Install	L.F.	51.50	8.59	4.54	64.63	
	Minimum Charge	Job		375		375	
W10 x 45							
	Demolish	L.F.		6.35	3.20	9.55	Includes material and labor to install W shaped steel beam / girder.
	Install	L.F.	34.50	2.14	1.28	37.92	
	Demolish and Install	L.F.	34.50	8.49	4.48	47.47	
	Minimum Charge	Job		375		375	
W10 x 68							
	Demolish	L.F.		6.35	3.20	9.55	Includes material and labor to install W shaped steel beam / girder.
	Install	L.F.	52.50	2.24	1.34	56.08	
	Demolish and Install	L.F.	52.50	8.59	4.54	65.63	
	Minimum Charge	Job		375		375	
W12 x 50							
	Demolish	L.F.		6.35	3.20	9.55	Includes material and labor to install W shaped steel beam / girder.
	Install	L.F.	38.50	2.14	1.28	41.92	
	Demolish and Install	L.F.	38.50	8.49	4.48	51.47	
	Minimum Charge	Job		375		375	
W12 x 87							
	Demolish	L.F.		9.30	4.71	14.01	Includes material and labor to install W shaped steel beam / girder.
	Install	L.F.	67	2.24	1.34	70.58	
	Demolish and Install	L.F.	67	11.54	6.05	84.59	
	Minimum Charge	Job		375		375	

Rough Frame / Structure

Beam / Girder	Unit	Material	Labor	Equip.	Total	Specification
W12 x 120						
Demolish	L.F.		9.30	4.71	14.01	Includes material and labor to install W shaped steel beam / girder.
Install	L.F.	92.50	2.30	1.38	96.18	
Demolish and Install	L.F.	92.50	11.60	6.09	110.19	
Minimum Charge	Job		375		375	
W14 x 74						
Demolish	L.F.		9.30	4.71	14.01	Includes material and labor to install W shaped steel beam / girder.
Install	L.F.	57	2.24	1.34	60.58	
Demolish and Install	L.F.	57	11.54	6.05	74.59	
Minimum Charge	Job		375		375	
W14 x 120						
Demolish	L.F.		9.30	4.71	14.01	Includes material and labor to install W shaped steel beam / girder.
Install	L.F.	92.50	2.30	1.38	96.18	
Demolish and Install	L.F.	92.50	11.60	6.09	110.19	
Minimum Charge	Job		375		375	

Column	Unit	Material	Labor	Equip.	Total	Specification
Concrete						
Small Diameter						
Demolish	L.F.		10.35	1.71	12.06	Includes material and labor to install small diameter concrete column.
Install	L.F.	62	9.40	2.24	73.64	
Demolish and Install	L.F.	62	19.75	3.95	85.70	
Minimum Charge	Job		470	112	582	
Large Diameter						
Demolish	L.F.		31	5.10	36.10	Includes material and labor to install large diameter concrete column.
Install	L.F.	182	11.05	2.64	195.69	
Demolish and Install	L.F.	182	42.05	7.74	231.79	
Minimum Charge	Job		470	112	582	

Floor Framing System	Unit	Material	Labor	Equip.	Total	Specification
12″ O.C.						
2″ x 6″ Joists						
Demolish	S.F.		.50		.50	Cost includes material and labor to install 2″ x 6″ joists, 12″ O.C. including box or band joist. Blocking or bridging not included.
Install	S.F.	.67	.24		.91	
Demolish and Install	S.F.	.67	.74		1.41	
Remove and Install	S.F.		.19		.19	
Minimum Charge	Job		150		150	
2″ x 8″ Joists						
Demolish	S.F.		.52		.52	Cost includes material and labor to install 2″ x 8″ joists, 12″ O.C. including box or band joist. Blocking or bridging not included.
Install	S.F.	1.03	.27		1.30	
Demolish and Install	S.F.	1.03	.79		1.82	
Remove and Install	S.F.		.22		.22	
Minimum Charge	Job		150		150	
2″ x 10″ Joists						
Demolish	S.F.		.53		.53	Cost includes material and labor to install 2″ x 10″ joists, 12″ O.C. including box or band joist. Blocking or bridging not included.
Install	S.F.	1.47	.33		1.80	
Demolish and Install	S.F.	1.47	.86		2.33	
Remove and Install	S.F.		.27		.27	
Minimum Charge	Job		150		150	

Floor Framing System		Unit	Material	Labor	Equip.	Total	Specification
2″ x 12″ Joists							
	Demolish	S.F.		.55		.55	Cost includes material and labor to install 2″ x 12″ joists, 12″ O.C. including box or band joist. Blocking or bridging not included.
	Install	S.F.	2.01	.34		2.35	
	Demolish and Install	S.F.	2.01	.89		2.90	
	Remove and Install	S.F.		.28		.28	
	Minimum Charge	Job		150		150	
Block / Bridge							
	Demolish	Ea.		.76		.76	Includes material and labor to install set of cross bridging or per block of solid bridging for 2″ x 10″ joists cut to size on site.
	Install	Ea.	1.69	1.69		3.38	
	Demolish and Install	Ea.	1.69	2.45		4.14	
	Minimum Charge	Job		150		150	
16″ O.C.							
2″ x 6″ Joists							
	Demolish	S.F.		.37		.37	Cost includes material and labor to install 2″ x 6″ joists including box or band joist installed 16″ O.C. Does not include beams, blocking or bridging.
	Install	S.F.	.52	.36		.88	
	Demolish and Install	S.F.	.52	.73		1.25	
	Remove and Install	S.F.		.29		.29	
	Minimum Charge	Job		150		150	
2″ x 8″ Joists							
	Demolish	S.F.		.39		.39	Cost includes material and labor to install 2″ x 8″ joists including box or band joist installed 16″ O.C. Does not include beams, blocking or bridging.
	Install	S.F.	.79	.41		1.20	
	Demolish and Install	S.F.	.79	.80		1.59	
	Remove and Install	S.F.		.33		.33	
	Minimum Charge	Job		150		150	
2″ x 10″ Joists							
	Demolish	S.F.		.40		.40	Cost includes material and labor to install 2″ x 10″ joists including box or band joist 16″ O.C. Does not include beams, blocking or bridging.
	Install	S.F.	1.13	.50		1.63	
	Demolish and Install	S.F.	1.13	.90		2.03	
	Remove and Install	S.F.		.40		.40	
	Minimum Charge	Job		150		150	
2″ x 12″ Joists							
	Demolish	S.F.		.41		.41	Cost includes material and labor to install 2″ x 12″ joists including box or band joist installed 16″ O.C. Does not include beams, blocking or bridging.
	Install	S.F.	1.55	.52		2.07	
	Demolish and Install	S.F.	1.55	.93		2.48	
	Remove and Install	S.F.		.41		.41	
	Minimum Charge	Job		150		150	
Block / Bridge							
	Demolish	Ea.		.76		.76	Includes material and labor to install set of cross bridging or per block of solid bridging for 2″ x 10″ joists cut to size on site.
	Install	Ea.	1.69	1.69		3.38	
	Demolish and Install	Ea.	1.69	2.45		4.14	
	Minimum Charge	Job		150		150	
24″ O.C.							
2″ x 6″ Joists							
	Demolish	S.F.		.25		.25	Cost includes material and labor to install 2″ x 6″ joists including box or band joist installed 24″ O.C. Does not include beams, blocking or bridging.
	Install	S.F.	.36	.12		.48	
	Demolish and Install	S.F.	.36	.37		.73	
	Remove and Install	S.F.		.10		.10	
	Minimum Charge	Job		150		150	
2″ x 8″ Joists							
	Demolish	S.F.		.26		.26	Cost includes material and labor to install 2″ x 8″ joists including box or band joist installed 24″ O.C. Does not include beams, blocking or bridging.
	Install	S.F.	.56	.14		.70	
	Demolish and Install	S.F.	.56	.40		.96	
	Remove and Install	S.F.		.11		.11	
	Minimum Charge	Job		150		150	
2″ x 10″ Joists							
	Demolish	S.F.		.27		.27	Cost includes material and labor to install 2″ x 10″ joists including box or band joist installed 24″ O.C. Does not include beams, blocking or bridging.
	Install	S.F.	.80	.17		.97	
	Demolish and Install	S.F.	.80	.44		1.24	
	Remove and Install	S.F.		.13		.13	
	Minimum Charge	Job		150		150	

Floor Framing System	Unit	Material	Labor	Equip.	Total	Specification
2″ x 12″ Joists						
Demolish	S.F.		.27		.27	Cost includes material and labor to install 2″ x 12″ joists including box or band joist installed 24″ O.C. Does not include beams, blocking or bridging.
Install	S.F.	1.10	.17		1.27	
Demolish and Install	S.F.	1.10	.44		1.54	
Remove and Install	S.F.		.14		.14	
Minimum Charge	Job		150		150	
Block / Bridge						
Demolish	Ea.		.76		.76	Includes material and labor to install set of cross bridging or per block of solid bridging for 2″ x 10″ joists cut to size on site.
Install	Ea.	1.69	1.69		3.38	
Demolish and Install	Ea.	1.69	2.45		4.14	
Minimum Charge	Job		150		150	
Engineered Lumber, Joist						
9-1/2″						
Demolish	S.F.		.40		.40	Includes material and labor to install engineered lumber truss / joists per S.F. of floor area based on joists 16″ O.C. Beams, supports and bridging are not included.
Install	S.F.	1.54	.30		1.84	
Demolish and Install	S.F.	1.54	.70		2.24	
Remove and Install	S.F.		.24		.24	
Minimum Charge	Job		150		150	
11-7/8″						
Demolish	S.F.		.41		.41	Includes material and labor to install engineered lumber truss / joists per S.F. of floor area based on joists 16″ O.C. Beams, supports and bridging are not included.
Install	S.F.	1.66	.31		1.97	
Demolish and Install	S.F.	1.66	.72		2.38	
Remove and Install	S.F.		.25		.25	
Minimum Charge	Job		150		150	
14″						
Demolish	S.F.		.43		.43	Includes material and labor to install engineered lumber truss / joists per S.F. of floor area based on joists 16″ O.C. Beams, supports and bridging are not included.
Install	S.F.	1.82	.33		2.15	
Demolish and Install	S.F.	1.82	.76		2.58	
Remove and Install	S.F.		.26		.26	
Minimum Charge	Job		150		150	
16″						
Demolish	S.F.		.44		.44	Includes material and labor to install engineered lumber truss / joists per S.F. of floor area based on joists 16″ O.C. Beams, supports and bridging are not included.
Install	S.F.	2.27	.35		2.62	
Demolish and Install	S.F.	2.27	.79		3.06	
Remove and Install	S.F.		.28		.28	
Minimum Charge	Job		150		150	
Block / Bridge						
Demolish	Pr.		.76		.76	Includes material and labor to install set of steel, one-nail type cross bridging for trusses placed 16″ O.C.
Install	Pr.	1.52	2.23		3.75	
Demolish and Install	Pr.	1.52	2.99		4.51	
Minimum Charge	Job		150		150	
Block / Bridge						
2″ x 6″						
Demolish	Ea.		.76		.76	Includes material and labor to install set of cross bridging or per block of solid bridging for 2″ x 6″ joists cut to size on site.
Install	Ea.	.77	1.36		2.13	
Demolish and Install	Ea.	.77	2.12		2.89	
Minimum Charge	Job		150		150	
2″ x 8″						
Demolish	Ea.		.76		.76	Includes material and labor to install set of cross bridging or per block of solid bridging for 2″ x 8″ joists cut to size on site.
Install	Ea.	1.18	1.50		2.68	
Demolish and Install	Ea.	1.18	2.26		3.44	
Minimum Charge	Job		150		150	

Floor Framing System	Unit	Material	Labor	Equip.	Total	Specification
2" x 10"						
Demolish	Ea.		.76		.76	Includes material and labor to install set of cross bridging or per block of solid bridging for 2" x 10" joists cut to size on site.
Install	Ea.	1.69	1.69		3.38	
Demolish and Install	Ea.	1.69	2.45		4.14	
Minimum Charge	Job		150		150	
2" x 12"						
Demolish	Ea.		.76		.76	Includes material and labor to install set of cross bridging or per block of solid bridging for 2" x 12" joists cut to size on site.
Install	Ea.	2.31	1.99		4.30	
Demolish and Install	Ea.	2.31	2.75		5.06	
Minimum Charge	Job		150		150	
Ledger Strips						
1" x 2"						
Demolish	L.F.		.20		.20	Cost includes material and labor to install 1" x 2" ledger strip nailed to the face of studs, beams or joist.
Install	L.F.	.22	.87		1.09	
Demolish and Install	L.F.	.22	1.07		1.29	
Minimum Charge	Job		150		150	
1" x 3"						
Demolish	L.F.		.20		.20	Includes material and labor to install up to 1" x 4" ledger strip nailed to the face of studs, beams or joist.
Install	L.F.	.66	1		1.66	
Demolish and Install	L.F.	.66	1.20		1.86	
Minimum Charge	Job		150		150	
1" x 4"						
Demolish	L.F.		.20		.20	Includes material and labor to install up to 1" x 4" ledger strip nailed to the face of studs, beams or joist.
Install	L.F.	.66	1		1.66	
Demolish and Install	L.F.	.66	1.20		1.86	
Minimum Charge	Job		150		150	
2" x 2"						
Demolish	L.F.		.22		.22	Cost includes material and labor to install 2" x 2" ledger strip nailed to the face of studs, beams or joist.
Install	L.F.	.28	.91		1.19	
Demolish and Install	L.F.	.28	1.13		1.41	
Minimum Charge	Job		150		150	
2" x 4"						
Demolish	L.F.		.24		.24	Cost includes material and labor to install 2" x 4" ledger strip nailed to the face of studs, beams or joist.
Install	L.F.	.36	1.20		1.56	
Demolish and Install	L.F.	.36	1.44		1.80	
Remove and Install	L.F.		.96		.96	
Minimum Charge	Job		150		150	
Ledger Boards						
2" x 6"						
Demolish	L.F.		.24		.24	Cost includes material and labor to install 2" x 6" ledger board fastened to a wall, joists or studs.
Install	L.F.	1.55	1.88		3.43	
Demolish and Install	L.F.	1.55	2.12		3.67	
Remove and Install	L.F.		1.50		1.50	
Minimum Charge	Job		150		150	
4" x 6"						
Demolish	L.F.		.40		.40	Cost includes material and labor to install 4" x 6" ledger board fastened to a wall, joists or studs.
Install	L.F.	4.07	1.88		5.95	
Demolish and Install	L.F.	4.07	2.28		6.35	
Remove and Install	L.F.		1.50		1.50	
Minimum Charge	Job		150		150	
4" x 8"						
Demolish	L.F.		.54		.54	Cost includes material and labor to install 4" x 8" ledger board fastened to a wall, joists or studs.
Install	L.F.	5.05	1.88		6.93	
Demolish and Install	L.F.	5.05	2.42		7.47	
Remove and Install	L.F.		1.50		1.50	
Minimum Charge	Job		150		150	

Floor Framing System	Unit	Material	Labor	Equip.	Total	Specification
Sill Plate Per L.F.						
2" x 4"						
Demolish	L.F.		.26		.26	Cost includes material and labor to install 2" x 4" pressure treated lumber, drilled and installed on foundation bolts 48" O.C. Bolts, nuts and washers are not included.
Install	L.F.	.46	1.09		1.55	
Demolish and Install	L.F.	.46	1.35		1.81	
Remove and Install	L.F.		.88		.88	
Minimum Charge	Job		150		150	
2" x 6"						
Demolish	L.F.		.26		.26	Cost includes material and labor to install 2" x 6" pressure treated lumber, drilled and installed on foundation bolts 48" O.C. Bolts, nuts and washers not included.
Install	L.F.	.84	1.20		2.04	
Demolish and Install	L.F.	.84	1.46		2.30	
Remove and Install	L.F.		.96		.96	
Minimum Charge	Job		150		150	
2" x 8"						
Demolish	L.F.		.26		.26	Cost includes material and labor to install 2" x 8" pressure treated lumber, drilled and installed on foundation bolts 48" O.C. Bolts, nuts and washers not included.
Install	L.F.	1.05	1.34		2.39	
Demolish and Install	L.F.	1.05	1.60		2.65	
Remove and Install	L.F.		1.07		1.07	
Minimum Charge	Job		150		150	
Earthquake Strapping						
Install	Ea.	1.68	1.88		3.56	Includes labor and material to replace an earthquake strap.
Minimum Charge	Job		150		150	

Subflooring	Unit	Material	Labor	Equip.	Total	Specification
Plywood						
1/4"						
Demolish	S.F.		.32		.32	Cost includes material and labor to install 1/4" lauan subfloor, standard interior grade, nailed every 6".
Install	S.F.	.50	.40		.90	
Demolish and Install	S.F.	.50	.72		1.22	
Minimum Charge	Job		150		150	
1/2"						
Demolish	S.F.		.32		.32	Cost includes material and labor to install 1/2" plywood subfloor, CD standard interior grade, plugged and touch sanded.
Install	SF Flr.	.41	.32		.73	
Demolish and Install	SF Flr.	.41	.64		1.05	
Minimum Charge	Job		150		150	
5/8"						
Demolish	S.F.		.32		.32	Cost includes material and labor to install 5/8" plywood subfloor, CD standard interior grade, plugged and touch sanded.
Install	SF Flr.	.48	.36		.84	
Demolish and Install	SF Flr.	.48	.68		1.16	
Remove and Install	SF Flr.		.29		.29	
Minimum Charge	Job		150		150	
3/4"						
Demolish	S.F.		.32		.32	Cost includes material and labor to install 3/4" plywood subfloor, CD standard interior grade, plugged and touch sanded.
Install	SF Flr.	.55	.39		.94	
Demolish and Install	SF Flr.	.55	.71		1.26	
Remove and Install	SF Flr.		.31		.31	
Minimum Charge	Job		150		150	
Particle Board						
3/8"						
Demolish	S.F.		.31		.31	Cost includes material and labor to install 3/8" particle board subfloor.
Install	S.F.	.42	.19		.61	
Demolish and Install	S.F.	.42	.50		.92	
Minimum Charge	Job		150		150	

Subflooring		Unit	Material	Labor	Equip.	Total	Specification
1/2"							
	Demolish	S.F.		.32		.32	Cost includes material and labor to install 1/2" particle board subfloor.
	Install	SF Flr.	.44	.40		.84	
	Demolish and Install	SF Flr.	.44	.72		1.16	
	Minimum Charge	Job		150		150	
5/8"							
	Demolish	S.F.		.32		.32	Cost includes material and labor to install 5/8" particle board subfloor.
	Install	S.F.	.59	.22		.81	
	Demolish and Install	S.F.	.59	.54		1.13	
	Minimum Charge	Job		150		150	
3/4"							
	Demolish	S.F.		.32		.32	Cost includes material and labor to install 3/4" particle board subfloor.
	Install	SF Flr.	.64	.48		1.12	
	Demolish and Install	SF Flr.	.64	.80		1.44	
	Minimum Charge	Job		150		150	
Plank Board							
	Demolish	S.F.		1.03		1.03	Cost includes material and labor to install 1" x 6" standard grade plank flooring.
	Install	S.F.	1.20	.33		1.53	
	Demolish and Install	S.F.	1.20	1.36		2.56	
	Remove and Install	S.F.		.33		.33	
	Minimum Charge	Job		150		150	
Felt Underlay							
	Demolish	S.F.		.06		.06	Includes material and labor to install #15 felt building paper.
	Install	S.F.	.02	.08		.10	
	Demolish and Install	S.F.	.02	.14		.16	
	Minimum Charge	Job		150		150	
Prep (for flooring)							
	Install	S.F.		.26		.26	Preparation of subflooring for installation of new finished flooring. Cost reflects average time to prep area.
	Minimum Charge	Job		138		138	

Wall Framing System		Unit	Material	Labor	Equip.	Total	Specification
2" x 4"							
8' High							
	Demolish	L.F.		3.23		3.23	Cost includes material and labor to install 2" x 4" x 8' high wall system, including studs, treated bottom plate, double top plate and one row of fire blocking.
	Install	L.F.	2.63	3.54		6.17	
	Demolish and Install	L.F.	2.63	6.77		9.40	
	Remove and Install	L.F.		2.83		2.83	
	Minimum Charge	Job		150		150	
9' High							
	Demolish	L.F.		3.62		3.62	Cost includes material and labor to install 2" x 4" x 9' wall system including studs, treated bottom plate, double top plate and one row of fire blocking.
	Install	L.F.	2.89	3.54		6.43	
	Demolish and Install	L.F.	2.89	7.16		10.05	
	Remove and Install	L.F.		2.83		2.83	
	Minimum Charge	Job		150		150	
10' High							
	Demolish	L.F.		4.04		4.04	Cost includes material and labor to install 2" x 4" x 10' wall system including studs, treated bottom plate, double top plate and one row of fire blocking.
	Install	L.F.	3.17	3.54		6.71	
	Demolish and Install	L.F.	3.17	7.58		10.75	
	Remove and Install	L.F.		2.83		2.83	
	Minimum Charge	Job		150		150	

Wall Framing System	Unit	Material	Labor	Equip.	Total	Specification
12' High						
Demolish	L.F.		4.84		4.84	Cost includes material and labor to install 2" x 4" x 12' wall system including studs, treated bottom plate, double top plate and one row of fire blocking.
Install	L.F.	3.71	4.30		8.01	
Demolish and Install	L.F.	3.71	9.14		12.85	
Remove and Install	L.F.		3.44		3.44	
Minimum Charge	Job		150		150	
2" x 6"						
8' High						
Demolish	L.F.		3.03		3.03	Cost includes material and labor to install 2" x 6" x 8' wall system including studs, treated bottom plate, double top plate and one row of fire blocking.
Install	L.F.	4.28	6.55		10.83	
Demolish and Install	L.F.	4.28	9.58		13.86	
Remove and Install	L.F.		5.23		5.23	
Minimum Charge	Job		150		150	
9' High						
Demolish	L.F.		3.41		3.41	Cost includes material and labor to install 2" x 6" x 9' wall system including studs, treated bottom plate, double top plate and one row of fire blocking.
Install	L.F.	4.71	6.55		11.26	
Demolish and Install	L.F.	4.71	9.96		14.67	
Remove and Install	L.F.		5.23		5.23	
Minimum Charge	Job		150		150	
10' High						
Demolish	L.F.		3.73		3.73	Cost includes material and labor to install 2" x 6" x 10' wall system including studs, treated bottom plate, double top plate and one row of fire blocking.
Install	L.F.	5.15	6.55		11.70	
Demolish and Install	L.F.	5.15	10.28		15.43	
Remove and Install	L.F.		5.23		5.23	
Minimum Charge	Job		150		150	
12' High						
Demolish	L.F.		4.40		4.40	Cost includes material and labor to install 2" x 6" x 12' wall system including studs, treated bottom plate, double top plate and one row of fire blocking.
Install	L.F.	6	8.05		14.05	
Demolish and Install	L.F.	6	12.45		18.45	
Remove and Install	L.F.		6.44		6.44	
Minimum Charge	Job		150		150	
Fireblock						
2" x 4", 16" O.C. System						
Demolish	L.F.		.30		.30	Cost includes material and labor to install 2" x 4" fireblocks in wood frame walls per L.F. of wall to be blocked.
Install	L.F.	.36	1		1.36	
Demolish and Install	L.F.	.36	1.30		1.66	
Remove and Install	L.F.		.80		.80	
Minimum Charge	Job		150		150	
2" x 4", 24" O.C. System						
Demolish	L.F.		.30		.30	Cost includes material and labor to install 2" x 4" fireblocks in wood frame walls per L.F. of wall to be blocked.
Install	L.F.	.36	1		1.36	
Demolish and Install	L.F.	.36	1.30		1.66	
Remove and Install	L.F.		.80		.80	
Minimum Charge	Job		150		150	
2" x 6", 16" O.C. System						
Demolish	L.F.		.30		.30	Cost includes material and labor to install 2" x 6" fireblocks in wood frame walls per L.F. of wall to be blocked.
Install	L.F.	.57	1		1.57	
Demolish and Install	L.F.	.57	1.30		1.87	
Remove and Install	L.F.		.80		.80	
Minimum Charge	Job		150		150	
2" x 6", 24" O.C. System						
Demolish	L.F.		.30		.30	Cost includes material and labor to install 2" x 6" fireblocks in wood frame walls per L.F. of wall to be blocked.
Install	L.F.	.57	1		1.57	
Demolish and Install	L.F.	.57	1.30		1.87	
Remove and Install	L.F.		.80		.80	
Minimum Charge	Job		150		150	

Wall Framing System	Unit	Material	Labor	Equip.	Total	Specification
Bracing						
1" x 3"						
Demolish	L.F.		.23		.23	Cost includes material and labor to install 1" x 3" let-in bracing.
Install	L.F.	.21	1.88		2.09	
Demolish and Install	L.F.	.21	2.11		2.32	
Minimum Charge	Job		150		150	
1" x 4"						
Demolish	L.F.		.23		.23	Cost includes material and labor to install 1" x 4" let-in bracing.
Install	L.F.	.39	1.88		2.27	
Demolish and Install	L.F.	.39	2.11		2.50	
Minimum Charge	Job		150		150	
1" x 6"						
Demolish	L.F.		.23		.23	Cost includes material and labor to install 1" x 6" let-in bracing.
Install	L.F.	.55	2.01		2.56	
Demolish and Install	L.F.	.55	2.24		2.79	
Minimum Charge	Job		150		150	
2" x 3"						
Demolish	L.F.		.30		.30	Cost includes material and labor to install 2" x 3" let-in bracing.
Install	L.F.	.28	2.01		2.29	
Demolish and Install	L.F.	.28	2.31		2.59	
Remove and Install	L.F.		1.60		1.60	
Minimum Charge	Job		150		150	
2" x 4"						
Demolish	L.F.		.30		.30	Cost includes material and labor to install 2" x 4" let-in bracing.
Install	L.F.	.36	2.01		2.37	
Demolish and Install	L.F.	.36	2.31		2.67	
Remove and Install	L.F.		1.60		1.60	
Minimum Charge	Job		150		150	
2" x 6"						
Demolish	L.F.		.30		.30	Cost includes material and labor to install 2" x 6" let-in bracing.
Install	L.F.	.57	2.15		2.72	
Demolish and Install	L.F.	.57	2.45		3.02	
Remove and Install	L.F.		1.72		1.72	
Minimum Charge	Job		150		150	
2" x 8"						
Demolish	L.F.		.30		.30	Cost includes material and labor to install 2" x 8" let-in bracing.
Install	L.F.	.88	2.15		3.03	
Demolish and Install	L.F.	.88	2.45		3.33	
Remove and Install	L.F.		1.72		1.72	
Minimum Charge	Job		150		150	
Earthquake Strapping						
Install	Ea.	1.68	1.88		3.56	Includes labor and material to replace an earthquake strap.
Minimum Charge	Job		150		150	
Hurricane Clips						
Install	Ea.	1.49	2.07		3.56	Includes labor and material to install a hurricane clip.
Minimum Charge	Job		150		150	
Stud						
2" x 4"						
Demolish	L.F.		.24		.24	Cost includes material and labor to install 2" x 4" wall stud per L.F. of stud.
Install	L.F.	.36	.55		.91	
Demolish and Install	L.F.	.36	.79		1.15	
Remove and Install	L.F.		.44		.44	
Minimum Charge	Job		150		150	

Wall Framing System	Unit	Material	Labor	Equip.	Total	Specification
2" x 6"						
Demolish	L.F.		.30		.30	Cost includes material and labor to install 2" x 6" stud per L.F. of stud.
Install	L.F.	.57	.60		1.17	
Demolish and Install	L.F.	.57	.90		1.47	
Remove and Install	L.F.		.48		.48	
Minimum Charge	Job		150		150	
Plates						
2" x 4"						
Demolish	L.F.		.22		.22	Cost includes material and labor to install 2" x 4" plate per L.F. of plate.
Install	L.F.	.36	.75		1.11	
Demolish and Install	L.F.	.36	.97		1.33	
Remove and Install	L.F.		.60		.60	
Minimum Charge	Job		150		150	
2" x 6"						
Demolish	L.F.		.23		.23	Cost includes material and labor to install 2" x 6" plate per L.F. of plate.
Install	L.F.	.57	.80		1.37	
Demolish and Install	L.F.	.57	1.03		1.60	
Remove and Install	L.F.		.64		.64	
Minimum Charge	Job		150		150	
Headers						
Demolish	L.F.		.22		.22	Includes material and labor to install header over wall openings and around floor, ceiling and roof openings or flush beam.
Install	L.F.	.57	1.67		2.24	
Demolish and Install	L.F.	.57	1.89		2.46	
Clean	L.F.	.01	.27		.28	
Paint	L.F.	.09	1.06		1.15	
Minimum Charge	Job		150		150	
2" x 8"						
Demolish	L.F.		.23		.23	Includes material and labor to install header over wall openings and around floor, ceiling and roof openings or flush beam.
Install	L.F.	.88	1.77		2.65	
Demolish and Install	L.F.	.88	2		2.88	
Clean	L.F.	.02	.28		.30	
Paint	L.F.	.12	1.43		1.55	
Minimum Charge	Job		150		150	
2" x 10"						
Demolish	L.F.		.23		.23	Includes material and labor to install header over wall openings and around floor, ceiling and roof openings or flush beam.
Install	L.F.	1.27	1.88		3.15	
Demolish and Install	L.F.	1.27	2.11		3.38	
Clean	L.F.	.02	.28		.30	
Paint	L.F.	.14	1.76		1.90	
Minimum Charge	Job		150		150	
2" x 12"						
Demolish	L.F.		.24		.24	Includes material and labor to install header over wall openings and around floor, ceiling and roof openings or flush beam.
Install	L.F.	1.72	2.01		3.73	
Demolish and Install	L.F.	1.72	2.25		3.97	
Clean	L.F.	.02	.28		.30	
Paint	L.F.	.18	2.11		2.29	
Minimum Charge	Job		150		150	
4" x 8"						
Demolish	L.F.		.43		.43	Includes material and labor to install header over wall openings and around floor, ceiling and roof openings or flush beam.
Install	L.F.	3.83	2.31		6.14	
Demolish and Install	L.F.	3.83	2.74		6.57	
Clean	L.F.	.02	.28		.30	
Paint	L.F.	.12	1.43		1.55	
Minimum Charge	Job		150		150	

Wall Framing System	Unit	Material	Labor	Equip.	Total	Specification
4" x 10"						
Demolish	L.F.		.46		.46	Includes material and labor to install header over wall openings and around floor, ceiling and roof openings or flush beam.
Install	L.F.	4.79	2.51		7.30	
Demolish and Install	L.F.	4.79	2.97		7.76	
Clean	L.F.	.02	.28		.30	
Paint	L.F.	.14	1.76		1.90	
Minimum Charge	Job		150		150	
4" x 12"						
Demolish	L.F.		.48		.48	Includes material and labor to install header over wall openings and around floor, ceiling and roof openings or flush beam.
Install	L.F.	5.70	3.17		8.87	
Demolish and Install	L.F.	5.70	3.65		9.35	
Clean	L.F.	.02	.28		.30	
Paint	L.F.	.18	2.11		2.29	
Minimum Charge	Job		150		150	
6" x 8"						
Demolish	L.F.		.43		.43	Includes material and labor to install header over wall openings and around floor, ceiling and roof openings or flush beam.
Install	L.F.	8.25	1.67		9.92	
Demolish and Install	L.F.	8.25	2.10		10.35	
Clean	L.F.	.02	.28		.30	
Paint	L.F.	.12	1.43		1.55	
Minimum Charge	Job		150		150	
6" x 10"						
Demolish	L.F.		.46		.46	Includes material and labor to install header over wall openings and around floor, ceiling and roof openings or flush beam.
Install	L.F.	10.60	3.65		14.25	
Demolish and Install	L.F.	10.60	4.11		14.71	
Clean	L.F.	.02	.28		.30	
Paint	L.F.	.14	1.76		1.90	
Minimum Charge	Job		150		150	
6" x 12"						
Demolish	L.F.		.48		.48	Includes material and labor to install header over wall openings and around floor, ceiling and roof openings or flush beam.
Install	L.F.	12.55	4.30		16.85	
Demolish and Install	L.F.	12.55	4.78		17.33	
Clean	L.F.	.02	.28		.30	
Paint	L.F.	.18	.73		.91	
Minimum Charge	Job		150		150	

Rough-in Opening	Unit	Material	Labor	Equip.	Total	Specification
Door w / 2" x 4" Lumber						
3' Wide						
Demolish	Ea.		7.55		7.55	Includes material and labor to install header, double studs each side, cripples, blocking and nails, up to 3' opening in 2" x 4" stud wall 8' high.
Install	Ea.	15.30	9.40		24.70	
Demolish and Install	Ea.	15.30	16.95		32.25	
Remove and Install	Ea.		7.52		7.52	
Minimum Charge	Job		150		150	
4' Wide						
Demolish	Ea.		7.55		7.55	Includes material and labor to install header, double studs each side, cripples, blocking and nails, up to 4' opening in 2" x 4" stud wall 8' high.
Install	Ea.	16.40	9.40		25.80	
Demolish and Install	Ea.	16.40	16.95		33.35	
Remove and Install	Ea.		7.52		7.52	
Minimum Charge	Job		150		150	
5' Wide						
Demolish	Ea.		7.55		7.55	Includes material and labor to install header, double studs each side, cripples, blocking and nails, up to 5' opening in 2" x 4" stud wall 8' high.
Install	Ea.	21	9.40		30.40	
Demolish and Install	Ea.	21	16.95		37.95	
Remove and Install	Ea.		7.52		7.52	
Minimum Charge	Job		150		150	

Rough-in Opening	Unit	Material	Labor	Equip.	Total	Specification
6' Wide						
Demolish	Ea.		7.55		7.55	Includes material and labor to install header, double studs each side, cripples, blocking and nails, up to 6' opening in 2" x 4" stud wall 8' high.
Install	Ea.	22.50	9.40		31.90	
Demolish and Install	Ea.	22.50	16.95		39.45	
Remove and Install	Ea.		7.52		7.52	
Minimum Charge	Job		150		150	
8' Wide						
Demolish	Ea.		8.05		8.05	Includes material and labor to install header, double studs each side, cripples, blocking and nails, up to 8' opening in 2" x 4" stud wall 8' high.
Install	Ea.	32.50	10.05		42.55	
Demolish and Install	Ea.	32.50	18.10		50.60	
Remove and Install	Ea.		8.02		8.02	
Minimum Charge	Job		150		150	
10' Wide						
Demolish	Ea.		8.05		8.05	Includes material and labor to install header, double studs each side, cripples, blocking and nails, up to 10' opening in 2" x 4" stud wall 8' high.
Install	Ea.	47	10.05		57.05	
Demolish and Install	Ea.	47	18.10		65.10	
Remove and Install	Ea.		8.02		8.02	
Minimum Charge	Job		150		150	
12' Wide						
Demolish	Ea.		8.05		8.05	Includes material and labor to install header, double studs each side, cripples, blocking and nails, up to 12' opening in 2" x 4" stud wall 8' high.
Install	Ea.	61.50	10.05		71.55	
Demolish and Install	Ea.	61.50	18.10		79.60	
Remove and Install	Ea.		8.02		8.02	
Minimum Charge	Job		150		150	
Door w / 2" x 6" Lumber						
3' Wide						
Demolish	Ea.		7.55		7.55	Includes material and labor to install header, double studs each side, cripples, blocking and nails, up to 3' opening in 2" x 6" stud wall 8' high.
Install	Ea.	22	9.40		31.40	
Demolish and Install	Ea.	22	16.95		38.95	
Remove and Install	Ea.		7.52		7.52	
Minimum Charge	Job		150		150	
4' Wide						
Demolish	Ea.		7.55		7.55	Includes material and labor to install header, double studs each side, cripples, blocking and nails, up to 4' opening in 2" x 6" stud wall 8' high.
Install	Ea.	23	9.40		32.40	
Demolish and Install	Ea.	23	16.95		39.95	
Remove and Install	Ea.		7.52		7.52	
Minimum Charge	Job		150		150	
5' Wide						
Demolish	Ea.		7.55		7.55	Includes material and labor to install header, double studs each side, cripples, blocking and nails up to 5' opening in 2" x 6" stud wall 8' high.
Install	Ea.	27.50	9.40		36.90	
Demolish and Install	Ea.	27.50	16.95		44.45	
Remove and Install	Ea.		7.52		7.52	
Minimum Charge	Job		150		150	
6' Wide						
Demolish	Ea.		7.55		7.55	Includes material and labor to install header, double studs each side, cripples, blocking and nails up to 6' opening in 2" x 6" stud wall 8' high.
Install	Ea.	29	9.40		38.40	
Demolish and Install	Ea.	29	16.95		45.95	
Remove and Install	Ea.		7.52		7.52	
Minimum Charge	Job		150		150	
8' Wide						
Demolish	Ea.		8.05		8.05	Includes material and labor to install header, double studs each side, cripples, blocking and nails up to 8' opening in 2" x 6" stud wall 8' high.
Install	Ea.	39	10.05		49.05	
Demolish and Install	Ea.	39	18.10		57.10	
Remove and Install	Ea.		8.02		8.02	
Minimum Charge	Job		150		150	
10' Wide						
Demolish	Ea.		8.05		8.05	Includes material and labor to install header, double studs each side, cripples, blocking and nails up to 10' opening in 2" x 6" stud wall 8' high.
Install	Ea.	54	10.05		64.05	
Demolish and Install	Ea.	54	18.10		72.10	
Remove and Install	Ea.		8.02		8.02	
Minimum Charge	Job		150		150	

Rough-in Opening	Unit	Material	Labor	Equip.	Total	Specification
12' Wide						
Demolish	Ea.		8.05		8.05	Includes material and labor to install header, double studs each side, cripples, blocking and nails, up to 12' opening in 2" x 6" stud wall 8' high.
Install	Ea.	69	10.05		79.05	
Demolish and Install	Ea.	69	18.10		87.10	
Remove and Install	Ea.		8.02		8.02	
Minimum Charge	Job		150		150	

Rough-in Opening	Unit	Material	Labor	Equip.	Total	Specification
Window w / 2" x 4" Lumber						
2' Wide						
Demolish	Ea.		10.10		10.10	Includes material and labor to install header, double studs each side of opening, cripples, blocking, nails and sub-sills for opening up to 2' in a 2" x 4" stud wall 8' high.
Install	Ea.	16.05	12.55		28.60	
Demolish and Install	Ea.	16.05	22.65		38.70	
Remove and Install	Ea.		10.03		10.03	
Minimum Charge	Job		150		150	
3' Wide						
Demolish	Ea.		10.10		10.10	Includes material and labor to install header, double studs each side of opening, cripples, blocking, nails and sub-sills for opening up to 3' in a 2" x 4" stud wall 8' high.
Install	Ea.	18.85	12.55		31.40	
Demolish and Install	Ea.	18.85	22.65		41.50	
Remove and Install	Ea.		10.03		10.03	
Minimum Charge	Job		150		150	
4' Wide						
Demolish	Ea.		10.10		10.10	Includes material and labor to install header, double studs each side of opening, cripples, blocking, nails and sub-sills for opening up to 4' in a 2" x 4" stud wall 8' high.
Install	Ea.	20.50	12.55		33.05	
Demolish and Install	Ea.	20.50	22.65		43.15	
Remove and Install	Ea.		10.03		10.03	
Minimum Charge	Job		150		150	
5' Wide						
Demolish	Ea.		10.10		10.10	Includes material and labor to install header, double studs each side of opening, cripples, blocking, nails and sub-sills for opening up to 5' in a 2" x 4" stud wall 8' high.
Install	Ea.	25.50	12.55		38.05	
Demolish and Install	Ea.	25.50	22.65		48.15	
Remove and Install	Ea.		10.03		10.03	
Minimum Charge	Job		150		150	
6' Wide						
Demolish	Ea.		10.10		10.10	Includes material and labor to install header, double studs each side of opening, cripples, blocking, nails and sub-sills for opening up to 6' in a 2" x 4" stud wall 8' high.
Install	Ea.	28	12.55		40.55	
Demolish and Install	Ea.	28	22.65		50.65	
Remove and Install	Ea.		10.03		10.03	
Minimum Charge	Job		150		150	
7' Wide						
Demolish	Ea.		10.10		10.10	Includes material and labor to install header, double studs each side of opening, cripples, blocking, nails and sub-sills for opening up to 7' in a 2" x 4" stud wall 8' high.
Install	Ea.	36.50	12.55		49.05	
Demolish and Install	Ea.	36.50	22.65		59.15	
Remove and Install	Ea.		10.03		10.03	
Minimum Charge	Job		150		150	
8' Wide						
Demolish	Ea.		11		11	Includes material and labor to install header, double studs each side of opening, cripples, blocking, nails and sub-sills for opening up to 8' in a 2" x 4" stud wall 8' high.
Install	Ea.	40	13.65		53.65	
Demolish and Install	Ea.	40	24.65		64.65	
Remove and Install	Ea.		10.94		10.94	
Minimum Charge	Job		150		150	

Rough-in Opening	Unit	Material	Labor	Equip.	Total	Specification
10' Wide						
Demolish	Ea.		11		11	Includes material and labor to install header, double studs each side of opening, cripples, blocking, nails and sub-sills for opening up to 10' in a 2" x 4" stud wall 8' high.
Install	Ea.	55.50	13.65		69.15	
Demolish and Install	Ea.	55.50	24.65		80.15	
Remove and Install	Ea.		10.94		10.94	
Minimum Charge	Job		150		150	
12' Wide						
Demolish	Ea.		11		11	Includes material and labor to install header, double studs each side of opening, cripples, blocking, nails and sub-sills for opening up to 12' in a 2" x 4" stud wall 8' high.
Install	Ea.	72.50	13.65		86.15	
Demolish and Install	Ea.	72.50	24.65		97.15	
Remove and Install	Ea.		10.94		10.94	
Minimum Charge	Job		150		150	
Window w / 2" x 6" Lumber						
2' Wide						
Demolish	Ea.		10.10		10.10	Includes material and labor to install header, double studs each side of opening, cripples, blocking, nails and sub-sills for opening up to 2' in a 2" x 6" stud wall 8' high.
Install	Ea.	25	12.55		37.55	
Demolish and Install	Ea.	25	22.65		47.65	
Remove and Install	Ea.		10.03		10.03	
Minimum Charge	Job		150		150	
3' Wide						
Demolish	Ea.		10.10		10.10	Includes material and labor to install header, double studs each side of opening, cripples, blocking, nails and sub-sills for opening up to 3' in a 2" x 6" stud wall 8' high.
Install	Ea.	28	12.55		40.55	
Demolish and Install	Ea.	28	22.65		50.65	
Remove and Install	Ea.		10.03		10.03	
Minimum Charge	Job		150		150	
4' Wide						
Demolish	Ea.		10.10		10.10	Includes material and labor to install header, double studs each side of opening, cripples, blocking, nails and sub-sills for opening up to 4' in a 2" x 6" stud wall 8' high.
Install	Ea.	29.50	12.55		42.05	
Demolish and Install	Ea.	29.50	22.65		52.15	
Remove and Install	Ea.		10.03		10.03	
Minimum Charge	Job		150		150	
5' Wide						
Demolish	Ea.		10.10		10.10	Includes material and labor to install header, double studs each side of opening, cripples, blocking, nails and sub-sills for opening up to 5' in a 2" x 6" stud wall 8' high.
Install	Ea.	34.50	12.55		47.05	
Demolish and Install	Ea.	34.50	22.65		57.15	
Remove and Install	Ea.		10.03		10.03	
Minimum Charge	Job		150		150	
6' Wide						
Demolish	Ea.		10.10		10.10	Includes material and labor to install header, double studs each side of opening, cripples, blocking, nails and sub-sills for opening up to 6' in a 2" x 6" stud wall 8' high.
Install	Ea.	38	12.55		50.55	
Demolish and Install	Ea.	38	22.65		60.65	
Remove and Install	Ea.		10.03		10.03	
Minimum Charge	Job		150		150	
7' Wide						
Demolish	Ea.		10.10		10.10	Includes material and labor to install header, double studs each side of opening, cripples, blocking, nails and sub-sills for opening up to 7' in a 2" x 6" stud wall 8' high.
Install	Ea.	47	12.55		59.55	
Demolish and Install	Ea.	47	22.65		69.65	
Remove and Install	Ea.		10.03		10.03	
Minimum Charge	Job		150		150	

Rough-in Opening	Unit	Material	Labor	Equip.	Total	Specification
8′ Wide						
Demolish	Ea.		11		11	Includes material and labor to install header, double studs each side of opening, cripples, blocking, nails and sub-sills for opening up to 8′ in a 2″ x 6″ stud wall 8′ high.
Install	Ea.	51	13.65		64.65	
Demolish and Install	Ea.	51	24.65		75.65	
Remove and Install	Ea.		10.94		10.94	
Minimum Charge	Job		150		150	
10′ Wide						
Demolish	Ea.		11		11	Includes material and labor to install header, double studs each side of opening, cripples, blocking, nails and sub-sills for opening up to 10′ in a 2″ x 6″ stud wall 8′ high.
Install	Ea.	67.50	13.65		81.15	
Demolish and Install	Ea.	67.50	24.65		92.15	
Remove and Install	Ea.		10.94		10.94	
Minimum Charge	Job		150		150	
12′ Wide						
Demolish	Ea.		11		11	Includes material and labor to install header, double studs each side of opening, cripples, blocking, nails and sub-sills for opening up to 12′ in a 2″ x 6″ stud wall 8′ high.
Install	Ea.	86	13.65		99.65	
Demolish and Install	Ea.	86	24.65		110.65	
Remove and Install	Ea.		10.94		10.94	
Minimum Charge	Job		150		150	

Glue-Laminated Beams	Unit	Material	Labor	Equip.	Total	Specification
3-1/2″ x 6″						
Demolish	L.F.		3.01		3.01	Includes material and labor to install glue-laminated wood beam.
Install	L.F.	4.48	2.26	1.17	7.91	
Demolish and Install	L.F.	4.48	5.27	1.17	10.92	
Remove and Install	L.F.		1.80	.93	2.73	
Clean	L.F.	.09	.27		.36	
Paint	L.F.	.28	.78		1.06	
Minimum Charge	Job		150		150	
3-1/2″ x 9″						
Demolish	L.F.		3.01		3.01	Includes material and labor to install glue-laminated wood beam.
Install	L.F.	6	2.26	1.17	9.43	
Demolish and Install	L.F.	6	5.27	1.17	12.44	
Remove and Install	L.F.		1.80	.93	2.73	
Clean	L.F.	.12	.28		.40	
Paint	L.F.	.35	1.01		1.36	
Minimum Charge	Job		150		150	
3-1/2″ x 12″						
Demolish	L.F.		4.51		4.51	Includes material and labor to install glue-laminated wood beam.
Install	L.F.	8	2.26	1.17	11.43	
Demolish and Install	L.F.	8	6.77	1.17	15.94	
Remove and Install	L.F.		1.80	.93	2.73	
Clean	L.F.	.14	.29		.43	
Paint	L.F.	.44	1.26		1.70	
Minimum Charge	Job		150		150	
3-1/2″ x 15″						
Demolish	L.F.		4.51		4.51	Includes material and labor to install glue-laminated wood beam.
Install	L.F.	10	2.33	1.21	13.54	
Demolish and Install	L.F.	10	6.84	1.21	18.05	
Remove and Install	L.F.		1.87	.96	2.83	
Clean	L.F.	.18	.30		.48	
Paint	L.F.	.52	1.47		1.99	
Minimum Charge	Job		150		150	

Glue-Laminated Beams	Unit	Material	Labor	Equip.	Total	Specification
3-1/2" x 18"						
Demolish	L.F.		5.65		5.65	Includes material and labor to install glue-laminated wood beam.
Install	L.F.	13.40	2.33	1.21	16.94	
Demolish and Install	L.F.	13.40	7.98	1.21	22.59	
Remove and Install	L.F.		1.87	.96	2.83	
Clean	L.F.	.20	.31		.51	
Paint	L.F.	.61	1.70		2.31	
Minimum Charge	Job		150		150	
5-1/8" x 6"						
Demolish	L.F.		4.51		4.51	Includes material and labor to install glue-laminated wood beam.
Install	L.F.	6.55	2.26	1.17	9.98	
Demolish and Install	L.F.	6.55	6.77	1.17	14.49	
Remove and Install	L.F.		1.80	.93	2.73	
Clean	L.F.	.10	.27		.37	
Paint	L.F.	.30	.85		1.15	
Minimum Charge	Job		150		150	
5-1/8" x 9"						
Demolish	L.F.		5.65		5.65	Includes material and labor to install glue-laminated wood beam.
Install	L.F.	9.85	2.26	1.17	13.28	
Demolish and Install	L.F.	9.85	7.91	1.17	18.93	
Remove and Install	L.F.		1.80	.93	2.73	
Clean	L.F.	.13	.28		.41	
Paint	L.F.	.39	1.09		1.48	
Minimum Charge	Job		150		150	
5-1/8" x 12"						
Demolish	L.F.		6.70		6.70	Includes material and labor to install glue-laminated wood beam.
Install	L.F.	13.10	2.26	1.17	16.53	
Demolish and Install	L.F.	13.10	8.96	1.17	23.23	
Remove and Install	L.F.		1.80	.93	2.73	
Clean	L.F.	.15	.29		.44	
Paint	L.F.	.46	1.32		1.78	
Minimum Charge	Job		150		150	
5-1/8" x 18"						
Demolish	L.F.		11.30		11.30	Includes material and labor to install glue-laminated wood beam.
Install	L.F.	19.65	2.33	1.21	23.19	
Demolish and Install	L.F.	19.65	13.63	1.21	34.49	
Remove and Install	L.F.		1.87	.96	2.83	
Clean	L.F.	.21	.31		.52	
Paint	L.F.	.54	1.78		2.32	
Minimum Charge	Job		150		150	
6-3/4" x 12"						
Demolish	L.F.		6.70		6.70	Includes material and labor to install glue-laminated wood beam.
Install	L.F.	17.25	2.33	1.21	20.79	
Demolish and Install	L.F.	17.25	9.03	1.21	27.49	
Remove and Install	L.F.		1.87	.96	2.83	
Clean	L.F.	.18	.30		.48	
Paint	L.F.	.52	1.47		1.99	
Minimum Charge	Job		150		150	
6-3/4" x 15"						
Demolish	L.F.		11.30		11.30	Includes material and labor to install glue-laminated wood beam.
Install	L.F.	21.50	2.33	1.21	25.04	
Demolish and Install	L.F.	21.50	13.63	1.21	36.34	
Remove and Install	L.F.		1.87	.96	2.83	
Clean	L.F.	.20	.31		.51	
Paint	L.F.	.61	1.70		2.31	
Minimum Charge	Job		150		150	

Glue-Laminated Beams	Unit	Material	Labor	Equip.	Total	Specification
6-3/4" x 18"						
Demolish	L.F.		11.30		11.30	Includes material and labor to install glue-laminated wood beam.
Install	L.F.	26	2.42	1.25	29.67	
Demolish and Install	L.F.	26	13.72	1.25	40.97	
Remove and Install	L.F.		1.93	1	2.93	
Clean	L.F.	.23	.32		.55	
Paint	L.F.	.68	1.94		2.62	
Minimum Charge	Job		150		150	
Hardware						
5-1/4" Glue-Lam Seat						
Install	Ea.	49	1.67		50.67	Includes labor and material to install beam hangers.
Minimum Charge	Job		150		150	
6-3/4" Glue-Lam Seat						
Install	Ea.	50.50	1.67		52.17	Includes labor and material to install beam hangers.
Minimum Charge	Job		150		150	
8-3/4" Glue-Lam Seat						
Install	Ea.	58	1.67		59.67	Includes labor and material to install beam hangers.
Minimum Charge	Job		150		150	
Earthquake Strapping						
Install	Ea.	1.68	1.88		3.56	Includes labor and material to replace an earthquake strap.
Minimum Charge	Job		150		150	
Hurricane Clips						
Install	Ea.	1.49	2.07		3.56	Includes labor and material to install a hurricane clip.
Minimum Charge	Job		150		150	

Metal Stud Framing	Unit	Material	Labor	Equip.	Total	Specification
16" O.C. System						
4", 16 Ga.						
Demolish	S.F.		.90		.90	Includes material and labor to install load bearing cold rolled metal stud walls, to 10' high, including studs, top and bottom track and screws.
Install	S.F.	.95	.93		1.88	
Demolish and Install	S.F.	.95	1.83		2.78	
Remove and Install	S.F.		.74		.74	
Minimum Charge	Job		150		150	
6", 16 Ga.						
Demolish	S.F.		.90		.90	Includes material and labor to install load bearing cold rolled metal stud walls, to 10' high, including studs, top and bottom track and screws.
Install	S.F.	1.19	.94		2.13	
Demolish and Install	S.F.	1.19	1.84		3.03	
Remove and Install	S.F.		.75		.75	
Minimum Charge	Job		150		150	
4", 25 Ga.						
Demolish	S.F.		.63		.63	Includes material and labor to install cold rolled metal stud walls, to 10' high, including studs, top and bottom track and screws.
Install	S.F.	.30	.63		.93	
Demolish and Install	S.F.	.30	1.26		1.56	
Remove and Install	S.F.		.51		.51	
Minimum Charge	Job		150		150	
6", 25 Ga.						
Demolish	S.F.		.63		.63	Includes material and labor to install cold rolled metal stud walls, to 10' high, including studs, top and bottom track and screws.
Install	S.F.	.44	.64		1.08	
Demolish and Install	S.F.	.44	1.27		1.71	
Remove and Install	S.F.		.51		.51	
Minimum Charge	Job		150		150	

Metal Stud Framing

	Unit	Material	Labor	Equip.	Total	Specification
24" O.C. System						
4", 25 Ga.						
Demolish	S.F.		.53		.53	Includes material and labor to install cold rolled metal stud walls, to 10' high, including studs, top and bottom track and screws.
Install	S.F.	.22	.41		.63	
Demolish and Install	S.F.	.22	.94		1.16	
Remove and Install	S.F.		.33		.33	
Minimum Charge	Job		150		150	
6", 25 Ga.						
Demolish	S.F.		.53		.53	Includes material and labor to install cold rolled metal stud walls, to 10' high, including studs, top and bottom track and screws.
Install	S.F.	.33	.41		.74	
Demolish and Install	S.F.	.33	.94		1.27	
Remove and Install	S.F.		.33		.33	
Minimum Charge	Job		150		150	
4", 16 Ga.						
Demolish	L.F.		6.05		6.05	Includes material and labor to install load bearing cold rolled metal stud walls, to 10' high, including studs, top and bottom track and screws.
Install	S.F.	.68	.67		1.35	
Demolish and Install	S.F.	.68	6.72		7.40	
Remove and Install	S.F.		.53		.53	
Minimum Charge	Job		150		150	
6", 16 Ga.						
Demolish	L.F.		6.05		6.05	Includes material and labor to install load bearing cold rolled metal stud walls, to 10' high, including studs, top and bottom track and screws.
Install	S.F.	.86	.68		1.54	
Demolish and Install	S.F.	.86	6.73		7.59	
Remove and Install	S.F.		.55		.55	
Minimum Charge	Job		150		150	

Metal Joist

	Unit	Material	Labor	Equip.	Total	Specification
Bar Joist						
18K9						
Demolish	L.F.		.72	.36	1.08	Includes material, labor and equipment to install open web joist.
Install	L.F.	4.71	1.67	.75	7.13	
Demolish and Install	L.F.	4.71	2.39	1.11	8.21	
Remove and Install	L.F.		1.33	.60	1.93	
Minimum Charge	Job		1350	700	2050	
16K6						
Demolish	L.F.		.72	.36	1.08	Includes material, labor and equipment to install open web joist.
Install	L.F.	3.67	1.13	.06	4.86	
Demolish and Install	L.F.	3.67	1.85	.42	5.94	
Remove and Install	L.F.		.91	.05	.96	
Minimum Charge	Job		187		187	

Exterior Sheathing

	Unit	Material	Labor	Equip.	Total	Specification
CDX Plywood						
5/16"						
Demolish	S.F.		.21		.21	Cost includes material and labor to install 5/16" CDX plywood sheathing.
Install	S.F.	.47	.38		.85	
Demolish and Install	S.F.	.47	.59		1.06	
Minimum Charge	Job		150		150	
3/8"						
Demolish	S.F.		.21		.21	Cost includes material and labor to install 3/8" CDX plywood sheathing.
Install	S.F.	.48	.50		.98	
Demolish and Install	S.F.	.48	.71		1.19	
Minimum Charge	Job		150		150	

Exterior Sheathing		Unit	Material	Labor	Equip.	Total	Specification
1/2"							
	Demolish	S.F.		.22		.22	Cost includes material and labor to install 1/2" CDX plywood sheathing.
	Install	S.F.	.54	.53		1.07	
	Demolish and Install	S.F.	.54	.75		1.29	
	Minimum Charge	Job		150		150	
5/8"							
	Demolish	S.F.		.22		.22	Cost includes material and labor to install 5/8" CDX plywood sheathing.
	Install	S.F.	.63	.57		1.20	
	Demolish and Install	S.F.	.63	.79		1.42	
	Minimum Charge	Job		150		150	
3/4"							
	Demolish	S.F.		.23		.23	Cost includes material and labor to install 3/4" CDX plywood sheathing.
	Install	S.F.	.73	.62		1.35	
	Demolish and Install	S.F.	.73	.85		1.58	
	Minimum Charge	Job		150		150	
OSB							
1/2"							
	Demolish	S.F.		.22		.22	Cost includes material and labor to install 4' x 8' x 1/2" OSB sheathing.
	Install	S.F.	.34	.23		.57	
	Demolish and Install	S.F.	.34	.45		.79	
	Minimum Charge	Job		150		150	
5/8"							
	Demolish	S.F.		.22		.22	Cost includes material and labor to install 4' x 8' x 5/8" OSB sheathing.
	Install	S.F.	.52	.24		.76	
	Demolish and Install	S.F.	.52	.46		.98	
	Minimum Charge	Job		150		150	
Vapor Barrier							
Black Paper							
	Install	S.F.	.02	.08		.10	Includes material and labor to install #15 felt building paper.
	Minimum Charge	Job		150		150	

Plywood Sheathing		Unit	Material	Labor	Equip.	Total	Specification
Finish Plywood							
5/16"							
	Demolish	S.F.		.21		.21	Includes material and labor to install exterior 5/16" AC plywood on walls.
	Install	S.F.	.55	.19		.74	
	Demolish and Install	S.F.	.55	.40		.95	
	Minimum Charge	Job		150		150	
3/8"							
	Demolish	S.F.		.21		.21	Includes material and labor to install exterior 3/8" AC plywood on walls.
	Install	S.F.	.55	.25		.80	
	Demolish and Install	S.F.	.55	.46		1.01	
	Minimum Charge	Job		150		150	
1/2"							
	Demolish	S.F.		.22		.22	Includes material and labor to install exterior 1/2" AC plywood on walls.
	Install	S.F.	.68	.27		.95	
	Demolish and Install	S.F.	.68	.49		1.17	
	Minimum Charge	Job		150		150	
5/8"							
	Demolish	S.F.		.22		.22	Includes material and labor to install exterior 5/8" AC plywood on walls.
	Install	S.F.	.77	.27		1.04	
	Demolish and Install	S.F.	.77	.49		1.26	
	Remove and Install	S.F.		.27		.27	
	Minimum Charge	Job		150		150	

Plywood Sheathing	Unit	Material	Labor	Equip.	Total	Specification
3/4"						
Demolish	S.F.		.23		.23	Includes material and labor to install exterior 3/4" AC plywood on walls.
Install	S.F.	1.06	.27		1.33	
Demolish and Install	S.F.	1.06	.50		1.56	
Remove and Install	S.F.		.27		.27	
Minimum Charge	Job		150		150	

Stairs	Unit	Material	Labor	Equip.	Total	Specification
Job-built						
Treads and Risers						
Demolish	Ea.		30.50		30.50	Includes material and labor to install three 2" x 12" stringers, treads and risers of 3/4" CDX plywood, installed in a straight or "L" shaped run including carpet.
Install	Ea.	415	37.50		452.50	
Demolish and Install	Ea.	415	68		483	
Clean	Flight	2.73	18.90		21.63	
Paint	Ea.	.08	53		53.08	
Minimum Charge	Job		150		150	
Landing						
Demolish	S.F.		.24		.24	Includes material and labor to install landing framing, 3/4" CDX plywood surface and carpet.
Install	S.F.	44	2.09		46.09	
Demolish and Install	S.F.	44	2.33		46.33	
Clean	S.F.		.27		.27	
Paint	S.F.	.25	.65		.90	
Minimum Charge	Job		150		150	

Ceiling Framing	Unit	Material	Labor	Equip.	Total	Specification
16" O.C. System						
2" x 6" Joists						
Demolish	S.F.		.37		.37	Cost includes material and labor to install 2" x 6" joists including end and header joist installed 16" O.C. Does not include beams, ledger strips, blocking or bridging.
Install	S.F.	.44	.49		.93	
Demolish and Install	S.F.	.44	.86		1.30	
Remove and Install	S.F.		.40		.40	
Minimum Charge	Job		150		150	
2" x 8" Joists						
Demolish	S.F.		.39		.39	Cost includes material and labor to install 2" x 8" joists including end and header joist installed 16" O.C. Does not include beams, ledger strips, blocking or bridging.
Install	S.F.	.67	.61		1.28	
Demolish and Install	S.F.	.67	1		1.67	
Remove and Install	S.F.		.49		.49	
Minimum Charge	Job		150		150	
2" x 10" Joists						
Demolish	S.F.		.40		.40	Cost includes material and labor to install 2" x 10" joists including end and header joist installed 16" O.C. Does not include beams, ledger strips, blocking or bridging.
Install	S.F.	.96	.72		1.68	
Demolish and Install	S.F.	.96	1.12		2.08	
Remove and Install	S.F.		.58		.58	
Minimum Charge	Job		150		150	
2" x 12" Joists						
Demolish	S.F.		.41		.41	Cost includes material and labor to install 2" x 12" joists including end and header joist installed 16" O.C. Does not include beams, ledger strips, blocking or bridging.
Install	S.F.	1.31	.88		2.19	
Demolish and Install	S.F.	1.31	1.29		2.60	
Remove and Install	S.F.		.70		.70	
Minimum Charge	Job		150		150	

Ceiling Framing	Unit	Material	Labor	Equip.	Total	Specification
Block / Bridge						
Demolish	Ea.		.76		.76	Includes material and labor to install set of cross bridging or per block of solid bridging for 2" x 10" joists cut to size on site.
Install	Ea.	1.69	1.69		3.38	
Demolish and Install	Ea.	1.69	2.45		4.14	
Minimum Charge	Job		150		150	
24" O.C. System						
2" x 6" Joists						
Demolish	S.F.		.25		.25	Cost includes material and labor to install 2" x 6" joists including end and header joist installed 24" O.C. Does not include beams, ledger strips, blocking or bridging.
Install	S.F.	.30	.33		.63	
Demolish and Install	S.F.	.30	.58		.88	
Remove and Install	S.F.		.26		.26	
Minimum Charge	Job		150		150	
2" x 8" Joists						
Demolish	S.F.		.26		.26	Cost includes material and labor to install 2" x 8" joists including end and header joist installed 24" O.C. Does not include beams, ledger strips, blocking or bridging.
Install	S.F.	.45	.41		.86	
Demolish and Install	S.F.	.45	.67		1.12	
Remove and Install	S.F.		.32		.32	
Minimum Charge	Job		150		150	
2" x 10" Joists						
Demolish	S.F.		.27		.27	Cost includes material and labor to install 2" x 10" joists including end and header joist installed 24" O.C. Does not include beams, ledger strips, blocking or bridging.
Install	S.F.	.64	.48		1.12	
Demolish and Install	S.F.	.64	.75		1.39	
Remove and Install	S.F.		.38		.38	
Minimum Charge	Job		150		150	
2" x 12" Joists						
Demolish	S.F.		.27		.27	Cost includes material and labor to install 2" x 12" joists including end and header joist installed 24" O.C. Does not include beams, ledger strips, blocking or bridging.
Install	S.F.	.88	.59		1.47	
Demolish and Install	S.F.	.88	.86		1.74	
Remove and Install	S.F.		.47		.47	
Minimum Charge	Job		150		150	
Block / Bridge						
Demolish	Ea.		.76		.76	Includes material and labor to install set of cross bridging or per block of solid bridging for 2" x 10" joists cut to size on site.
Install	Ea.	1.69	1.69		3.38	
Demolish and Install	Ea.	1.69	2.45		4.14	
Minimum Charge	Job		150		150	
Ledger Strips						
1" x 2"						
Demolish	L.F.		.20		.20	Cost includes material and labor to install 1" x 2" ledger strip nailed to the face of studs, beams or joist.
Install	L.F.	.22	.87		1.09	
Demolish and Install	L.F.	.22	1.07		1.29	
Minimum Charge	Job		150		150	
1" x 3"						
Demolish	L.F.		.20		.20	Includes material and labor to install up to 1" x 4" ledger strip nailed to the face of studs, beams or joist.
Install	L.F.	.66	1		1.66	
Demolish and Install	L.F.	.66	1.20		1.86	
Minimum Charge	Job		150		150	
1" x 4"						
Demolish	L.F.		.20		.20	Includes material and labor to install up to 1" x 4" ledger strip nailed to the face of studs, beams or joist.
Install	L.F.	.66	1		1.66	
Demolish and Install	L.F.	.66	1.20		1.86	
Minimum Charge	Job		150		150	
2" x 2"						
Demolish	L.F.		.22		.22	Cost includes material and labor to install 2" x 2" ledger strip nailed to the face of studs, beams or joist.
Install	L.F.	.28	.91		1.19	
Demolish and Install	L.F.	.28	1.13		1.41	
Minimum Charge	Job		150		150	

Ceiling Framing	Unit	Material	Labor	Equip.	Total	Specification
2" x 4"						
Demolish	L.F.		.24		.24	Cost includes material and labor to install 2" x 4" ledger strip nailed to the face of studs, beams or joist.
Install	L.F.	.36	1.20		1.56	
Demolish and Install	L.F.	.36	1.44		1.80	
Remove and Install	L.F.		.96		.96	
Minimum Charge	Job		150		150	
Ledger Boards						
Demolish	L.F.		.24		.24	Cost includes material and labor to install 2" x 4" ledger board fastened to a wall, joists or studs.
Install	L.F.	1.52	1.67		3.19	
Demolish and Install	L.F.	1.52	1.91		3.43	
Remove and Install	L.F.		1.34		1.34	
Minimum Charge	Job		150		150	
4" x 6"						
Demolish	L.F.		.40		.40	Cost includes material and labor to install 4" x 6" ledger board fastened to a wall, joists or studs.
Install	L.F.	4.07	1.88		5.95	
Demolish and Install	L.F.	4.07	2.28		6.35	
Remove and Install	L.F.		1.50		1.50	
Minimum Charge	Job		150		150	
4" x 8"						
Demolish	L.F.		.54		.54	Cost includes material and labor to install 4" x 8" ledger board fastened to a wall, joists or studs.
Install	L.F.	5.05	1.88		6.93	
Demolish and Install	L.F.	5.05	2.42		7.47	
Remove and Install	L.F.		1.50		1.50	
Minimum Charge	Job		150		150	
Earthquake Strapping						
Install	Ea.	1.68	1.88		3.56	Includes labor and material to replace an earthquake strap.
Minimum Charge	Job		150		150	
Hurricane Clips						
Install	Ea.	1.49	2.07		3.56	Includes labor and material to install a hurricane clip.
Minimum Charge	Job		150		150	

Roof Framing	Unit	Material	Labor	Equip.	Total	Specification
16" O.C. System						
2" x 4" Rafters						
Demolish	S.F.		.55		.55	Cost includes material and labor to install 2" x 4", 16" O.C. rafter framing for flat, shed or gable roofs, 25' span, up to 5/12 slope, per S.F. of roof.
Install	S.F.	.28	.27		.55	
Demolish and Install	S.F.	.28	.82		1.10	
Remove and Install	S.F.		.21		.21	
Minimum Charge	Job		150		150	
2" x 6" Rafters						
Demolish	S.F.		.58		.58	Cost includes material and labor to install 2" x 6", 16" O.C. rafter framing for flat, shed or gable roofs, 25' span, up to 5/12 slope, per S.F. of roof.
Install	S.F.	.43	.28		.71	
Demolish and Install	S.F.	.43	.86		1.29	
Remove and Install	S.F.		.23		.23	
Minimum Charge	Job		150		150	
2" x 8" Rafters						
Demolish	S.F.		.59		.59	Cost includes material and labor to install 2" x 8", 16" O.C. rafter framing for flat, shed or gable roofs, 25' span, up to 5/12 slope, per S.F. of roof.
Install	S.F.	.66	.30		.96	
Demolish and Install	S.F.	.66	.89		1.55	
Remove and Install	S.F.		.24		.24	
Minimum Charge	Job		150		150	

Roof Framing		Unit	Material	Labor	Equip.	Total	Specification
2" x 10" Rafters							
	Demolish	S.F.		.59		.59	Cost includes material and labor to install 2" x 10", 16" O.C. rafter framing for flat, shed or gable roofs, 25' span, up to 5/12 slope, per S.F. of roof.
	Install	S.F.	.95	.46		1.41	
	Demolish and Install	S.F.	.95	1.05		2	
	Remove and Install	S.F.		.36		.36	
	Minimum Charge	Job		150		150	
2" x 12" Rafters							
	Demolish	S.F.		.60		.60	Cost includes material and labor to install 2" x 12", 16" O.C. rafter framing for flat, shed or gable roofs, 25' span, up to 5/12 slope, per S.F. of roof.
	Install	S.F.	1.29	.50		1.79	
	Demolish and Install	S.F.	1.29	1.10		2.39	
	Remove and Install	S.F.		.40		.40	
	Minimum Charge	Job		150		150	
24" O.C. System							
2" x 4" Rafters							
	Demolish	S.F.		.41		.41	Cost includes material and labor to install 2" x 4", 24" O.C. rafter framing for flat, shed or gable roofs, 25' span, up to 5/12 slope, per S.F. of roof.
	Install	S.F.	.18	.20		.38	
	Demolish and Install	S.F.	.18	.61		.79	
	Remove and Install	S.F.		.16		.16	
	Minimum Charge	Job		150		150	
2" x 6" Rafters							
	Demolish	S.F.		.43		.43	Cost includes material and labor to install 2" x 6", 24" O.C. rafter framing for flat, shed or gable roofs, 25' span, up to 5/12 slope, per S.F. of roof.
	Install	S.F.	.29	.21		.50	
	Demolish and Install	S.F.	.29	.64		.93	
	Remove and Install	S.F.		.17		.17	
	Minimum Charge	Job		150		150	
2" x 8" Rafters							
	Demolish	S.F.		.44		.44	Cost includes material and labor to install 2" x 8", 24" O.C. rafter framing for flat, shed or gable roofs, 25' span, up to 5/12 slope, per S.F. of roof.
	Install	S.F.	.44	.23		.67	
	Demolish and Install	S.F.	.44	.67		1.11	
	Remove and Install	S.F.		.18		.18	
	Minimum Charge	Job		150		150	
2" x 10" Rafters							
	Demolish	S.F.		.44		.44	Cost includes material and labor to install 2" x 10", 24" O.C. rafter framing for flat, shed or gable roofs, 25' span, up to 5/12 slope, per S.F. of roof.
	Install	S.F.	.63	.34		.97	
	Demolish and Install	S.F.	.63	.78		1.41	
	Remove and Install	S.F.		.27		.27	
	Minimum Charge	Job		150		150	
2" x 12" Rafters							
	Demolish	S.F.		.45		.45	Cost includes material and labor to install 2" x 12", 24" O.C. rafter framing for flat, shed or gable roofs, 25' span, up to 5/12 slope, per S.F. of roof.
	Install	S.F.	.86	.37		1.23	
	Demolish and Install	S.F.	.86	.82		1.68	
	Remove and Install	S.F.		.30		.30	
	Minimum Charge	Job		150		150	
Rafter							
2" x 4"							
	Demolish	L.F.		.56		.56	Cost includes material and labor to install 2" x 4" rafters for flat, shed or gable roofs, up to 5/12 slope, 25' span, per L.F.
	Install	L.F.	.36	.87		1.23	
	Demolish and Install	L.F.	.36	1.43		1.79	
	Remove and Install	L.F.		.70		.70	
	Minimum Charge	Job		150		150	
2" x 6"							
	Demolish	L.F.		.57		.57	Cost includes material and labor to install 2" x 6" rafters for flat, shed or gable roofs, up to 5/12 slope, 25' span, per L.F.
	Install	L.F.	.57	.75		1.32	
	Demolish and Install	L.F.	.57	1.32		1.89	
	Remove and Install	L.F.		.60		.60	
	Minimum Charge	Job		150		150	

Roof Framing	Unit	Material	Labor	Equip.	Total	Specification
2" x 8"						
Demolish	L.F.		.58		.58	Cost includes material and labor to install 2" x 8" rafters for flat, shed or gable roofs, up to 5/12 slope, 25' span, per L.F.
Install	L.F.	.88	.80		1.68	
Demolish and Install	L.F.	.88	1.38		2.26	
Remove and Install	L.F.		.64		.64	
Minimum Charge	Job		150		150	
2" x 10"						
Demolish	L.F.		.59		.59	Cost includes material and labor to install 2" x 10" rafters for flat, shed or gable roofs, up to 5/12 slope, 25' span, per L.F.
Install	L.F.	1.27	1.22		2.49	
Demolish and Install	L.F.	1.27	1.81		3.08	
Remove and Install	L.F.		.97		.97	
Minimum Charge	Job		150		150	
2" x 12"						
Demolish	L.F.		.60		.60	Cost includes material and labor to install 2" x 12" rafters for flat, shed or gable roofs, up to 5/12 slope, 25' span, per L.F.
Install	L.F.	1.72	1.32		3.04	
Demolish and Install	L.F.	1.72	1.92		3.64	
Remove and Install	L.F.		1.06		1.06	
Minimum Charge	Job		150		150	
2"x 4" Valley / Jack						
Demolish	L.F.		.56		.56	Includes material and labor to install up to 2" x 6" valley/jack rafters for flat, shed or gable roofs, up to 5/12 slope, 25' span, per L.F.
Install	L.F.	.57	1.27		1.84	
Demolish and Install	L.F.	.57	1.83		2.40	
Remove and Install	L.F.		1.01		1.01	
Minimum Charge	Job		150		150	
2"x 6" Valley / Jack						
Demolish	L.F.		.57		.57	Includes material and labor to install up to 2" x 6" valley/jack rafters for flat, shed or gable roofs, up to 5/12 slope, 25' span, per L.F.
Install	L.F.	.57	1.27		1.84	
Demolish and Install	L.F.	.57	1.84		2.41	
Remove and Install	L.F.		1.01		1.01	
Minimum Charge	Job		150		150	
Ridgeboard						
2" x 4"						
Demolish	L.F.		.54		.54	Cost includes material and labor to install 2" x 4" ridgeboard for flat, shed or gable roofs, up to 5/12 slope, 25' span, per L.F.
Install	L.F.	.36	.55		.91	
Demolish and Install	L.F.	.36	1.09		1.45	
Remove and Install	L.F.		.44		.44	
Minimum Charge	Job		150		150	
2" x 6"						
Demolish	L.F.		.55		.55	Cost includes material and labor to install 2" x 6" ridgeboard for flat, shed or gable roofs, up to 5/12 slope, 25' span, per L.F.
Install	L.F.	.57	1.20		1.77	
Demolish and Install	L.F.	.57	1.75		2.32	
Remove and Install	L.F.		.96		.96	
Minimum Charge	Job		150		150	
2" x 8"						
Demolish	L.F.		.57		.57	Cost includes material and labor to install 2" x 8" ridgeboard for flat, shed or gable roofs, up to 5/12 slope, 25' span, per L.F.
Install	L.F.	.88	1.34		2.22	
Demolish and Install	L.F.	.88	1.91		2.79	
Remove and Install	L.F.		1.07		1.07	
Minimum Charge	Job		150		150	
2" x 10"						
Demolish	L.F.		.59		.59	Cost includes material and labor to install 2" x 10" ridgeboard for flat, shed or gable roofs, up to 5/12 slope, 25' span, per L.F.
Install	L.F.	1.27	1.50		2.77	
Demolish and Install	L.F.	1.27	2.09		3.36	
Remove and Install	L.F.		1.20		1.20	
Minimum Charge	Job		150		150	

Roof Framing		Unit	Material	Labor	Equip.	Total	Specification
2" x 12"							
	Demolish	L.F.		.61		.61	Cost includes material and labor to install 2" x 12" ridgeboard for flat, shed or gable roofs, up to 5/12 slope, 25' span, per L.F.
	Install	L.F.	1.72	.86		2.58	
	Demolish and Install	L.F.	1.72	1.47		3.19	
	Remove and Install	L.F.		.69		.69	
	Minimum Charge	Job		150		150	
Collar Beam							
1" x 6"							
	Demolish	L.F.		.43		.43	Cost includes material and labor to install 1" x 6" collar beam or tie for roof framing.
	Install	L.F.	.55	.38		.93	
	Demolish and Install	L.F.	.55	.81		1.36	
	Remove and Install	L.F.		.30		.30	
	Minimum Charge	Job		150		150	
2" x 6"							
	Demolish	L.F.		.50		.50	Cost includes material and labor to install 2" x 6" collar beam or tie for roof framing.
	Install	L.F.	.57	.38		.95	
	Demolish and Install	L.F.	.57	.88		1.45	
	Remove and Install	L.F.		.30		.30	
	Minimum Charge	Job		150		150	
Purlins							
	Demolish	L.F.		.48		.48	Cost includes material and labor to install 2" x 6" purlins below roof rafters.
	Install	L.F.	.57	.33		.90	
	Demolish and Install	L.F.	.57	.81		1.38	
	Remove and Install	L.F.		.27		.27	
	Minimum Charge	Job		150		150	
2" x 8"							
	Demolish	L.F.		.49		.49	Cost includes material and labor to install 2" x 8" purlins below roof rafters.
	Install	L.F.	.88	.34		1.22	
	Demolish and Install	L.F.	.88	.83		1.71	
	Remove and Install	L.F.		.27		.27	
	Minimum Charge	Job		150		150	
2" x 10"							
	Demolish	L.F.		.49		.49	Cost includes material and labor to install 2" x 10" purlins below roof rafters.
	Install	L.F.	1.27	.34		1.61	
	Demolish and Install	L.F.	1.27	.83		2.10	
	Remove and Install	L.F.		.27		.27	
	Minimum Charge	Job		150		150	
2" x 12"							
	Demolish	L.F.		.50		.50	Cost includes material and labor to install 2" x 12" purlins below roof rafters.
	Install	L.F.	1.72	.35		2.07	
	Demolish and Install	L.F.	1.72	.85		2.57	
	Remove and Install	L.F.		.28		.28	
	Minimum Charge	Job		150		150	
4" x 6"							
	Demolish	L.F.		.50		.50	Cost includes material and labor to install 4" x 6" purlins below roof rafters.
	Install	L.F.	2.87	.35		3.22	
	Demolish and Install	L.F.	2.87	.85		3.72	
	Remove and Install	L.F.		.28		.28	
	Minimum Charge	Job		150		150	
4" x 8"							
	Demolish	L.F.		.50		.50	Cost includes material and labor to install 4" x 8" purlins below roof rafters.
	Install	L.F.	3.83	.35		4.18	
	Demolish and Install	L.F.	3.83	.85		4.68	
	Remove and Install	L.F.		.28		.28	
	Minimum Charge	Job		150		150	

Roof Framing	Unit	Material	Labor	Equip.	Total	Specification
Ledger Board						
2" x 6"						
Demolish	L.F.		.24		.24	Cost includes material and labor to install 2" x 6" ledger board nailed to the face of a rafter.
Install	L.F.	.57	1.36		1.93	
Demolish and Install	L.F.	.57	1.60		2.17	
Remove and Install	L.F.		1.08		1.08	
Minimum Charge	Job		150		150	
4" x 6"						
Demolish	L.F.		.40		.40	Cost includes material and labor to install 4" x 6" ledger board nailed to the face of a rafter.
Install	L.F.	2.87	1.99		4.86	
Demolish and Install	L.F.	2.87	2.39		5.26	
Remove and Install	L.F.		1.59		1.59	
Minimum Charge	Job		150		150	
4" x 8"						
Demolish	L.F.		.54		.54	Cost includes material and labor to install 4" x 8" ledger board nailed to the face of a rafter.
Install	L.F.	3.83	2.30		6.13	
Demolish and Install	L.F.	3.83	2.84		6.67	
Remove and Install	L.F.		1.84		1.84	
Minimum Charge	Job		150		150	
Outriggers						
2" x 4"						
Demolish	L.F.		.17		.17	Cost includes material and labor to install 2" x 4" outrigger rafters for flat, shed or gabled roofs, up to 5/12 slope, per L.F.
Install	L.F.	.36	1.20		1.56	
Demolish and Install	L.F.	.36	1.37		1.73	
Remove and Install	L.F.		.96		.96	
Minimum Charge	Job		150		150	
2" x 6"						
Demolish	L.F.		.25		.25	Cost includes material and labor to install 2" x 6" outrigger rafters for flat, shed or gabled roofs, up to 5/12 slope, per L.F.
Install	L.F.	.57	1.20		1.77	
Demolish and Install	L.F.	.57	1.45		2.02	
Remove and Install	L.F.		.96		.96	
Minimum Charge	Job		150		150	
Lookout Rafter						
2" x 4"						
Demolish	L.F.		.56		.56	Cost includes material and labor to install 2" x 4" lookout rafters for flat, shed or gabled roofs, up to 5/12 slope, per L.F.
Install	L.F.	.36	.50		.86	
Demolish and Install	L.F.	.36	1.06		1.42	
Remove and Install	L.F.		.40		.40	
Minimum Charge	Job		150		150	
2" x 6"						
Demolish	L.F.		.57		.57	Cost includes material and labor to install 2" x 6" lookout rafters for flat, shed or gabled roofs, up to 5/12 slope, per L.F.
Install	L.F.	.57	.51		1.08	
Demolish and Install	L.F.	.57	1.08		1.65	
Remove and Install	L.F.		.41		.41	
Minimum Charge	Job		150		150	
Fly Rafter						
2" x 4"						
Demolish	L.F.		.56		.56	Cost includes material and labor to install 2" x 4" fly rafters for flat, shed or gabled roofs, up to 5/12 slope, per L.F.
Install	L.F.	.36	.50		.86	
Demolish and Install	L.F.	.36	1.06		1.42	
Remove and Install	L.F.		.40		.40	
Minimum Charge	Job		150		150	

Roof Framing	Unit	Material	Labor	Equip.	Total	Specification
2" x 6"						
Demolish	L.F.		.57		.57	Cost includes material and labor to install 2" x 6" fly rafters for flat, shed or gabled roofs, up to 5/12 slope, per L.F.
Install	L.F.	.57	.51		1.08	
Demolish and Install	L.F.	.57	1.08		1.65	
Remove and Install	L.F.		.41		.41	
Minimum Charge	Job		150		150	

Residential Trusses	Unit	Material	Labor	Equip.	Total	Specification
W or Fink Truss						
24' Span						
Demolish	Ea.		7.35		7.35	Includes material, labor and equipment to install wood gang-nailed residential truss, up to 24' span.
Install	Ea.	51	22.50	11.65	85.15	
Demolish and Install	Ea.	51	29.85	11.65	92.50	
Remove and Install	Ea.		22.42	11.66	34.08	
Minimum Charge	Job		150		150	
28' Span						
Demolish	Ea.		7.80		7.80	Includes material, labor and equipment to install wood gang-nailed residential truss, up to 28' span.
Install	Ea.	62	25.50	13.20	100.70	
Demolish and Install	Ea.	62	33.30	13.20	108.50	
Remove and Install	Ea.		25.38	13.20	38.58	
Minimum Charge	Job		150		150	
32' Span						
Demolish	Ea.		8.65		8.65	Includes material, labor and equipment to install wood gang-nailed residential truss, up to 32' span.
Install	Ea.	87.50	27	14	128.50	
Demolish and Install	Ea.	87.50	35.65	14	137.15	
Remove and Install	Ea.		26.90	13.99	40.89	
Minimum Charge	Job		150		150	
36' Span						
Demolish	Ea.		9.30		9.30	Includes material, labor and equipment to install wood gang-nailed residential truss, up to 36' span.
Install	Ea.	109	29	15.20	153.20	
Demolish and Install	Ea.	109	38.30	15.20	162.50	
Remove and Install	Ea.		29.24	15.21	44.45	
Minimum Charge	Job		150		150	
40' Span						
Demolish	Ea.		10.10		10.10	Includes material, labor and equipment to install wood gang-nailed residential truss, up to 40' span.
Install	Ea.	161	31.50	16.25	208.75	
Demolish and Install	Ea.	161	41.60	16.25	218.85	
Remove and Install	Ea.		31.28	16.27	47.55	
Minimum Charge	Job		150		150	
Gable End						
24' Span						
Demolish	Ea.		7.35		7.35	Includes material, labor and equipment to install wood gang-nailed residential gable end truss, up to 24' span.
Install	Ea.	72.50	24	12.50	109	
Demolish and Install	Ea.	72.50	31.35	12.50	116.35	
Remove and Install	Ea.		24.02	12.49	36.51	
Clean	Ea.	1.14	29		30.14	
Paint	Ea.	3.42	35		38.42	
Minimum Charge	Job		150		150	
28' Span						
Demolish	Ea.		7.80		7.80	Includes material, labor and equipment to install wood gang-nailed residential gable end truss, up to 30' span.
Install	Ea.	88	28.50	14.90	131.40	
Demolish and Install	Ea.	88	36.30	14.90	139.20	
Remove and Install	Ea.		28.62	14.89	43.51	
Clean	Ea.	1.19	39.50		40.69	
Paint	Ea.	3.56	48		51.56	
Minimum Charge	Job		150		150	

Rough Frame / Structure

Residential Trusses	Unit	Material	Labor	Equip.	Total	Specification
32′ Span						
Demolish	Ea.		8.65		8.65	Includes material, labor and equipment to install wood gang-nailed residential gable end truss, up to 32′ span.
Install	Ea.	98.50	29	15.20	142.70	
Demolish and Install	Ea.	98.50	37.65	15.20	151.35	
Remove and Install	Ea.		29.24	15.21	44.45	
Clean	Ea.	1.24	52		53.24	
Paint	Ea.	3.73	63		66.73	
Minimum Charge	Job		150		150	
36′ Span						
Demolish	Ea.		9.30		9.30	Includes material, labor and equipment to install wood gang-nailed residential gable end truss, up to 36′ span.
Install	Ea.	117	33.50	17.50	168	
Demolish and Install	Ea.	117	42.80	17.50	177.30	
Remove and Install	Ea.		33.63	17.49	51.12	
Clean	Ea.	1.30	65.50		66.80	
Paint	Ea.	3.91	79.50		83.41	
Minimum Charge	Job		150		150	
40′ Span						
Demolish	Ea.		10.10		10.10	Includes material, labor and equipment to install wood gang-nailed residential gable end truss, up to 40′ span.
Install	Ea.	130	35.50	18.40	183.90	
Demolish and Install	Ea.	130	45.60	18.40	194	
Remove and Install	Ea.		35.40	18.41	53.81	
Clean	Ea.	4.10	80.50		84.60	
Paint	Ea.	4.10	98		102.10	
Minimum Charge	Job		150		150	
Truss Hardware						
5-1/4″ Glue-Lam Seat						
Install	Ea.	49	1.67		50.67	Includes labor and material to install beam hangers.
Minimum Charge	Job		150		150	
6-3/4″ Glue-Lam Seat						
Install	Ea.	50.50	1.67		52.17	Includes labor and material to install beam hangers.
Minimum Charge	Job		150		150	
8-3/4″ Glue-Lam Seat						
Install	Ea.	58	1.67		59.67	Includes labor and material to install beam hangers.
Minimum Charge	Job		150		150	
2″ x 4″ Joist Hanger						
Install	Ea.	.59	1.72		2.31	Includes labor and material to install joist and beam hangers.
Minimum Charge	Job		150		150	
2″ x 12″ Joist Hanger						
Install	Ea.	.51	1.82		2.33	Cost includes labor and material to install 2″ x 12″ joist hanger.
Minimum Charge	Job		150		150	

Commercial Trusses	Unit	Material	Labor	Equip.	Total	Specification
Engineered Lumber, Truss / Joist						
9-1/2″						
Demolish	S.F.		.40		.40	Includes material and labor to install engineered lumber truss / joists per S.F. of floor area based on joists 16″ O.C. Beams, supports and bridging are not included.
Install	S.F.	1.54	.30		1.84	
Demolish and Install	S.F.	1.54	.70		2.24	
Remove and Install	S.F.		.24		.24	
Minimum Charge	Job		150		150	
11-7/8″						
Demolish	S.F.		.41		.41	Includes material and labor to install engineered lumber truss / joists per S.F. of floor area based on joists 16″ O.C. Beams, supports and bridging are not included.
Install	S.F.	1.66	.31		1.97	
Demolish and Install	S.F.	1.66	.72		2.38	
Remove and Install	S.F.		.25		.25	
Minimum Charge	Job		150		150	

Commercial Trusses	Unit	Material	Labor	Equip.	Total	Specification
14"						
Demolish	S.F.		.43		.43	Includes material and labor to install engineered lumber truss / joists per S.F. of floor area based on joists 16" O.C. Beams, supports and bridging are not included.
Install	S.F.	1.82	.33		2.15	
Demolish and Install	S.F.	1.82	.76		2.58	
Remove and Install	S.F.		.26		.26	
Minimum Charge	Job		150		150	
16"						
Demolish	S.F.		.44		.44	Includes material and labor to install engineered lumber truss / joists per S.F. of floor area based on joists 16" O.C. Beams, supports and bridging are not included.
Install	S.F.	2.27	.35		2.62	
Demolish and Install	S.F.	2.27	.79		3.06	
Remove and Install	S.F.		.28		.28	
Minimum Charge	Job		150		150	
Bowstring Truss						
100' Clear Span						
Demolish	SF Flr.		.41	.33	.74	Includes material, labor and equipment to install bow string truss system up to 100' clear span per S.F. of floor area.
Install	SF Flr.	5.90	.34	.32	6.56	
Demolish and Install	SF Flr.	5.90	.75	.65	7.30	
Minimum Charge	Job		150		150	
120' Clear Span						
Demolish	SF Flr.		.45	.37	.82	Includes material, labor and equipment to install bow string truss system up to 120' clear span per S.F. of floor area.
Install	SF Flr.	6.35	.38	.35	7.08	
Demolish and Install	SF Flr.	6.35	.83	.72	7.90	
Minimum Charge	Job		150		150	

Roof Sheathing	Unit	Material	Labor	Equip.	Total	Specification
Plywood						
3/8"						
Demolish	S.F.		.32		.32	Cost includes material and labor to install 3/8" CDX plywood roof sheathing.
Install	S.F.	.48	.39		.87	
Demolish and Install	S.F.	.48	.71		1.19	
Remove and Install	S.F.		.39		.39	
Minimum Charge	Job		150		150	
1/2"						
Demolish	S.F.		.32		.32	Cost includes material and labor to install 1/2" CDX plywood roof sheathing.
Install	S.F.	.54	.43		.97	
Demolish and Install	S.F.	.54	.75		1.29	
Minimum Charge	Job		150		150	
5/8"						
Demolish	S.F.		.32		.32	Cost includes material and labor to install 5/8" CDX plywood roof sheathing.
Install	S.F.	.63	.46		1.09	
Demolish and Install	S.F.	.63	.78		1.41	
Minimum Charge	Job		150		150	
3/4"						
Demolish	S.F.		.32		.32	Cost includes material and labor to install 3/4" CDX plywood roof sheathing.
Install	S.F.	.73	.50		1.23	
Demolish and Install	S.F.	.73	.82		1.55	
Minimum Charge	Job		150		150	
OSB						
1/2"						
Demolish	S.F.		.32		.32	Cost includes material and labor to install 4' x 8' x 1/2" OSB sheathing.
Install	S.F.	.34	.23		.57	
Demolish and Install	S.F.	.34	.55		.89	
Minimum Charge	Job		150		150	

Roof Sheathing	Unit	Material	Labor	Equip.	Total	Specification
5/8″						
Demolish	S.F.		.32		.32	Cost includes material and labor to install 4′ x 8′ x 5/8″ OSB sheathing.
Install	S.F.	.52	.24		.76	
Demolish and Install	S.F.	.52	.56		1.08	
Minimum Charge	Job		150		150	
Plank						
Demolish	S.F.		.32		.32	Cost includes material and labor to install 1″ x 6″ or 1″ x 8″ utility T&G board sheathing laid diagonal.
Install	S.F.	1.23	.83		2.06	
Demolish and Install	S.F.	1.23	1.15		2.38	
Remove and Install	S.F.		.83		.83	
Minimum Charge	Job		150		150	
Skip-type						
1″ x 4″						
Demolish	S.F.		.48		.48	Includes material and labor to install sheathing board material 1″ x 4″, 7″ O.C.
Install	S.F.	.48	.25		.73	
Demolish and Install	S.F.	.48	.73		1.21	
Minimum Charge	Job		150		150	
1″ x 6″						
Demolish	S.F.		.48		.48	Includes material and labor to install sheathing board material 1″ x 6″, 7″ or 9″ O.C.
Install	S.F.	.55	.21		.76	
Demolish and Install	S.F.	.55	.69		1.24	
Minimum Charge	Job		150		150	

Wood Deck	Unit	Material	Labor	Equip.	Total	Specification
Treated						
Demolish	S.F.		1.01		1.01	Includes material and labor to install unfinished deck using 2″ x 4″ treated decking, with 2″ x 6″ double beams 24″ O.C., 4″ x 4″ posts 5′ O.C. set in concrete.
Install	S.F.	4.21	7.50		11.71	
Demolish and Install	S.F.	4.21	8.51		12.72	
Remove and Install	S.F.		6.02		6.02	
Clean	S.F.	.06	.22		.28	
Paint	S.F.	.06	.20		.26	
Minimum Charge	Job		150		150	
Redwood						
Demolish	S.F.		1.01		1.01	Includes material and labor to install unfinished deck using 2″ x 4″ select redwood decking, with 2″ x 6″ double beams 24″ O.C., 4″ x 4″ posts 5′ O.C. set in concrete.
Install	S.F.	16.35	7.50		23.85	
Demolish and Install	S.F.	16.35	8.51		24.86	
Remove and Install	S.F.		6.02		6.02	
Clean	S.F.	.06	.22		.28	
Paint	S.F.	.06	.20		.26	
Minimum Charge	Job		150		150	
Cedar						
Demolish	S.F.		1.01		1.01	Includes material and labor to install unfinished deck using 2″ x 4″ select cedar decking, with 2″ x 6″ double beams 24″ O.C., 4″ x 4″ posts 5′ O.C. set in concrete.
Install	S.F.	7.15	7.50		14.65	
Demolish and Install	S.F.	7.15	8.51		15.66	
Remove and Install	S.F.		6.02		6.02	
Clean	S.F.	.06	.22		.28	
Paint	S.F.	.06	.20		.26	
Minimum Charge	Job		150		150	
Stairs						
Demolish	Ea.		290		290	Includes material and labor to install wood steps.
Install	Ea.	146	520		666	
Demolish and Install	Ea.	146	810		956	
Remove and Install	Ea.		416.64		416.64	
Clean	Flight	2.73	18.90		21.63	
Paint	Flight	10	35		45	
Minimum Charge	Job		150		150	

Wood Deck

		Unit	Material	Labor	Equip.	Total	Specification
Railing							
	Demolish	L.F.		3.73		3.73	Includes material and labor to install porch rail with balusters.
	Install	L.F.	13.95	13.65		27.60	
	Demolish and Install	L.F.	13.95	17.38		31.33	
	Clean	L.F.	.55	1.36		1.91	
	Paint	L.F.	.47	2.20		2.67	
	Minimum Charge	Job		150		150	
Bench Seating							
	Demolish	L.F.		4.04		4.04	Includes material and labor to install pressure treated bench seating.
	Install	L.F.	2.15	15.05		17.20	
	Demolish and Install	L.F.	2.15	19.09		21.24	
	Remove and Install	L.F.		12.03		12.03	
	Clean	L.F.	.11	1.09		1.20	
	Paint	L.F.	.48	2.11		2.59	
	Minimum Charge	Job		150		150	
Privacy Wall							
	Demolish	SF Flr.		.97		.97	Includes material, labor and equipment to install privacy wall.
	Install	SF Flr.	.42	.51		.93	
	Demolish and Install	SF Flr.	.42	1.48		1.90	
	Remove and Install	SF Flr.		.41		.41	
	Clean	SF Flr.	.06	.22		.28	
	Paint	S.F.	.06	.20		.26	
	Minimum Charge	Job		150		150	

Porch

		Unit	Material	Labor	Equip.	Total	Specification
Treated							
	Demolish	S.F.		2.02		2.02	Includes material and labor to install floor and roof frame and decking using 2″ x 4″ treated decking, with 2″ x 6″ double beams 24″ O.C., 4″ x 4″ posts 5′ O.C. set in concrete.
	Install	S.F.	5.10	15.05		20.15	
	Demolish and Install	S.F.	5.10	17.07		22.17	
	Remove and Install	S.F.		12.03		12.03	
	Clean	S.F.	.11	.45		.56	
	Paint	S.F.	.17	.53		.70	
	Minimum Charge	Job		150		150	
Redwood							
	Demolish	S.F.		2.02		2.02	Includes material and labor to install floor and roof frame and decking using 2″ x 4″ redwood decking, with 2″ x 6″ double beams 24″ O.C., 4″ x 4″ posts 5′ O.C. set in concrete.
	Install	S.F.	17.50	15.05		32.55	
	Demolish and Install	S.F.	17.50	17.07		34.57	
	Remove and Install	S.F.		12.03		12.03	
	Clean	S.F.	.11	.45		.56	
	Paint	S.F.	.17	.53		.70	
	Minimum Charge	Job		150		150	
Cedar							
	Demolish	S.F.		2.02		2.02	Includes material and labor to install floor and roof frame and cedar decking using 2″ x 4″ select cedar, with 2″ x 6″ double beams 24″ O.C., 4″ x 4″ posts 5′ O.C. set in concrete.
	Install	S.F.	8.30	15.05		23.35	
	Demolish and Install	S.F.	8.30	17.07		25.37	
	Remove and Install	S.F.		12.03		12.03	
	Clean	S.F.	.11	.45		.56	
	Paint	S.F.	.17	.53		.70	
	Minimum Charge	Job		150		150	
Stairs							
	Demolish	Ea.		290		290	Includes material and labor to install wood steps.
	Install	Ea.	146	520		666	
	Demolish and Install	Ea.	146	810		956	
	Remove and Install	Ea.		416.64		416.64	
	Clean	Flight	2.73	18.90		21.63	
	Paint	Flight	10	35		45	
	Minimum Charge	Job		150		150	

Porch	Unit	Material	Labor	Equip.	Total	Specification
Railing						
Demolish	L.F.		3.73		3.73	Includes material and labor to install porch rail with balusters.
Install	L.F.	13.95	13.65		27.60	
Demolish and Install	L.F.	13.95	17.38		31.33	
Clean	L.F.	.55	1.70		2.25	
Paint	L.F.	.47	2.20		2.67	
Minimum Charge	Job		150		150	
Bench Seating						
Demolish	L.F.		4.04		4.04	Includes material and labor to install pressure treated bench seating.
Install	L.F.	2.15	15.05		17.20	
Demolish and Install	L.F.	2.15	19.09		21.24	
Remove and Install	L.F.		12.03		12.03	
Clean	L.F.	.11	1.09		1.20	
Paint	L.F.	.48	2.11		2.59	
Minimum Charge	Job		150		150	
Privacy Wall						
Demolish	SF Flr.		.97		.97	Includes material, labor and equipment to install privacy wall.
Install	SF Flr.	.42	.51		.93	
Demolish and Install	SF Flr.	.42	1.48		1.90	
Remove and Install	SF Flr.		.41		.41	
Clean	SF Flr.	.06	.22		.28	
Paint	S.F.	.06	.20		.26	
Minimum Charge	Job		150		150	
Wood Board Ceiling						
Demolish	S.F.		.40		.40	Includes material and labor to install wood board on ceiling.
Install	S.F.	1.77	1.34		3.11	
Demolish and Install	S.F.	1.77	1.74		3.51	
Clean	S.F.	.02	.13		.15	
Paint	S.F.	.15	.33		.48	
Minimum Charge	Job		150		150	
Re-screen						
Install	S.F.	.52	1.20		1.72	Includes labor and material to re-screen wood frame.
Minimum Charge	Job		150		150	

Exterior Trim

Exterior Trim	Unit	Material	Labor	Equip.	Total	Specification
Stock Lumber						
1″ x 2″						
Demolish	L.F.		.20		.20	Cost includes material and labor to install 1″ x 2″ pine trim.
Install	L.F.	.33	.91		1.24	
Demolish and Install	L.F.	.33	1.11		1.44	
Clean	L.F.		.15		.15	
Paint	L.F.	.02	.41		.43	
Minimum Charge	Job		150		150	
1″ x 3″						
Demolish	L.F.		.20		.20	Cost includes material and labor to install 1″ x 3″ pine trim.
Install	L.F.	.21	1		1.21	
Demolish and Install	L.F.	.21	1.20		1.41	
Clean	L.F.	.01	.15		.16	
Paint	L.F.	.02	.41		.43	
Minimum Charge	Job		150		150	
1″ x 4″						
Demolish	L.F.		.20		.20	Cost includes material and labor to install 1″ x 4″ pine trim.
Install	L.F.	.67	1.20		1.87	
Demolish and Install	L.F.	.67	1.40		2.07	
Clean	L.F.	.01	.15		.16	
Paint	L.F.	.02	.41		.43	
Minimum Charge	Job		150		150	
1″ x 6″						
Demolish	L.F.		.20		.20	Cost includes material and labor to install 1″ x 6″ pine trim.
Install	L.F.	1.18	1.20		2.38	
Demolish and Install	L.F.	1.18	1.40		2.58	
Clean	L.F.	.01	.15		.16	
Paint	L.F.	.07	.41		.48	
Minimum Charge	Job		150		150	
1″ x 8″						
Demolish	L.F.		.22		.22	Cost includes material and labor to install 1″ x 8″ pine trim.
Install	L.F.	1.45	1.50		2.95	
Demolish and Install	L.F.	1.45	1.72		3.17	
Remove and Install	L.F.		1.20		1.20	
Clean	L.F.	.02	.16		.18	
Paint	L.F.	.07	.41		.48	
Minimum Charge	Job		150		150	

Exterior Trim

Exterior Trim

Exterior Trim	Unit	Material	Labor	Equip.	Total	Specification
1" x 10"						
Demolish	L.F.		.34		.34	Cost includes material and labor to install 1" x 10" pine trim.
Install	L.F.	1.60	1.67		3.27	
Demolish and Install	L.F.	1.60	2.01		3.61	
Minimum Charge	Job		150		150	
1" x 12"						
Demolish	L.F.		.34		.34	Cost includes material and labor to install 1" x 12" pine trim.
Install	L.F.	1.75	1.67		3.42	
Demolish and Install	L.F.	1.75	2.01		3.76	
Minimum Charge	Job		150		150	

Fascia

Fascia	Unit	Material	Labor	Equip.	Total	Specification
Aluminum						
Demolish	L.F.		.42		.42	Includes material and labor to install aluminum fascia.
Install	S.F.	1.11	1.43		2.54	
Demolish and Install	S.F.	1.11	1.85		2.96	
Remove and Install	S.F.		1.15		1.15	
Clean	L.F.		.27		.27	
Minimum Charge	Job		150		150	
Vinyl						
Demolish	L.F.		.42		.42	Cost includes material and labor to install 6" vinyl fascia with 12" vinyl soffit and J-channel.
Install	L.F.	2.34	1.72		4.06	
Demolish and Install	L.F.	2.34	2.14		4.48	
Remove and Install	L.F.		1.38		1.38	
Clean	L.F.		.27		.27	
Minimum Charge	Job		150		150	
Plywood						
Demolish	L.F.		.42		.42	Cost includes material and labor to install 12" wide plywood fascia.
Install	L.F.	1.02	.67		1.69	
Demolish and Install	L.F.	1.02	1.09		2.11	
Remove and Install	L.F.		.53		.53	
Clean	L.F.	.03	.18		.21	
Paint	L.F.	.14	.66		.80	
Minimum Charge	Job		150		150	
Cedar						
Demolish	L.F.		.42		.42	Cost includes material and labor to install 12" wide cedar fascia.
Install	L.F.	1.58	.67		2.25	
Demolish and Install	L.F.	1.58	1.09		2.67	
Remove and Install	L.F.		.53		.53	
Clean	L.F.	.03	.18		.21	
Paint	L.F.	.14	.66		.80	
Minimum Charge	Job		150		150	
Redwood						
1" x 6"						
Demolish	L.F.		.32		.32	Cost includes material and labor to install 1" x 6" redwood fascia board.
Install	L.F.	2.17	1.20		3.37	
Demolish and Install	L.F.	2.17	1.52		3.69	
Remove and Install	L.F.		.96		.96	
Minimum Charge	Job		150		150	
1" x 8"						
Demolish	L.F.		.37		.37	Cost includes material and labor to install 1" x 8" redwood fascia board.
Install	L.F.	2.97	1.34		4.31	
Demolish and Install	L.F.	2.97	1.71		4.68	
Remove and Install	L.F.		1.07		1.07	
Minimum Charge	Job		150		150	

Exterior Trim

Fascia	Unit	Material	Labor	Equip.	Total	Specification
2" x 6"						
Demolish	L.F.		.32		.32	Cost includes material and labor to install 2" x 6" redwood fascia board.
Install	L.F.	2.10	1.34		3.44	
Demolish and Install	L.F.	2.10	1.66		3.76	
Remove and Install	L.F.		1.07		1.07	
Minimum Charge	Job		150		150	
2" x 8"						
Demolish	L.F.		.37		.37	Cost includes material and labor to install 2" x 8" redwood fascia board.
Install	L.F.	2.84	1.50		4.34	
Demolish and Install	L.F.	2.84	1.87		4.71	
Remove and Install	L.F.		1.20		1.20	
Minimum Charge	Job		150		150	
Hem-fir Std & Better						
2" x 6"						
Demolish	L.F.		.32		.32	Cost includes material and labor to install 2" x 6" hem-fir exterior trim board.
Install	L.F.	.57	2.41		2.98	
Demolish and Install	L.F.	.57	2.73		3.30	
Remove and Install	L.F.		1.93		1.93	
Clean	L.F.	.01	.18		.19	
Paint	L.F.	.14	.66		.80	
Minimum Charge	Job		150		150	
2" x 8"						
Demolish	L.F.		.37		.37	Cost includes material and labor to install 2" x 8" hem-fir exterior trim board.
Install	L.F.	.88	2.67		3.55	
Demolish and Install	L.F.	.88	3.04		3.92	
Remove and Install	L.F.		2.14		2.14	
Clean	L.F.	.02	.18		.20	
Paint	L.F.	.14	.66		.80	
Minimum Charge	Job		150		150	
2" x 10"						
Demolish	L.F.		.42		.42	Cost includes material and labor to install 2" x 10" hem-fir exterior trim board.
Install	L.F.	1.27	3.34		4.61	
Demolish and Install	L.F.	1.27	3.76		5.03	
Remove and Install	L.F.		2.67		2.67	
Clean	L.F.	.03	.20		.23	
Paint	L.F.	.14	.66		.80	
Minimum Charge	Job		150		150	
Barge Rafter						
2" x 6"						
Demolish	L.F.		.65		.65	Cost includes material and labor to install 2" x 6" barge rafters.
Install	L.F.	.57	.33		.90	
Demolish and Install	L.F.	.57	.98		1.55	
Remove and Install	L.F.		.27		.27	
Clean	L.F.	.01	.18		.19	
Paint	L.F.	.14	.66		.80	
Minimum Charge	Job		150		150	
2" x 8"						
Demolish	L.F.		.66		.66	Cost includes material and labor to install 2" x 8" barge rafters.
Install	L.F.	.88	.35		1.23	
Demolish and Install	L.F.	.88	1.01		1.89	
Remove and Install	L.F.		.28		.28	
Clean	L.F.	.02	.18		.20	
Paint	L.F.	.14	.66		.80	
Minimum Charge	Job		150		150	

Fascia	Unit	Material	Labor	Equip.	Total	Specification
2″ x 10″						
Demolish	L.F.		.67		.67	Cost includes material and labor to install 2″ x 10″ barge rafters.
Install	L.F.	1.27	.36		1.63	
Demolish and Install	L.F.	1.27	1.03		2.30	
Remove and Install	L.F.		.29		.29	
Clean	L.F.	.03	.20		.23	
Paint	L.F.	.14	.66		.80	
Minimum Charge	Job		150		150	

Soffit	Unit	Material	Labor	Equip.	Total	Specification
Aluminum						
12″ w / Fascia						
Demolish	L.F.		.42		.42	Cost includes material and labor to install 12″ aluminum soffit with 6″ fascia and J-channel.
Install	L.F.	2.70	1.72		4.42	
Demolish and Install	L.F.	2.70	2.14		4.84	
Remove and Install	L.F.		1.38		1.38	
Clean	L.F.		.27		.27	
Paint	L.F.	.18	.66		.84	
Minimum Charge	Job		150		150	
18″ w / Fascia						
Demolish	L.F.		.42		.42	Cost includes material and labor to install 18″ aluminum soffit with 6″ fascia and J-channel.
Install	L.F.	3.21	2.01		5.22	
Demolish and Install	L.F.	3.21	2.43		5.64	
Remove and Install	L.F.		1.60		1.60	
Clean	L.F.		.27		.27	
Paint	L.F.	.18	.66		.84	
Minimum Charge	Job		150		150	
24″ w / Fascia						
Demolish	L.F.		.42		.42	Cost includes material and labor to install 24″ aluminum soffit with 6″ fascia and J-channel.
Install	L.F.	3.76	2.23		5.99	
Demolish and Install	L.F.	3.76	2.65		6.41	
Remove and Install	L.F.		1.78		1.78	
Clean	L.F.		.40		.40	
Paint	L.F.	.18	.66		.84	
Minimum Charge	Job		150		150	
Vinyl						
Demolish	L.F.		.42		.42	Cost includes material and labor to install 6″ vinyl fascia with 12″ vinyl soffit and J-channel.
Install	L.F.	2.34	1.72		4.06	
Demolish and Install	L.F.	2.34	2.14		4.48	
Remove and Install	L.F.		1.38		1.38	
Clean	L.F.		.27		.27	
Paint	L.F.	.18	.66		.84	
Minimum Charge	Job		150		150	
Plywood						
Demolish	L.F.		.47		.47	Cost includes material and labor to install 12″ wide plywood soffit including 6″ fascia.
Install	L.F.	1.02	.86		1.88	
Demolish and Install	L.F.	1.02	1.33		2.35	
Remove and Install	L.F.		.69		.69	
Clean	L.F.	.03	.18		.21	
Paint	L.F.	.14	.66		.80	
Minimum Charge	Job		150		150	

Soffit	Unit	Material	Labor	Equip.	Total	Specification
Redwood						
Demolish	L.F.		.47		.47	Cost includes material and labor to install 12″ wide redwood soffit including 6″ fascia.
Install	L.F.	4.33	.86		5.19	
Demolish and Install	L.F.	4.33	1.33		5.66	
Remove and Install	L.F.		.69		.69	
Clean	L.F.	.03	.18		.21	
Paint	L.F.	.14	.66		.80	
Minimum Charge	Job		150		150	
Cedar						
Demolish	L.F.		.47		.47	Cost includes material and labor to install 12″ wide cedar soffit including 6″ fascia.
Install	L.F.	1.58	.86		2.44	
Demolish and Install	L.F.	1.58	1.33		2.91	
Remove and Install	L.F.		.69		.69	
Clean	L.F.	.03	.18		.21	
Paint	L.F.	.14	.66		.80	
Minimum Charge	Job		150		150	
Douglas Fir						
Demolish	L.F.		.47		.47	Cost includes material and labor to install 12″ wide Douglas fir soffit including 6″ fascia.
Install	L.F.	1.75	.86		2.61	
Demolish and Install	L.F.	1.75	1.33		3.08	
Remove and Install	L.F.		.69		.69	
Clean	L.F.	.03	.18		.21	
Paint	L.F.	.14	.66		.80	
Minimum Charge	Job		150		150	

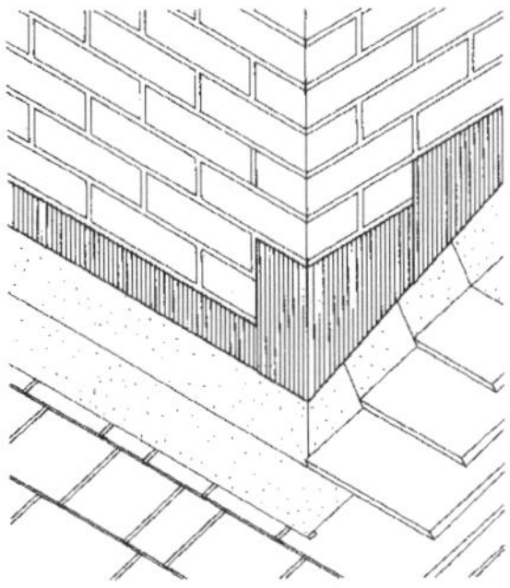

Roofing and Flashing

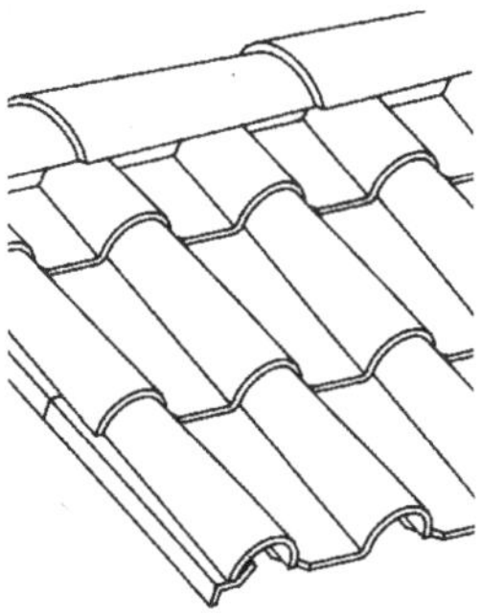

Tile Roofing

Metal Flashing	Unit	Material	Labor	Equip.	Total	Specification
Galvanized						
6"						
Demolish	L.F.		.16		.16	Includes material and labor to install flat galvanized steel flashing, 6" wide.
Install	L.F.	.41	.71		1.12	
Demolish and Install	L.F.	.41	.87		1.28	
Paint	L.F.	.10	.33		.43	
Minimum Charge	Job		141		141	
14"						
Demolish	L.F.		.16		.16	Includes material and labor to install flat galvanized steel flashing, 14" wide.
Install	L.F.	.95	.71		1.66	
Demolish and Install	L.F.	.95	.87		1.82	
Paint	L.F.	.10	.33		.43	
Minimum Charge	Job		141		141	
20"						
Demolish	L.F.		.16		.16	Includes material and labor to install flat galvanized steel flashing, 20" wide.
Install	L.F.	1.34	.71		2.05	
Demolish and Install	L.F.	1.34	.87		2.21	
Paint	L.F.	.10	.33		.43	
Minimum Charge	Job		141		141	
Aluminum						
6"						
Demolish	L.F.		.16		.16	Includes material and labor to install flat aluminum flashing, 6" wide.
Install	L.F.	.25	.71		.96	
Demolish and Install	L.F.	.25	.87		1.12	
Paint	L.F.	.10	.33		.43	
Minimum Charge	Job		141		141	
14"						
Demolish	L.F.		.16		.16	Includes material and labor to install flat aluminum flashing, 14" wide.
Install	L.F.	.48	.71		1.19	
Demolish and Install	L.F.	.48	.87		1.35	
Paint	L.F.	.10	.33		.43	
Minimum Charge	Job		141		141	
20"						
Demolish	L.F.		.16		.16	Includes material and labor to install flat aluminum flashing, 20" wide.
Install	L.F.	.64	.71		1.35	
Demolish and Install	L.F.	.64	.87		1.51	
Paint	L.F.	.10	.33		.43	
Minimum Charge	Job		141		141	

Metal Flashing		Unit	Material	Labor	Equip.	Total	Specification
Copper							
6″							
	Demolish	L.F.		.16		.16	Includes material and labor to install flat copper flashing, 6″ wide.
	Install	L.F.	1.72	.94		2.66	
	Demolish and Install	L.F.	1.72	1.10		2.82	
	Minimum Charge	Job		164		164	
14″							
	Demolish	L.F.		.16		.16	Includes material and labor to install flat copper flashing, 14″ wide.
	Install	L.F.	3.98	.94		4.92	
	Demolish and Install	L.F.	3.98	1.10		5.08	
	Minimum Charge	Job		164		164	
18″							
	Demolish	L.F.		.16		.16	Includes material and labor to install flat copper flashing, 18″ wide.
	Install	L.F.	5.70	.94		6.64	
	Demolish and Install	L.F.	5.70	1.10		6.80	
	Minimum Charge	Job		164		164	
Aluminum Valley							
	Demolish	L.F.		.16		.16	Includes material and labor to install aluminum valley flashing, .024″ thick.
	Install	L.F.	2.29	1.77		4.06	
	Demolish and Install	L.F.	2.29	1.93		4.22	
	Paint	L.F.	.10	.33		.43	
	Minimum Charge	Job		150		150	
Gravel Stop							
	Demolish	L.F.		.15		.15	Includes material and labor to install aluminum gravel stop, .050″ thick, 4″ high, mill finish.
	Install	L.F.	3.54	2.27		5.81	
	Demolish and Install	L.F.	3.54	2.42		5.96	
	Remove and Install	L.F.		1.81		1.81	
	Paint	L.F.	.15	.33		.48	
	Minimum Charge	Job		150		150	
Drip Edge							
	Demolish	L.F.		.15		.15	Includes material and labor to install aluminum drip edge, .016″ thick, 5″ wide, mill finish.
	Install	L.F.	.22	.75		.97	
	Demolish and Install	L.F.	.22	.90		1.12	
	Remove and Install	L.F.		.60		.60	
	Paint	L.F.	.15	.33		.48	
	Minimum Charge	Job		150		150	
End Wall							
	Demolish	L.F.		.75		.75	Includes material and labor to install aluminum end wall flashing, .024″ thick.
	Install	L.F.	1.42	1.77		3.19	
	Demolish and Install	L.F.	1.42	2.52		3.94	
	Minimum Charge	Job		150		150	
Side Wall							
	Demolish	L.F.		.14		.14	Includes material and labor to install aluminum side wall flashing, .024″ thick.
	Install	L.F.	1.41	1.77		3.18	
	Demolish and Install	L.F.	1.41	1.91		3.32	
	Minimum Charge	Job		150		150	
Coping and Wall Cap							
	Demolish	L.F.		1.01		1.01	Includes material and labor to install aluminum flashing, mill finish, 0.40 thick, 12″ wide.
	Install	L.F.	1.60	1.95		3.55	
	Demolish and Install	L.F.	1.60	2.96		4.56	
	Clean	L.F.	.07	.68		.75	
	Minimum Charge	Job		141		141	

Roof Shingles

Roof Shingles	Unit	Material	Labor	Equip.	Total	Specification
Underlayment						
15#						
Demolish	Sq.		8.05		8.05	Includes material and labor to install #15 felt underlayment.
Install	Sq.	2.45	4.41		6.86	
Demolish and Install	Sq.	2.45	12.46		14.91	
Minimum Charge	Job		141		141	
30#						
Demolish	Sq.		8.05		8.05	Includes material and labor to install #30 felt underlayment.
Install	Sq.	5.55	4.87		10.42	
Demolish and Install	Sq.	5.55	12.92		18.47	
Minimum Charge	Job		141		141	
Self Adhering						
Demolish	Sq.		20		20	Includes material and labor to install self adhering ice barrier roofing underlayment.
Install	Sq.	42	12.85		54.85	
Demolish and Install	Sq.	42	32.85		74.85	
Minimum Charge	Job		141		141	

Rolled Roofing

Rolled Roofing	Unit	Material	Labor	Equip.	Total	Specification
90 lb						
Demolish	Sq.		20		20	Cost includes material and labor to install 90 lb mineral surface rolled roofing.
Install	Sq.	16.90	18.85		35.75	
Demolish and Install	Sq.	16.90	38.85		55.75	
Minimum Charge	Job		141		141	
110 lb						
Demolish	Sq.		20		20	Cost includes material and labor to install 110 lb mineral surface rolled roofing.
Install	Sq.	44.50	18.85		63.35	
Demolish and Install	Sq.	44.50	38.85		83.35	
Minimum Charge	Job		141		141	
140 lb						
Demolish	Sq.		20		20	Cost includes material and labor to install 140 lb mineral surface rolled roofing.
Install	Sq.	44.50	28		72.50	
Demolish and Install	Sq.	44.50	48		92.50	
Minimum Charge	Job		141		141	

Composition Shingles

Composition Shingles	Unit	Material	Labor	Equip.	Total	Specification
Fiberglass (3-tab)						
225 lb, 20 Year						
Demolish	Sq.		22		22	Cost includes material and labor to install 225 lb, 20 year fiberglass shingles.
Install	Sq.	26.50	55		81.50	
Demolish and Install	Sq.	26.50	77		103.50	
Minimum Charge	Job		141		141	
300 lb, 25 Year						
Demolish	Sq.		22		22	Cost includes material and labor to install 300 lb, 25 year fiberglass shingles.
Install	Sq.	27.50	55		82.50	
Demolish and Install	Sq.	27.50	77		104.50	
Minimum Charge	Job		141		141	
Emergency Repairs						
Install	Ea.	72	141		213	Includes labor and material for emergency repairs.
Minimum Charge	Job		141		141	

Composition Shingles	Unit	Material	Labor	Equip.	Total	Specification
Architectural						
Laminated, 25 Year						
Demolish	Sq.		22		22	Cost includes material and labor to install 25 year fiberglass roof shingles.
Install	Sq.	51	55		106	
Demolish and Install	Sq.	51	77		128	
Minimum Charge	Job		141		141	
Laminated, 30 Year						
Demolish	Sq.		22		22	Cost includes material and labor to install 30 year fiberglass roof shingles.
Install	Sq.	33.50	55		88.50	
Demolish and Install	Sq.	33.50	77		110.50	
Minimum Charge	Job		141		141	
Laminated, 40 Year						
Demolish	Sq.		22		22	Cost includes material and labor to install 40 year fiberglass roof shingles.
Install	Sq.	69	55		124	
Demolish and Install	Sq.	69	77		146	
Minimum Charge	Job		141		141	
Laminated Shake, 40 Year						
Demolish	Sq.		22		22	Cost includes material and labor to install 40 year fiberglass roof shakes.
Install	Sq.	98.50	73.50		172	
Demolish and Install	Sq.	98.50	95.50		194	
Minimum Charge	Job		141		141	
Emergency Repairs						
Install	Ea.	72	141		213	Includes labor and material for emergency repairs.
Minimum Charge	Job		141		141	
Roof Jack						
Demolish	Ea.		4.40		4.40	Includes material and labor to install residential roof jack, w/bird screen, backdraft damper, 3" & 4" dia. round duct.
Install	Ea.	13.20	30		43.20	
Demolish and Install	Ea.	13.20	34.40		47.60	
Paint	Ea.	.06	3.30		3.36	
Minimum Charge	Job		165		165	
Ridge Vent						
Demolish	L.F.		.78		.78	Includes material and labor to install a mill finish aluminum ridge vent strip.
Install	L.F.	2.35	2.12		4.47	
Demolish and Install	L.F.	2.35	2.90		5.25	
Minimum Charge	Job		150		150	
Shingle Molding						
1" x 2"						
Demolish	L.F.		.15		.15	Cost includes material and labor to install 1" x 2" pine trim.
Install	L.F.	.33	.91		1.24	
Demolish and Install	L.F.	.33	1.06		1.39	
Paint	L.F.	.02	.41		.43	
Minimum Charge	Job		150		150	
1" x 3"						
Demolish	L.F.		.15		.15	Cost includes material and labor to install 1" x 3" pine trim.
Install	L.F.	.40	1.04		1.44	
Demolish and Install	L.F.	.40	1.19		1.59	
Paint	L.F.	.02	.41		.43	
Minimum Charge	Job		150		150	
1" x 4"						
Demolish	L.F.		.15		.15	Cost includes material and labor to install 1" x 4" pine trim.
Install	L.F.	.67	1.20		1.87	
Demolish and Install	L.F.	.67	1.35		2.02	
Paint	L.F.	.02	.41		.43	
Minimum Charge	Job		150		150	
Add for Steep Pitch						
Install	Sq.		17.65		17.65	Includes labor and material added for 2nd story or steep roofs.

Composition Shingles	Unit	Material	Labor	Equip.	Total	Specification
Add for Additional Story						
Install	Sq.		17.65		17.65	Includes labor and material added for 2nd story or steep roofs.

Wood Shingles	Unit	Material	Labor	Equip.	Total	Specification
Cedar Wood Shingles						
16" Red Label (#2)						
Demolish	Sq.		54		54	Includes material and labor to install wood shingles/shakes.
Install	Sq.	127	120		247	
Demolish and Install	Sq.	127	174		301	
Minimum Charge	Job		150		150	
16" Blue Label (#1)						
Demolish	Sq.		54		54	Includes material and labor to install wood shingles/shakes.
Install	Sq.	166	92.50		258.50	
Demolish and Install	Sq.	166	146.50		312.50	
Minimum Charge	Job		150		150	
18" Red Label (#2)						
Demolish	Sq.		54		54	Includes material and labor to install wood shingles/shakes.
Install	Sq.	219	77		296	
Demolish and Install	Sq.	219	131		350	
Minimum Charge	Job		150		150	
18" Blue Label (#1)						
Demolish	Sq.		54		54	Includes material and labor to install wood shingles/shakes.
Install	Sq.	176	84.50		260.50	
Demolish and Install	Sq.	176	138.50		314.50	
Minimum Charge	Job		150		150	
Replace Shingle						
Install	Ea.	13.20	7.50		20.70	Includes labor and material to replace individual wood shingle/shake. Cost is per shingle.
Minimum Charge	Job		150		150	
Cedar Shake						
18" Medium Handsplit						
Demolish	Sq.		54		54	Includes material and labor to install wood shingles/shakes.
Install	Sq.	107	116		223	
Demolish and Install	Sq.	107	170		277	
Minimum Charge	Job		150		150	
18" Heavy Handsplit						
Demolish	Sq.		54		54	Includes material and labor to install wood shingles/shakes.
Install	Sq.	107	129		236	
Demolish and Install	Sq.	107	183		290	
Minimum Charge	Job		150		150	
24" Heavy Handsplit						
Demolish	Sq.		54		54	Includes material and labor to install wood shingles/shakes.
Install	Sq.	152	103		255	
Demolish and Install	Sq.	152	157		309	
Minimum Charge	Job		150		150	
24" Medium Handsplit						
Demolish	Sq.		54		54	Includes material and labor to install wood shingles/shakes.
Install	Sq.	152	92.50		244.50	
Demolish and Install	Sq.	152	146.50		298.50	
Minimum Charge	Job		150		150	
24" Straight Split						
Demolish	Sq.		54		54	Includes material and labor to install wood shingles/shakes.
Install	Sq.	163	137		300	
Demolish and Install	Sq.	163	191		354	
Minimum Charge	Job		150		150	

Wood Shingles	Unit	Material	Labor	Equip.	Total	Specification
24" Tapersplit						
Demolish	Sq.		54		54	Includes material and labor to install wood shingles/shakes.
Install	Sq.	163	137		300	
Demolish and Install	Sq.	163	191		354	
Minimum Charge	Job		150		150	
18" Straight Split						
Demolish	Sq.		54		54	Includes material and labor to install wood shingles/shakes.
Install	Sq.	121	137		258	
Demolish and Install	Sq.	121	191		312	
Minimum Charge	Job		150		150	
Replace Shingle						
Install	Ea.	13.20	7.50		20.70	Includes labor and material to replace individual wood shingle/shake. Cost is per shingle.
Minimum Charge	Job		150		150	

Slate Tile Roofing	Unit	Material	Labor	Equip.	Total	Specification
Unfading Green						
Demolish	Sq.		48.50		48.50	Includes material and labor to install clear Vermont slate tile.
Install	Sq.	490	163		653	
Demolish and Install	Sq.	490	211.50		701.50	
Remove and Install	Sq.		130.01		130.01	
Minimum Charge	Job		142		142	
Unfading Purple						
Demolish	Sq.		48.50		48.50	Includes material and labor to install clear Vermont slate tile.
Install	Sq.	430	163		593	
Demolish and Install	Sq.	430	211.50		641.50	
Remove and Install	Sq.		130.01		130.01	
Minimum Charge	Job		142		142	
Replace Slate Tiles						
Install	Ea.	5.95	14.95		20.90	Includes labor and material to replace individual slate tiles. Cost is per tile.
Minimum Charge	Job		142		142	
Variegated Purple						
Demolish	Sq.		48.50		48.50	Includes material and labor to install clear Vermont slate tile.
Install	Sq.	415	163		578	
Demolish and Install	Sq.	415	211.50		626.50	
Remove and Install	Sq.		130.01		130.01	
Minimum Charge	Job		142		142	
Unfading Grey / Black						
Demolish	Sq.		48.50		48.50	Includes material and labor to install clear Vermont slate tile.
Install	Sq.	380	163		543	
Demolish and Install	Sq.	380	211.50		591.50	
Remove and Install	Sq.		130.01		130.01	
Minimum Charge	Job		142		142	
Unfading Red						
Demolish	Sq.		48.50		48.50	Includes material and labor to install clear Vermont slate tile.
Install	Sq.	1250	163		1413	
Demolish and Install	Sq.	1250	211.50		1461.50	
Remove and Install	Sq.		130.01		130.01	
Minimum Charge	Job		142		142	
Weathering Black						
Demolish	Sq.		48.50		48.50	Includes material and labor to install clear Pennsylvania slate tile.
Install	Sq.	480	163		643	
Demolish and Install	Sq.	480	211.50		691.50	
Remove and Install	Sq.		130.01		130.01	
Minimum Charge	Job		142		142	

Slate Tile Roofing	Unit	Material	Labor	Equip.	Total	Specification
Weathering Green						
Demolish	Sq.		48.50		48.50	Includes material and labor to install clear Vermont slate tile.
Install	Sq.	340	163		503	
Demolish and Install	Sq.	340	211.50		551.50	
Remove and Install	Sq.		130.01		130.01	
Minimum Charge	Job		142		142	

Flat Clay Tile Roofing	Unit	Material	Labor	Equip.	Total	Specification
Glazed						
Demolish	Sq.		54		54	Includes material and labor to install tile roofing. Cost based on terra cotta red flat clay tile and includes ridge and rake tiles.
Install	Sq.	495	114		609	
Demolish and Install	Sq.	495	168		663	
Remove and Install	Sq.		91.01		91.01	
Minimum Charge	Job		142		142	
Terra Cotta Red						
Demolish	Sq.		54		54	Includes material and labor to install tile roofing. Cost based on terra cotta red flat clay tile and includes ridge and rake tiles.
Install	Sq.	495	114		609	
Demolish and Install	Sq.	495	168		663	
Remove and Install	Sq.		91.01		91.01	
Minimum Charge	Job		142		142	

Mission Tile Roofing	Unit	Material	Labor	Equip.	Total	Specification
Glazed Red						
Demolish	Sq.		54		54	Includes material and labor to install tile roofing. Cost based on glazed red tile and includes birdstop, ridge and rake tiles.
Install	Sq.	715	190		905	
Demolish and Install	Sq.	715	244		959	
Remove and Install	Sq.		151.68		151.68	
Minimum Charge	Job		142		142	
Unglazed Red						
Demolish	Sq.		54		54	Includes material and labor to install tile roofing. Cost based on unglazed red tile and includes birdstop, ridge and rake tiles.
Install	Sq.	715	190		905	
Demolish and Install	Sq.	715	244		959	
Remove and Install	Sq.		151.68		151.68	
Minimum Charge	Job		142		142	
Glazed Blue						
Demolish	Sq.		54		54	Includes material and labor to install tile roofing. Cost based on glazed blue tile and includes birdstop, ridge and rake tiles.
Install	Sq.	715	190		905	
Demolish and Install	Sq.	715	244		959	
Remove and Install	Sq.		151.68		151.68	
Minimum Charge	Job		142		142	
Glazed White						
Demolish	Sq.		54		54	Includes material and labor to install tile roofing. Cost based on glazed white tile and includes birdstop, ridge and rake tiles.
Install	Sq.	715	190		905	
Demolish and Install	Sq.	715	244		959	
Remove and Install	Sq.		151.68		151.68	
Minimum Charge	Job		142		142	
Unglazed White						
Demolish	Sq.		54		54	Includes material and labor to install tile roofing. Cost based on unglazed white tile and includes birdstop, ridge and rake tiles.
Install	Sq.	715	190		905	
Demolish and Install	Sq.	715	244		959	
Remove and Install	Sq.		151.68		151.68	
Minimum Charge	Job		142		142	

Mission Tile Roofing	Unit	Material	Labor	Equip.	Total	Specification
Color Blend						
Demolish	Sq.		54		54	Includes material and labor to install tile roofing. Cost based on color blend tile and includes birdstop, ridge and rake tiles.
Install	Sq.	715	190		905	
Demolish and Install	Sq.	715	244		959	
Remove and Install	Sq.		151.68		151.68	
Minimum Charge	Job		142		142	
Accessories / Extras						
Add for Cap Furring Strips						
Install	Sq.	78	178		256	Includes labor and material to install vertically placed wood furring strips under cap tiles.
Add for Tile Adhesive						
Install	Sq.	13.35	10.40		23.75	Includes material and labor for adhesive material placed between tiles.
Add for Wire Attachment						
Install	Sq.	13.20	124		137.20	Includes labor and material to install single wire anchors for attaching tile.
Add for Braided Runners						
Install	Sq.	23.50	119		142.50	Includes labor and material to install braided wire runners for secure attachment in high wind areas.
Add for Stainless Steel Nails						
Install	Sq.	18.20			18.20	Includes material only for stainless steel nails.
Add for Brass Nails						
Install	Sq.	24			24	Includes material only for brass nails.
Add for Hurricane Clips						
Install	Sq.	15.95	167		182.95	Includes material and labor for Hurricane or wind clips for high wind areas.
Add for Copper Nails						
Install	Sq.	24			24	Includes material only for copper nails.
Add for Vertical Furring Strips						
Install	Sq.	15.40	33		48.40	Includes labor and material to install vertically placed wood furring strips under tiles.
Add for Horizontal Furring Strips						
Install	Sq.	9.90	14.80		24.70	Includes labor and material to install horizontally placed wood furring strips under tiles.

Spanish Tile Roofing	Unit	Material	Labor	Equip.	Total	Specification
Glazed Red						
Demolish	Sq.		54		54	Includes material and labor to install tile roofing. Cost based on glazed red tile and includes birdstop, ridge and rake tiles.
Install	Sq.	495	119		614	
Demolish and Install	Sq.	495	173		668	
Remove and Install	Sq.		94.80		94.80	
Minimum Charge	Job		142		142	

Spanish Tile Roofing	Unit	Material	Labor	Equip.	Total	Specification
Glazed White						
Demolish	Sq.		54		54	Includes material and labor to install tile roofing. Cost based on glazed white tile and includes birdstop, ridge and rake tiles.
Install	Sq.	495	119		614	
Demolish and Install	Sq.	495	173		668	
Remove and Install	Sq.		94.80		94.80	
Minimum Charge	Job		142		142	
Glazed Blue						
Demolish	Sq.		54		54	Includes material and labor to install tile roofing. Cost based on glazed blue tile and includes birdstop, ridge and rake tiles.
Install	Sq.	495	119		614	
Demolish and Install	Sq.	495	173		668	
Remove and Install	Sq.		94.80		94.80	
Minimum Charge	Job		142		142	
Unglazed Red						
Demolish	Sq.		54		54	Includes material and labor to install tile roofing. Cost based on unglazed red tile and includes birdstop, ridge and rake tiles.
Install	Sq.	495	119		614	
Demolish and Install	Sq.	495	173		668	
Remove and Install	Sq.		94.80		94.80	
Minimum Charge	Job		142		142	
Unglazed White						
Demolish	Sq.		54		54	Includes material and labor to install tile roofing. Cost based on unglazed white tile and includes birdstop, ridge and rake tiles.
Install	Sq.	495	119		614	
Demolish and Install	Sq.	495	173		668	
Remove and Install	Sq.		94.80		94.80	
Minimum Charge	Job		142		142	
Color Blend						
Demolish	Sq.		54		54	Includes material and labor to install tile roofing. Cost based on color blend tile and includes birdstop, ridge and rake tiles.
Install	Sq.	495	119		614	
Demolish and Install	Sq.	495	173		668	
Remove and Install	Sq.		94.80		94.80	
Minimum Charge	Job		142		142	
Accessories / Extras						
Add for Vertical Furring Strips						
Install	Sq.	15.40	33		48.40	Includes labor and material to install vertically placed wood furring strips under tiles.
Add for Wire Attachment						
Install	Sq.	11	95		106	Includes labor and material to install single wire ties.
Add for Braided Runners						
Install	Sq.	19.45	81.50		100.95	Includes labor and material to install braided wire runners for secure attachment in high wind areas.
Add for Stainless Steel Nails						
Install	Sq.	15.40			15.40	Includes material only for stainless steel nails.
Add for Copper Nails						
Install	Sq.	20.50			20.50	Includes material only for copper nails.
Add for Brass Nails						
Install	Sq.	20.50			20.50	Includes material only for brass nails.

Spanish Tile Roofing		Unit	Material	Labor	Equip.	Total	Specification
Add for Hurricane Clips							
	Install	Sq.	12.20	98		110.20	Includes material and labor to install hurricane or wind clips for high wind areas.
Add for Mortar Set Tiles							
	Install	Sq.	38	142		180	Includes material and labor to install setting tiles by using mortar material placed between tiles.
Add for Tile Adhesive							
	Install	Sq.	12.40	9.25		21.65	Includes material and labor for adhesive placed between tiles.
Add for Cap Furring Strips							
	Install	Sq.	78	178		256	Includes labor and material to install vertically placed wood furring strips under cap tiles.
Add for Horizontal Furring Strips							
	Install	Sq.	9.90	14.80		24.70	Includes labor and material to install horizontally placed wood furring strips under tiles.

Concrete Tile Roofing		Unit	Material	Labor	Equip.	Total	Specification
Corrugated Red / Brown							
	Demolish	Sq.		54		54	Includes material and labor to install tile roofing. Cost includes birdstop, booster, ridge and rake tiles.
	Install	Sq.	121	211		332	
	Demolish and Install	Sq.	121	265		386	
	Remove and Install	Sq.		168.53		168.53	
	Minimum Charge	Job		142		142	
Corrugated Gray							
	Demolish	Sq.		54		54	Includes material and labor to install tile roofing. Cost includes birdstop, booster, ridge and rake tiles.
	Install	Sq.	121	211		332	
	Demolish and Install	Sq.	121	265		386	
	Remove and Install	Sq.		168.53		168.53	
	Minimum Charge	Job		142		142	
Corrugated Bright Red							
	Demolish	Sq.		54		54	Includes material and labor to install tile roofing. Cost includes birdstop, booster, ridge and rake tiles.
	Install	Sq.	121	211		332	
	Demolish and Install	Sq.	121	265		386	
	Remove and Install	Sq.		168.53		168.53	
	Minimum Charge	Job		142		142	
Corrugated Black							
	Demolish	Sq.		54		54	Includes material and labor to install tile roofing. Cost includes birdstop, booster, ridge and rake tiles.
	Install	Sq.	121	211		332	
	Demolish and Install	Sq.	121	265		386	
	Remove and Install	Sq.		168.53		168.53	
	Minimum Charge	Job		142		142	
Corrugated Green							
	Demolish	Sq.		54		54	Includes material and labor to install tile roofing. Cost includes birdstop, booster, ridge and rake tiles.
	Install	Sq.	121	211		332	
	Demolish and Install	Sq.	121	265		386	
	Remove and Install	Sq.		168.53		168.53	
	Minimum Charge	Job		142		142	

Concrete Tile Roofing

Concrete Tile Roofing		Unit	Material	Labor	Equip.	Total	Specification
Corrugated Blue							
	Demolish	Sq.		54		54	Includes material and labor to install tile roofing. Cost includes birdstop, booster, ridge and rake tiles.
	Install	Sq.	121	211		332	
	Demolish and Install	Sq.	121	265		386	
	Remove and Install	Sq.		168.53		168.53	
	Minimum Charge	Job		142		142	
Flat Natural Gray							
	Demolish	Sq.		54		54	Includes material and labor to install tile roofing. Cost includes birdstop, booster, ridge and rake tiles.
	Install	Sq.	105	211		316	
	Demolish and Install	Sq.	105	265		370	
	Remove and Install	Sq.		168.53		168.53	
	Minimum Charge	Job		142		142	
Flat Red / Brown							
	Demolish	Sq.		54		54	Includes material and labor to install tile roofing. Cost includes birdstop, booster, ridge and rake tiles.
	Install	Sq.	105	211		316	
	Demolish and Install	Sq.	105	265		370	
	Remove and Install	Sq.		168.53		168.53	
	Minimum Charge	Job		142		142	
Flat Bright Red							
	Demolish	Sq.		54		54	Includes material and labor to install tile roofing. Cost includes birdstop, booster, ridge and rake tiles.
	Install	Sq.	105	211		316	
	Demolish and Install	Sq.	105	265		370	
	Remove and Install	Sq.		168.53		168.53	
	Minimum Charge	Job		142		142	
Flat Green							
	Demolish	Sq.		54		54	Includes material and labor to install tile roofing. Cost includes birdstop, booster, ridge and rake tiles.
	Install	Sq.	105	211		316	
	Demolish and Install	Sq.	105	265		370	
	Minimum Charge	Job		142		142	
Flat Blue							
	Demolish	Sq.		54		54	Includes material and labor to install tile roofing. Cost includes birdstop, booster, ridge and rake tiles.
	Install	Sq.	105	211		316	
	Demolish and Install	Sq.	105	265		370	
	Remove and Install	Sq.		168.53		168.53	
	Minimum Charge	Job		142		142	
Flat Black							
	Demolish	Sq.		54		54	Includes material and labor to install tile roofing. Cost includes birdstop, booster, ridge and rake tiles.
	Install	Sq.	105	211		316	
	Demolish and Install	Sq.	105	265		370	
	Remove and Install	Sq.		168.53		168.53	
	Minimum Charge	Job		142		142	

Aluminum Sheet

Aluminum Sheet		Unit	Material	Labor	Equip.	Total	Specification
Corrugated							
.019" Thick Natural							
	Demolish	S.F.		.58		.58	Includes material and labor to install aluminum sheet metal roofing.
	Install	S.F.	.88	1.37		2.25	
	Demolish and Install	S.F.	.88	1.95		2.83	
	Remove and Install	S.F.		1.10		1.10	
	Minimum Charge	Job		330		330	
.016" Thick Natural							
	Demolish	S.F.		.58		.58	Includes material and labor to install aluminum sheet metal roofing.
	Install	S.F.	.68	1.37		2.05	
	Demolish and Install	S.F.	.68	1.95		2.63	
	Remove and Install	S.F.		1.10		1.10	
	Minimum Charge	Job		330		330	

Aluminum Sheet	Unit	Material	Labor	Equip.	Total	Specification
.016" Colored Finish						
Demolish	S.F.		.58		.58	Includes material and labor to install aluminum sheet metal roofing.
Install	S.F.	.98	1.37		2.35	
Demolish and Install	S.F.	.98	1.95		2.93	
Remove and Install	S.F.		1.10		1.10	
Minimum Charge	Job		330		330	
.019" Colored Finish						
Demolish	S.F.		.58		.58	Includes material and labor to install aluminum sheet metal roofing.
Install	S.F.	1.09	1.37		2.46	
Demolish and Install	S.F.	1.09	1.95		3.04	
Remove and Install	S.F.		1.10		1.10	
Minimum Charge	Job		330		330	
Ribbed						
.019" Thick Natural						
Demolish	S.F.		.58		.58	Includes material and labor to install aluminum sheet metal roofing.
Install	S.F.	.88	1.37		2.25	
Demolish and Install	S.F.	.88	1.95		2.83	
Remove and Install	S.F.		1.10		1.10	
Minimum Charge	Job		330		330	
.016" Thick Natural						
Demolish	S.F.		.58		.58	Includes material and labor to install aluminum sheet metal roofing.
Install	S.F.	.68	1.37		2.05	
Demolish and Install	S.F.	.68	1.95		2.63	
Remove and Install	S.F.		1.10		1.10	
Minimum Charge	Job		330		330	
.050" Colored Finish						
Demolish	S.F.		.58		.58	Includes material and labor to install aluminum sheet metal roofing.
Install	S.F.	3.01	1.37		4.38	
Demolish and Install	S.F.	3.01	1.95		4.96	
Remove and Install	S.F.		1.10		1.10	
Minimum Charge	Job		330		330	
.032" Thick Natural						
Demolish	S.F.		.58		.58	Includes material and labor to install aluminum sheet metal roofing.
Install	S.F.	1.67	1.37		3.04	
Demolish and Install	S.F.	1.67	1.95		3.62	
Remove and Install	S.F.		1.10		1.10	
Minimum Charge	Job		330		330	
.040" Thick Natural						
Demolish	S.F.		.58		.58	Includes material and labor to install aluminum sheet metal roofing.
Install	S.F.	2.04	1.37		3.41	
Demolish and Install	S.F.	2.04	1.95		3.99	
Remove and Install	S.F.		1.10		1.10	
Minimum Charge	Job		330		330	
.050" Thick Natural						
Demolish	S.F.		.58		.58	Includes material and labor to install aluminum sheet metal roofing.
Install	S.F.	2.49	1.37		3.86	
Demolish and Install	S.F.	2.49	1.95		4.44	
Remove and Install	S.F.		1.10		1.10	
Minimum Charge	Job		330		330	
.016" Colored Finish						
Demolish	S.F.		.58		.58	Includes material and labor to install aluminum sheet metal roofing.
Install	S.F.	.98	1.37		2.35	
Demolish and Install	S.F.	.98	1.95		2.93	
Remove and Install	S.F.		1.10		1.10	
Minimum Charge	Job		330		330	

Aluminum Sheet

Aluminum Sheet	Unit	Material	Labor	Equip.	Total	Specification
.019" Colored Finish						
Demolish	S.F.		.58		.58	Includes material and labor to install aluminum sheet metal roofing.
Install	S.F.	1.09	1.37		2.46	
Demolish and Install	S.F.	1.09	1.95		3.04	
Remove and Install	S.F.		1.10		1.10	
Minimum Charge	Job		330		330	
.032" Colored Finish						
Demolish	S.F.		.58		.58	Includes material and labor to install aluminum sheet metal roofing.
Install	S.F.	2.10	1.37		3.47	
Demolish and Install	S.F.	2.10	1.95		4.05	
Remove and Install	S.F.		1.10		1.10	
Minimum Charge	Job		330		330	
.040" Colored Finish						
Demolish	S.F.		.58		.58	Includes material and labor to install aluminum sheet metal roofing.
Install	S.F.	2.51	1.37		3.88	
Demolish and Install	S.F.	2.51	1.95		4.46	
Remove and Install	S.F.		1.10		1.10	
Minimum Charge	Job		330		330	

Fiberglass Sheet

Fiberglass Sheet	Unit	Material	Labor	Equip.	Total	Specification
12 Ounce Corrugated						
Demolish	S.F.		1.03		1.03	Cost includes material and labor to install 12 ounce corrugated fiberglass sheet roofing.
Install	S.F.	5.70	1.09		6.79	
Demolish and Install	S.F.	5.70	2.12		7.82	
Remove and Install	S.F.		.88		.88	
Minimum Charge	Job		141		141	
8 Ounce Corrugated						
Demolish	S.F.		1.03		1.03	Cost includes material and labor to install 8 ounce corrugated fiberglass roofing.
Install	S.F.	2.46	1.09		3.55	
Demolish and Install	S.F.	2.46	2.12		4.58	
Remove and Install	S.F.		.88		.88	
Minimum Charge	Job		141		141	

Galvanized Steel

Galvanized Steel	Unit	Material	Labor	Equip.	Total	Specification
Corrugated						
24 Gauge						
Demolish	S.F.		.58		.58	Includes material and labor to install galvanized steel sheet metal roofing. Cost based on 24 gauge corrugated or ribbed steel roofing.
Install	S.F.	1.16	1.15		2.31	
Demolish and Install	S.F.	1.16	1.73		2.89	
Remove and Install	S.F.		.92		.92	
Minimum Charge	Job		141		141	
26 Gauge						
Demolish	S.F.		.58		.58	Includes material and labor to install galvanized steel sheet metal roofing. Cost based on 26 gauge corrugated or ribbed steel roofing.
Install	S.F.	.99	1.09		2.08	
Demolish and Install	S.F.	.99	1.67		2.66	
Remove and Install	S.F.		.88		.88	
Minimum Charge	Job		141		141	
28 Gauge						
Demolish	S.F.		.58		.58	Includes material and labor to install galvanized steel sheet metal roofing. Cost based on 28 gauge corrugated or ribbed steel roofing.
Install	S.F.	.89	1.04		1.93	
Demolish and Install	S.F.	.89	1.62		2.51	
Remove and Install	S.F.		.83		.83	
Minimum Charge	Job		141		141	

Galvanized Steel	Unit	Material	Labor	Equip.	Total	Specification
30 Gauge						
Demolish	S.F.		.58		.58	Includes material and labor to install galvanized steel sheet metal roofing. Cost based on 30 gauge corrugated or ribbed steel roofing.
Install	S.F.	.85	.99		1.84	
Demolish and Install	S.F.	.85	1.57		2.42	
Remove and Install	S.F.		.80		.80	
Minimum Charge	Job		141		141	
Ribbed						
24 Gauge						
Demolish	S.F.		.58		.58	Includes material and labor to install galvanized steel sheet metal roofing. Cost based on 24 gauge corrugated or ribbed steel roofing.
Install	S.F.	1.16	1.15		2.31	
Demolish and Install	S.F.	1.16	1.73		2.89	
Remove and Install	S.F.		.92		.92	
Minimum Charge	Job		141		141	
26 Gauge						
Demolish	S.F.		.58		.58	Includes material and labor to install galvanized steel sheet metal roofing. Cost based on 26 gauge corrugated or ribbed steel roofing.
Install	S.F.	.99	1.09		2.08	
Demolish and Install	S.F.	.99	1.67		2.66	
Remove and Install	S.F.		.88		.88	
Minimum Charge	Job		141		141	
28 Gauge						
Demolish	S.F.		.58		.58	Includes material and labor to install galvanized steel sheet metal roofing. Cost based on 28 gauge corrugated or ribbed steel roofing.
Install	S.F.	.89	1.04		1.93	
Demolish and Install	S.F.	.89	1.62		2.51	
Remove and Install	S.F.		.83		.83	
Minimum Charge	Job		141		141	
30 Gauge						
Demolish	S.F.		.58		.58	Includes material and labor to install galvanized steel sheet metal roofing. Cost based on 30 gauge corrugated or ribbed steel roofing.
Install	S.F.	.85	.99		1.84	
Demolish and Install	S.F.	.85	1.57		2.42	
Remove and Install	S.F.		.80		.80	
Minimum Charge	Job		141		141	

Standing Seam	Unit	Material	Labor	Equip.	Total	Specification
Copper						
16 Ounce						
Demolish	Sq.		57.50		57.50	Includes material and labor to install copper standing seam metal roofing. Cost based on 16 ounce copper roofing.
Install	Sq.	430	253		683	
Demolish and Install	Sq.	430	310.50		740.50	
Remove and Install	Sq.		202.34		202.34	
Minimum Charge	Job		164		164	
18 Ounce						
Demolish	Sq.		57.50		57.50	Includes material and labor to install copper standing seam metal roofing. Cost based on 18 ounce copper roofing.
Install	Sq.	480	274		754	
Demolish and Install	Sq.	480	331.50		811.50	
Remove and Install	Sq.		219.20		219.20	
Minimum Charge	Job		164		164	
20 Ounce						
Demolish	Sq.		57.50		57.50	Includes material and labor to install copper standing seam metal roofing. Cost based on 20 ounce copper roofing.
Install	Sq.	510	299		809	
Demolish and Install	Sq.	510	356.50		866.50	
Remove and Install	Sq.		239.13		239.13	

Standing Seam	Unit	Material	Labor	Equip.	Total	Specification
Stainless Steel						
26 Gauge						
Demolish	Sq.		57.50		57.50	Cost includes material and labor to install 26 gauge stainless steel standing seam metal roofing.
Install	Sq.	420	286		706	
Demolish and Install	Sq.	420	343.50		763.50	
Remove and Install	Sq.		228.73		228.73	
Minimum Charge	Job		164		164	
28 Gauge						
Demolish	Sq.		57.50		57.50	Includes material and labor to install standing seam metal roofing. Cost based on 28 gauge stainless steel roofing.
Install	Sq.	335	274		609	
Demolish and Install	Sq.	335	331.50		666.50	
Remove and Install	Sq.		219.20		219.20	
Minimum Charge	Job		164		164	

Elastomeric Roofing	Unit	Material	Labor	Equip.	Total	Specification
Neoprene						
1/16″						
Demolish	Sq.		20		20	Includes material and labor to install neoprene membrane, fully adhered, 1/16″ thick.
Install	Sq.	190	135	22.50	347.50	
Demolish and Install	Sq.	190	155	22.50	367.50	
Minimum Charge	Job		650	81.50	731.50	
GRM Membrane						
Demolish	Sq.		20		20	Includes material and labor to install coal tar based elastomeric roofing membrane, polyester fiber reinforced.
Install	Sq.	239	216	27	482	
Demolish and Install	Sq.	239	236	27	502	
Minimum Charge	Job		650	81.50	731.50	
Cant Strips						
3″						
Demolish	L.F.		.62		.62	Includes material and labor to install cants 3″ x 3″, treated timber, cut diagonally.
Install	L.F.	.44	.87		1.31	
Demolish and Install	L.F.	.44	1.49		1.93	
Minimum Charge	Job		141		141	
4″						
Demolish	L.F.		.62		.62	Includes material and labor to install cants 4″ x 4″, treated timber, cut diagonally.
Install	L.F.	1.39	.87		2.26	
Demolish and Install	L.F.	1.39	1.49		2.88	
Minimum Charge	Job		141		141	
Polyvinyl Chloride (PVC)						
45 mil (loose laid)						
Demolish	Sq.		20		20	Cost includes material and labor to install 45 mil PVC elastomeric roofing, loose-laid with stone.
Install	Sq.	89	25.50	3.19	117.69	
Demolish and Install	Sq.	89	45.50	3.19	137.69	
Minimum Charge	Job		650	81.50	731.50	
48 mil (loose laid)						
Demolish	Sq.		20		20	Cost includes material and labor to install 48 mil PVC elastomeric roofing, loose-laid with stone.
Install	Sq.	103	25.50	3.19	131.69	
Demolish and Install	Sq.	103	45.50	3.19	151.69	
Minimum Charge	Job		650	81.50	731.50	
60 mil (loose laid)						
Demolish	Sq.		20		20	Cost includes material and labor to install 60 mil PVC elastomeric roofing, loose-laid with stone.
Install	Sq.	101	25.50	3.19	129.69	
Demolish and Install	Sq.	101	45.50	3.19	149.69	
Minimum Charge	Job		650	81.50	731.50	

Elastomeric Roofing	Unit	Material	Labor	Equip.	Total	Specification
45 mil (attached)						
Demolish	Sq.		29.50		29.50	Cost includes material and labor to install 45 mil PVC elastomeric roofing fully attached to roof deck.
Install	Sq.	77.50	50	6.25	133.75	
Demolish and Install	Sq.	77.50	79.50	6.25	163.25	
Minimum Charge	Job		650	81.50	731.50	
48 mil (attached)						
Demolish	Sq.		29.50		29.50	Cost includes material and labor to install 48 mil PVC elastomeric roofing fully attached to roof deck.
Install	Sq.	129	50	6.25	185.25	
Demolish and Install	Sq.	129	79.50	6.25	214.75	
Minimum Charge	Job		650	81.50	731.50	
60 mil (attached)						
Demolish	Sq.		29.50		29.50	Cost includes material and labor to install 60 mil PVC elastomeric roofing fully attached to roof deck.
Install	Sq.	127	50	6.25	183.25	
Demolish and Install	Sq.	127	79.50	6.25	212.75	
Minimum Charge	Job		650	81.50	731.50	
CSPE Type						
45 mil (loose laid w / stone)						
Demolish	Sq.		20		20	Cost includes material and labor to install 45 mil CSPE (chlorosulfonated polyethylene-hypalon) roofing material loose laid on deck and ballasted with stone.
Install	Sq.	143	25.50	3.19	171.69	
Demolish and Install	Sq.	143	45.50	3.19	191.69	
Minimum Charge	Job		650	81.50	731.50	
45 mil (attached at seams)						
Demolish	Sq.		47.50		47.50	Cost includes material and labor to install 45 mil CSPE (chlorosulfonated polyethylene-hypalon) roofing material attached to deck at seams with batten strips.
Install	Sq.	138	37	4.65	179.65	
Demolish and Install	Sq.	138	84.50	4.65	227.15	
Minimum Charge	Job		650	81.50	731.50	
45 mil (fully attached)						
Demolish	Sq.		29.50		29.50	Cost includes material and labor to install 45 mil CSPE (chlorosulfonated polyethylene-hypalon) roofing material fully attached to deck.
Install	Sq.	135	50	6.25	191.25	
Demolish and Install	Sq.	135	79.50	6.25	220.75	
Minimum Charge	Job		650	81.50	731.50	
EPDM Type						
45 mil (attached at seams)						
Demolish	Sq.		47.50		47.50	Cost includes material and labor to install 45 mil EPDM (ethylene propylene diene monomer) roofing material attached to deck at seams.
Install	Sq.	61	37	4.65	102.65	
Demolish and Install	Sq.	61	84.50	4.65	150.15	
Minimum Charge	Job		650	81.50	731.50	
45 mil (fully attached)						
Demolish	Sq.		29.50		29.50	Cost includes material and labor to install 45 mil EPDM (ethylene propylene diene monomer) roofing material fully attached to deck.
Install	Sq.	88.50	50	6.25	144.75	
Demolish and Install	Sq.	88.50	79.50	6.25	174.25	
Minimum Charge	Job		650	81.50	731.50	
45 mil (loose laid w / stone)						
Demolish	Sq.		20		20	Cost includes material and labor to install 45 mil EPDM (ethylene propylene diene monomer) roofing material loose laid on deck and ballasted with stones (approx. 1/2 ton of stone per square).
Install	Sq.	70	25.50	3.19	98.69	
Demolish and Install	Sq.	70	45.50	3.19	118.69	
Minimum Charge	Job		650	81.50	731.50	
60 mil (attached at seams)						
Demolish	Sq.		47.50		47.50	Cost includes material and labor to install 60 mil EPDM (ethylene propylene diene monomer) roofing material attached to deck at seams.
Install	Sq.	73.50	37	4.65	115.15	
Demolish and Install	Sq.	73.50	84.50	4.65	162.65	
Minimum Charge	Job		650	81.50	731.50	

Elastomeric Roofing	Unit	Material	Labor	Equip.	Total	Specification
60 mil (fully attached)						
Demolish	Sq.		29.50		29.50	Cost includes material and labor to install 60 mil EPDM (ethylene propylene diene monomer) roofing material fully attached to deck.
Install	Sq.	101	50	6.25	157.25	
Demolish and Install	Sq.	101	79.50	6.25	186.75	
Minimum Charge	Job		650	81.50	731.50	
60 mil (loose laid w / stone)						
Demolish	Sq.		20		20	Cost includes material and labor to install 60 mil EPDM (ethylene propylene diene monomer) roofing material loose laid on deck and ballasted with stones (approx. 1/2 ton of stone per square).
Install	Sq.	83.50	25.50	3.19	112.19	
Demolish and Install	Sq.	83.50	45.50	3.19	132.19	
Minimum Charge	Job		650	81.50	731.50	
Silicone-urethane Foam						
Prep Roof for Foam System						
Install	Sq.	5.25	113		118.25	Includes labor and material to removal of any gravel and prepare existing built-up roof for application of silicone-urethane foam system.
Minimum Charge	Job		650	81.50	731.50	

Modified Bitumen Roofing	Unit	Material	Labor	Equip.	Total	Specification
Fully Attached						
120 mil						
Demolish	Sq.		47.50		47.50	Cost includes material and labor to install 120 mil modified bitumen roofing, fully attached to deck by torch.
Install	Sq.	78	6.30	1.05	85.35	
Demolish and Install	Sq.	78	53.80	1.05	132.85	
Minimum Charge	Job		141		141	
150 mil						
Demolish	Sq.		47.50		47.50	Cost includes material and labor to install 150 mil modified bitumen roofing, fully attached to deck by torch.
Install	Sq.	112	15.50	2.59	130.09	
Demolish and Install	Sq.	112	63	2.59	177.59	
Minimum Charge	Job		930	155	1085	
160 mil						
Demolish	Sq.		47.50		47.50	Cost includes material and labor to install 160 mil modified bitumen roofing, fully attached to deck by torch.
Install	Sq.	89.50	16.20	2.70	108.40	
Demolish and Install	Sq.	89.50	63.70	2.70	155.90	
Minimum Charge	Job		930	155	1085	

Built-up Roofing	Unit	Material	Labor	Equip.	Total	Specification
Emergency Repairs						
Install	Ea.	142	1325	222	1689	Minimum charge for work requiring hot asphalt.
Minimum Charge	Job		930	155	1085	
Asphalt						
3-ply Asphalt						
Demolish	Sq.		49.50		49.50	Cost includes material and labor to install 3-ply asphalt built-up roof including one 30 lb and two 15 lb layers of felt.
Install	Sq.	31.50	77.50	12.95	121.95	
Demolish and Install	Sq.	31.50	127	12.95	171.45	
Minimum Charge	Job		930	155	1085	

Built-up Roofing	Unit	Material	Labor	Equip.	Total	Specification
4-ply Asphalt						
Demolish	Sq.		59		59	Cost includes material and labor to install 4-ply asphalt built-up roof including one 30 lb and three 15 lb layers of felt.
Install	Sq.	45	84.50	14.10	143.60	
Demolish and Install	Sq.	45	143.50	14.10	202.60	
Minimum Charge	Job		930	155	1085	
5-ply Asphalt						
Demolish	Sq.		77.50		77.50	Cost includes material and labor to install 5-ply asphalt built-up roof including one 30 lb and four 15 lb layers of felt.
Install	Sq.	64.50	93	15.50	173	
Demolish and Install	Sq.	64.50	170.50	15.50	250.50	
Minimum Charge	Job		930	155	1085	
Components / Accessories						
Flood Coat						
Install	Sq.	11.40	43.50		54.90	Includes labor and material to install flood coat.
R / R Gravel and Flood Coat						
Install	Sq.	17	141		158	Includes labor and material to install flood coat and gravel.
Aluminum UV Coat						
Install	Sq.	14.15	162	20.50	196.65	Includes labor and material to install aluminum-based UV coating.
Gravel Coat						
Demolish	Sq.		25		25	Includes labor and material to install gravel coat.
Install	Sq.	5.65	23.50		29.15	
Demolish and Install	Sq.	5.65	48.50		54.15	
Mineral Surface Cap Sheet						
Install	Sq.	18.15	4.87		23.02	Cost includes labor and material to install 50 # mineral surfaced cap sheet.
Fibrated Aluminum UV Coating						
Install	Sq.	16.10	43.50		59.60	Includes labor and material to install fibered aluminum-based UV coating.
Impregnated Walk						
1″						
Install	S.F.	2.37	1.51		3.88	Cost includes labor and material to install 1″ layer of asphalt coating.
1/2″						
Install	S.F.	1.60	1.19		2.79	Cost includes labor and material to install 1/2″ layer of asphalt coating.
3/4″						
Install	S.F.	1.98	1.26		3.24	Cost includes labor and material to install 3/4″ layer of asphalt coating.

Doors

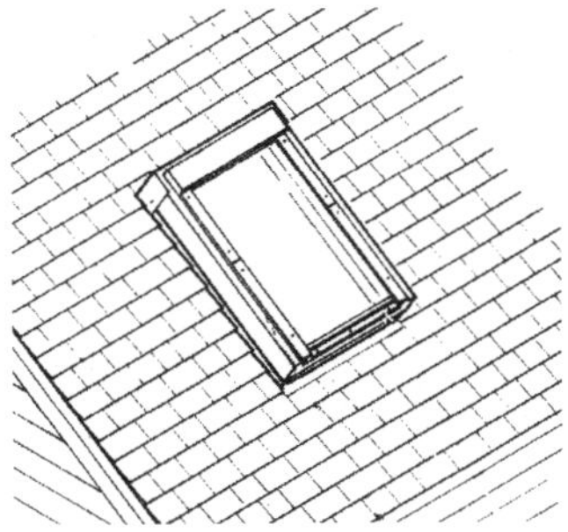

Roof Window

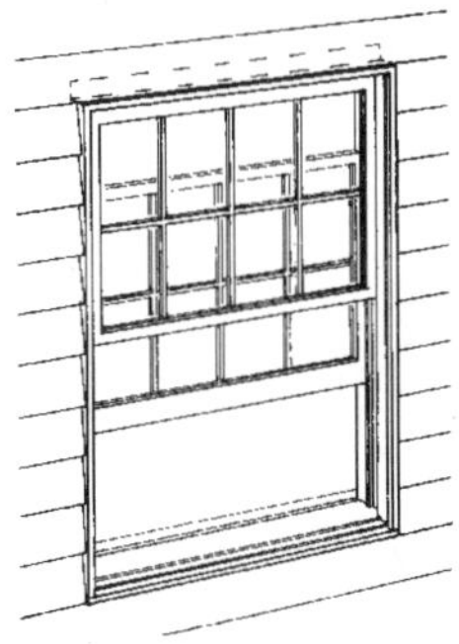

Double Hung Window

Exterior Type Door	Unit	Material	Labor	Equip.	Total	Specification
Entry						
Flush (1-3/4")						
Demolish	Ea.		28		28	Cost includes material and labor to install 3' x 6' 8" pre-hung 1-3/4" door, including casing and stop, hinges, jamb, aluminum sill, weatherstripped.
Install	Ea.	173	15.85		188.85	
Demolish and Install	Ea.	173	43.85		216.85	
Remove and Install	Ea.		15.83		15.83	
Clean	Ea.	2.73	9.10		11.83	
Paint	Ea.	7.55	37.50		45.05	
Minimum Charge	Job		150		150	
Raised Panel						
Demolish	Ea.		28		28	Cost includes material and labor to install 3' x 6' 8" pre-hung 1-3/4" door, jamb, including casing and stop, hinges, aluminum sill, weatherstripping.
Install	Ea.	340	16.70		356.70	
Demolish and Install	Ea.	340	44.70		384.70	
Remove and Install	Ea.		16.71		16.71	
Clean	Ea.	2.73	9.10		11.83	
Paint	Ea.	7.55	37.50		45.05	
Minimum Charge	Job		150		150	
Entry w / Lights						
Good Quality						
Demolish	Ea.		28		28	Cost includes material and labor to install 3' x 6' 8" pre-hung 1-3/4" door, solid wood lower panel, with up to 9-light glass panels, casing, sill, weatherstripped.
Install	Ea.	335	17.70		352.70	
Demolish and Install	Ea.	335	45.70		380.70	
Remove and Install	Ea.		17.69		17.69	
Clean	Ea.	2.73	9.10		11.83	
Paint	Ea.	5.55	17.60		23.15	
Minimum Charge	Job		150		150	
Better Quality						
Demolish	Ea.		28		28	Cost includes material and labor to install 3' x 6' 8" pre-hung 1-3/4" Ash door, solid wood, raised panel door without glass panels.
Install	Ea.	375	17.70		392.70	
Demolish and Install	Ea.	375	45.70		420.70	
Remove and Install	Ea.		17.69		17.69	
Clean	Ea.	2.73	9.10		11.83	
Paint	Ea.	7.55	37.50		45.05	
Minimum Charge	Job		150		150	
Premium Quality						
Demolish	Ea.		28		28	Cost includes material and labor to install 3' x 6' 8" pre-hung 1-3/4" Oak door, solid wood, raised panel door without glass panels.
Install	Ea.	580	17.70		597.70	
Demolish and Install	Ea.	580	45.70		625.70	
Remove and Install	Ea.		17.69		17.69	
Clean	Ea.	2.73	9.10		11.83	
Paint	Ea.	7.55	37.50		45.05	
Minimum Charge	Job		150		150	

Exterior Type Door	Unit	Material	Labor	Equip.	Total	Specification
Custom Quality						
Demolish	Ea.		28		28	Cost includes material and labor to install 3′ x 6′ 8″ pre-hung 1-3/4″ hand-made, solid wood, raised panel door with custom stained glass.
Install	Ea.	1200	17.70		1217.70	
Demolish and Install	Ea.	1200	45.70		1245.70	
Remove and Install	Ea.		17.69		17.69	
Clean	Ea.	.19	9.10		9.29	
Paint	Ea.	2.08	88		90.08	
Minimum Charge	Job		150		150	
Miami Style						
Demolish	Ea.		28		28	Cost includes material and labor to install 3′ x 6′ 8″ pre-hung 1-3/4″ lauan door of solid wood.
Install	Ea.	162	15.05		177.05	
Demolish and Install	Ea.	162	43.05		205.05	
Remove and Install	Ea.		15.04		15.04	
Clean	Ea.	.19	9.10		9.29	
Paint	Ea.	7.55	37.50		45.05	
Minimum Charge	Job		150		150	
French Style						
1-Light						
Demolish	Ea.		28		28	Cost includes material and labor to install 3′ x 6′ 8″ pre-hung 1-3/4″ french style door with 1 light.
Install	Ea.	320	21.50		341.50	
Demolish and Install	Ea.	320	49.50		369.50	
Remove and Install	Ea.		21.49		21.49	
Clean	Ea.	1.71	13.65		15.36	
Paint	Ea.	8.85	66		74.85	
Minimum Charge	Job		150		150	
5-Light						
Demolish	Ea.		28		28	Cost includes material and labor to install 3′ x 6′ 8″ pre-hung 1-3/4″ french style door with 5 lights.
Install	Ea.	320	21.50		341.50	
Demolish and Install	Ea.	320	49.50		369.50	
Remove and Install	Ea.		21.49		21.49	
Clean	Ea.	.37	18.15		18.52	
Paint	Ea.	8.85	66		74.85	
Minimum Charge	Job		150		150	
9-Light						
Demolish	Ea.		28		28	Cost includes material and labor to install 3′ x 6′ 8″ pre-hung 1-3/4″ french style door with 9 lights.
Install	Ea.	325	21.50		346.50	
Demolish and Install	Ea.	325	49.50		374.50	
Remove and Install	Ea.		21.49		21.49	
Clean	Ea.	.37	18.15		18.52	
Paint	Ea.	8.85	66		74.85	
Minimum Charge	Job		150		150	
10-Light						
Demolish	Ea.		30.50		30.50	Cost includes material and labor to install 3′ x 6′ 8″ pre-hung 1-3/4″ french style door with 10 lights.
Install	Ea.	535	37.50		572.50	
Demolish and Install	Ea.	535	68		603	
Remove and Install	Ea.		37.60		37.60	
Clean	Ea.	.37	18.15		18.52	
Paint	Ea.	8.85	66		74.85	
Minimum Charge	Job		150		150	
15-Light						
Demolish	Ea.		30.50		30.50	Cost includes material and labor to install 3′ x 6′ 8″ pre-hung 1-3/4″ french style door with 15 lights.
Install	Ea.	595	43		638	
Demolish and Install	Ea.	595	73.50		668.50	
Remove and Install	Ea.		42.97		42.97	
Clean	Ea.	.37	18.15		18.52	
Paint	Ea.	8.85	66		74.85	
Minimum Charge	Job		150		150	

Exterior Type Door	Unit	Material	Labor	Equip.	Total	Specification
Jamb						
Wood						
Demolish	Ea.		16.70		16.70	Includes material and labor to install flat pine jamb with square cut heads and rabbeted sides for 6′ 8″ high and 3-9/16″ door including trim sets for both sides.
Install	Ea.	77	8.15		85.15	
Demolish and Install	Ea.	77	24.85		101.85	
Clean	Ea.	.09	5.45		5.54	
Paint	Ea.	5.55	17.60		23.15	
Minimum Charge	Job		150		150	
Metal 6′ 8″ High						
Demolish	Ea.		16.70		16.70	Cost includes material and labor to install 18 gauge hollow metal door frame to fit a 4-1/2″ jamb 6′ 8″ high and up to 3′ 6″ opening.
Install	Ea.	77	18.80		95.80	
Demolish and Install	Ea.	77	35.50		112.50	
Remove and Install	Ea.		18.80		18.80	
Clean	Ea.	.57	4.54		5.11	
Paint	Ea.	1.03	5.30		6.33	
Minimum Charge	Job		150		150	
Metal 7′ High						
Demolish	Ea.		16.70		16.70	Cost includes material and labor to install 18 gauge hollow metal door frame to fit a 4-1/2″ jamb 7′ high and up to 3′ 6″ opening.
Install	Ea.	84	20		104	
Demolish and Install	Ea.	84	36.70		120.70	
Remove and Install	Ea.		20.05		20.05	
Clean	Ea.	.57	4.54		5.11	
Paint	Ea.	1.03	5.30		6.33	
Minimum Charge	Job		150		150	
Metal 8′ High						
Demolish	Ea.		16.70		16.70	Cost includes material and labor to install 18 gauge hollow metal door frame to fit a 4-1/2″ jamb 8′ high and up to 3′ 6″ opening.
Install	Ea.	100	25		125	
Demolish and Install	Ea.	100	41.70		141.70	
Remove and Install	Ea.		25.07		25.07	
Clean	Ea.	.80	5.45		6.25	
Paint	Ea.	1.16	5.85		7.01	
Minimum Charge	Job		150		150	
Metal 9′ High						
Demolish	Ea.		16.70		16.70	Cost includes material and labor to install 18 gauge hollow metal door frame to fit a 4-1/2″ jamb 9′ high and up to 3′ 6″ opening.
Install	Ea.	110	30		140	
Demolish and Install	Ea.	110	46.70		156.70	
Remove and Install	Ea.		30.08		30.08	
Clean	Ea.	.80	5.45		6.25	
Paint	Ea.	1.16	5.85		7.01	
Minimum Charge	Job		150		150	
Casing Trim						
Single Width						
Demolish	Ea.		5.30		5.30	Cost includes material and labor to install 11/16″ x 2-1/2″ pine ranch style casing for one side of a standard door opening.
Install	Opng.	15.90	51		66.90	
Demolish and Install	Opng.	15.90	56.30		72.20	
Remove and Install	Opng.		40.79		40.79	
Clean	Opng.	.39	2.18		2.57	
Paint	Ea.	3.06	5.30		8.36	
Minimum Charge	Job		150		150	
Double Width						
Demolish	Ea.		6.30		6.30	Cost includes material and labor to install 11/16″ x 2-1/2″ pine ranch style door casing for one side of a double door opening.
Install	Opng.	18.70	60		78.70	
Demolish and Install	Opng.	18.70	66.30		85	
Remove and Install	Opng.		48.13		48.13	
Clean	Opng.	.45	4.36		4.81	
Paint	Ea.	3.06	6		9.06	
Minimum Charge	Job		150		150	

Residential Metal Door

Residential Metal Door	Unit	Material	Labor	Equip.	Total	Specification
Pre-hung, Steel-clad						
2' 8" x 6' 8"						
Demolish	Ea.		28		28	Cost includes material and labor to install 2' 6" x 6' 8" pre-hung 1-3/4" 24 gauge steel clad door with 4-1/2" wood frame, trims, hinges, aluminum sill, weatherstripped.
Install	Ea.	191	37.50		228.50	
Demolish and Install	Ea.	191	65.50		256.50	
Remove and Install	Ea.		37.60		37.60	
Clean	Ea.	2.73	9.10		11.83	
Paint	Ea.	8.85	37.50		46.35	
Minimum Charge	Job		150		150	
3' x 6' 8"						
Demolish	Ea.		28		28	Cost includes material and labor to install 3' x 6' 8" pre-hung 1-3/4" 24 gauge steel clad door with 4-1/2" wood frame, trims, hinges, aluminum sill, weatherstripped.
Install	Ea.	195	40		235	
Demolish and Install	Ea.	195	68		263	
Remove and Install	Ea.		40.11		40.11	
Clean	Ea.	2.73	9.10		11.83	
Paint	Ea.	8.85	37.50		46.35	
Minimum Charge	Job		150		150	

Metal Door

Metal Door	Unit	Material	Labor	Equip.	Total	Specification
2' 8" x 6' 8", 18 Gauge						
Demolish	Ea.		5.95		5.95	Cost includes material and labor to install 18 gauge 2' 8" x 6' 8" steel door only, 1-3/4" thick, hinges.
Install	Ea.	222	18.80		240.80	
Demolish and Install	Ea.	222	24.75		246.75	
Remove and Install	Ea.		18.80		18.80	
Clean	Ea.	1.14	6.80		7.94	
Paint	Ea.	3.76	22		25.76	
Minimum Charge	Job		150		150	
3' x 6' 8", 18 Gauge						
Demolish	Ea.		5.95		5.95	Cost includes material and labor to install 18 gauge 3' x 6' 8" steel door only, 1-3/4" thick, hinges.
Install	Ea.	216	20		236	
Demolish and Install	Ea.	216	25.95		241.95	
Remove and Install	Ea.		20.05		20.05	
Clean	Ea.	1.14	6.80		7.94	
Paint	Ea.	3.76	22		25.76	
Minimum Charge	Job		150		150	
2' 8" x 7', 18 Gauge						
Demolish	Ea.		5.95		5.95	Cost includes material and labor to install 18 gauge 2' 8" x 7' steel door only, 1-3/4" thick, hinges.
Install	Ea.	211	20		231	
Demolish and Install	Ea.	211	25.95		236.95	
Remove and Install	Ea.		20.05		20.05	
Clean	Ea.	1.14	6.80		7.94	
Paint	Ea.	3.76	22		25.76	
Minimum Charge	Job		150		150	
3' x 7', 18 Gauge						
Demolish	Ea.		5.95		5.95	Cost includes material and labor to install 18 gauge 3' x 7' steel door only, 1-3/4" thick, hinges.
Install	Ea.	205	21.50		226.50	
Demolish and Install	Ea.	205	27.45		232.45	
Remove and Install	Ea.		21.49		21.49	
Clean	Ea.	1.14	6.80		7.94	
Paint	Ea.	3.76	22		25.76	
Minimum Charge	Job		150		150	
3' 6" x 7', 18 Gauge						
Demolish	Ea.		5.95		5.95	Cost includes material and labor to install 18 gauge 3' 6" x 7' steel door only, 1-3/4" thick, hinges.
Install	Ea.	267	23		290	
Demolish and Install	Ea.	267	28.95		295.95	
Remove and Install	Ea.		23.14		23.14	
Clean	Ea.	1.14	6.80		7.94	
Paint	Ea.	3.76	22		25.76	
Minimum Charge	Job		150		150	

Metal Door	Unit	Material	Labor	Equip.	Total	Specification
Casing Trim						
Single Width						
Demolish	Ea.		5.30		5.30	Cost includes material and labor to install 11/16" x 2-1/2" pine ranch style casing for one side of a standard door opening.
Install	Opng.	15.90	51		66.90	
Demolish and Install	Opng.	15.90	56.30		72.20	
Remove and Install	Opng.		40.79		40.79	
Clean	Opng.	.39	2.18		2.57	
Paint	Ea.	3.06	5.30		8.36	
Minimum Charge	Job		150		150	
Double Width						
Demolish	Ea.		6.30		6.30	Cost includes material and labor to install 11/16" x 2-1/2" pine ranch style door casing for one side of a double door opening.
Install	Opng.	18.70	60		78.70	
Demolish and Install	Opng.	18.70	66.30		85	
Remove and Install	Opng.		48.13		48.13	
Clean	Opng.	.45	4.36		4.81	
Paint	Ea.	3.06	6		9.06	
Minimum Charge	Job		150		150	

Solid Core Door	Unit	Material	Labor	Equip.	Total	Specification
Exterior Type						
Flush (1-3/4")						
Demolish	Ea.		28		28	Cost includes material and labor to install 3' x 6' 8" pre-hung 1-3/4" door, including casing and stop, hinges, jamb, aluminum sill, weatherstripped.
Install	Ea.	210	37.50		247.50	
Demolish and Install	Ea.	210	65.50		275.50	
Remove and Install	Ea.		37.60		37.60	
Clean	Ea.	2.73	9.10		11.83	
Paint	Ea.	8.85	37.50		46.35	
Minimum Charge	Job		150		150	
Raised Panel						
Demolish	Ea.		28		28	Cost includes material and labor to install 3' x 6' 8" pre-hung 1-3/4" door, jamb, including casing and stop, hinges, aluminum sill, weatherstripping.
Install	Ea.	340	16.70		356.70	
Demolish and Install	Ea.	340	44.70		384.70	
Remove and Install	Ea.		16.71		16.71	
Clean	Ea.	2.73	9.10		11.83	
Paint	Ea.	8.85	88		96.85	
Minimum Charge	Job		150		150	
Entry w / Lights						
Demolish	Ea.		28		28	Cost includes material and labor to install 3' x 6' 8" pre-hung 1-3/4" door, solid wood lower panel, with up to 9-light glass panels, casing, sill, weatherstripped.
Install	Ea.	335	17.70		352.70	
Demolish and Install	Ea.	335	45.70		380.70	
Remove and Install	Ea.		17.69		17.69	
Clean	Ea.	2.73	9.10		11.83	
Paint	Ea.	2.08	88		90.08	
Minimum Charge	Job		150		150	
Custom Entry						
Demolish	Ea.		28		28	Includes material and labor to install prehung, fir, single glazed with lead caming, 1-3/4" x 6' 8" x 3'.
Install	Ea.	910	17.70		927.70	
Demolish and Install	Ea.	910	45.70		955.70	
Remove and Install	Ea.		17.69		17.69	
Clean	Ea.	2.73	9.10		11.83	
Paint	Ea.	3.76	22		25.76	
Minimum Charge	Job		150		150	

Solid Core Door	Unit	Material	Labor	Equip.	Total	Specification
Dutch Style						
Demolish	Ea.		28		28	Includes material and labor to install dutch door, pine, 1-3/4" x 6' 8" x 2' 8" wide.
Install	Ea.	635	50		685	
Demolish and Install	Ea.	635	78		713	
Remove and Install	Ea.		50.13		50.13	
Clean	Ea.	2.73	9.10		11.83	
Paint	Ea.	2.08	88		90.08	
Minimum Charge	Job		150		150	
Miami Style						
Demolish	Ea.		28		28	Cost includes material and labor to install 3' x 6' 8" pre-hung 1-3/4" lauan door of solid wood.
Install	Ea.	162	15.05		177.05	
Demolish and Install	Ea.	162	43.05		205.05	
Remove and Install	Ea.		15.04		15.04	
Clean	Ea.	2.73	9.10		11.83	
Paint	Ea.	7.55	37.50		45.05	
Minimum Charge	Job		150		150	
French Style						
Demolish	Ea.		30.50		30.50	Cost includes material and labor to install 3' x 6' 8" pre-hung 1-3/4" french style door with 15 lights.
Install	Ea.	595	43		638	
Demolish and Install	Ea.	595	73.50		668.50	
Remove and Install	Ea.		42.97		42.97	
Clean	Ea.	.19	9.10		9.29	
Paint	Ea.	5.60	132		137.60	
Minimum Charge	Job		150		150	
Double						
Demolish	Ea.		20		20	Includes material and labor to install a pre-hung double entry door including frame and exterior trim.
Install	Ea.	1000	75		1075	
Demolish and Install	Ea.	1000	95		1095	
Clean	Ea.	.10	31.50		31.60	
Paint	Ea.	17.70	132		149.70	
Minimum Charge	Job		75		75	
Commercial Grade						
3' x 6' 8"						
Demolish	Ea.		28		28	Includes material and labor to install flush solid core 1-3/4" hardwood veneer 3' x 6' 8' door only.
Install	Ea.	264	17.70		281.70	
Demolish and Install	Ea.	264	45.70		309.70	
Remove and Install	Ea.		17.69		17.69	
Clean	Ea.	2.73	9.10		11.83	
Paint	Ea.	6.25	22		28.25	
Minimum Charge	Job		150		150	
3' 6" x 6' 8"						
Demolish	Ea.		28		28	Includes material and labor to install flush solid core 1-3/4" hardwood veneer 3' 6" x 6' 8' door only.
Install	Ea.	360	17.70		377.70	
Demolish and Install	Ea.	360	45.70		405.70	
Remove and Install	Ea.		17.69		17.69	
Clean	Ea.	2.73	9.10		11.83	
Paint	Ea.	6.25	22		28.25	
Minimum Charge	Job		150		150	
3' x 7'						
Demolish	Ea.		28		28	Includes material and labor to install flush solid core 1-3/4" hardwood veneer 3' x 7' door only.
Install	Ea.	279	43		322	
Demolish and Install	Ea.	279	71		350	
Remove and Install	Ea.		42.97		42.97	
Clean	Ea.	2.73	9.10		11.83	
Paint	Ea.	6.25	22		28.25	
Minimum Charge	Job		150		150	

Solid Core Door	Unit	Material	Labor	Equip.	Total	Specification
3′ 6″ x 7′						
Demolish	Ea.		28		28	Includes material and labor to install flush solid core 1-3/4″ hardwood veneer 3′ 6″ x 7′ door only.
Install	Ea.	385	21.50		406.50	
Demolish and Install	Ea.	385	49.50		434.50	
Remove and Install	Ea.		21.49		21.49	
Clean	Ea.	2.73	9.10		11.83	
Paint	Ea.	6.25	22		28.25	
Minimum Charge	Job		150		150	
3′ x 8′						
Demolish	Ea.		28		28	Includes material and labor to install flush solid core 1-3/4″ hardwood veneer 3′ x 8′ door only.
Install	Ea.	330	37.50		367.50	
Demolish and Install	Ea.	330	65.50		395.50	
Remove and Install	Ea.		37.60		37.60	
Clean	Ea.	3.28	13.65		16.93	
Paint	Ea.	6.25	22		28.25	
Minimum Charge	Job		150		150	
3′ 6″ x 8′						
Demolish	Ea.		28		28	Includes material and labor to install flush solid core 1-3/4″ hardwood veneer 3′ 6″ x 8′ door only.
Install	Ea.	405	37.50		442.50	
Demolish and Install	Ea.	405	65.50		470.50	
Remove and Install	Ea.		37.60		37.60	
Clean	Ea.	3.28	13.65		16.93	
Paint	Ea.	6.25	22		28.25	
Minimum Charge	Job		150		150	
3′ x 9′						
Demolish	Ea.		28		28	Includes material and labor to install flush solid core 1-3/4″ hardwood veneer 3′ x 9′ door only.
Install	Ea.	430	43		473	
Demolish and Install	Ea.	430	71		501	
Remove and Install	Ea.		42.97		42.97	
Clean	Ea.	3.28	13.65		16.93	
Paint	Ea.	6.25	22		28.25	
Minimum Charge	Job		150		150	
Stain						
Install	Ea.	5.65	29.50		35.15	Includes labor and material to stain exterior type door and trim on both sides.
Minimum Charge	Job		132		132	
Door Only (SC)						
2′ 4″ x 6′ 8″						
Demolish	Ea.		5.95		5.95	Cost includes material and labor to install 2′ 4″ x 6′ 8″ lauan solid core door slab.
Install	Ea.	176	16.70		192.70	
Demolish and Install	Ea.	176	22.65		198.65	
Remove and Install	Ea.		16.71		16.71	
Clean	Ea.	1.14	6.80		7.94	
Paint	Ea.	2.33	33		35.33	
Minimum Charge	Job		150		150	
2′ 8″ x 6′ 8″						
Demolish	Ea.		5.95		5.95	Cost includes material and labor to install 2′ 8″ x 6′ 8″ lauan solid core door slab.
Install	Ea.	180	16.70		196.70	
Demolish and Install	Ea.	180	22.65		202.65	
Remove and Install	Ea.		16.71		16.71	
Clean	Ea.	1.14	6.80		7.94	
Paint	Ea.	2.33	33		35.33	
Minimum Charge	Job		150		150	

Solid Core Door	Unit	Material	Labor	Equip.	Total	Specification
3′ x 6′ 8″						
Demolish	Ea.		5.95		5.95	Cost includes material and labor to install 3′ x 6′ 8″ lauan solid core door slab.
Install	Ea.	187	17.70		204.70	
Demolish and Install	Ea.	187	23.65		210.65	
Remove and Install	Ea.		17.69		17.69	
Clean	Ea.	1.14	6.80		7.94	
Paint	Ea.	2.33	33		35.33	
Minimum Charge	Job		150		150	
3′ 6″ x 6′ 8″						
Demolish	Ea.		5.95		5.95	Cost includes material and labor to install 3′ 6″ x 6′ 8″ lauan solid core door slab.
Install	Ea.	212	18.80		230.80	
Demolish and Install	Ea.	212	24.75		236.75	
Remove and Install	Ea.		18.80		18.80	
Clean	Ea.	1.14	6.80		7.94	
Paint	Ea.	2.33	33		35.33	
Minimum Charge	Job		150		150	
Panel						
Demolish	Ea.		5.95		5.95	Cost includes material and labor to install 3′ x 6′ 8″ six (6) panel interior door.
Install	Ea.	158	35.50		193.50	
Demolish and Install	Ea.	158	41.45		199.45	
Remove and Install	Ea.		35.39		35.39	
Clean	Ea.	1.36	6.80		8.16	
Paint	Ea.	2.33	33		35.33	
Minimum Charge	Job		150		150	
Casing Trim						
Single Width						
Demolish	Ea.		5.30		5.30	Cost includes material and labor to install 11/16″ x 2-1/2″ pine ranch style casing for one side of a standard door opening.
Install	Opng.	15.90	51		66.90	
Demolish and Install	Opng.	15.90	56.30		72.20	
Remove and Install	Opng.		40.79		40.79	
Clean	Opng.	.39	2.18		2.57	
Paint	Ea.	3.06	5.30		8.36	
Minimum Charge	Job		150		150	
Double Width						
Demolish	Ea.		6.30		6.30	Cost includes material and labor to install 11/16″ x 2-1/2″ pine ranch style door casing for one side of a double door opening.
Install	Opng.	18.70	60		78.70	
Demolish and Install	Opng.	18.70	66.30		85	
Remove and Install	Opng.		48.13		48.13	
Clean	Opng.	.45	4.36		4.81	
Paint	Ea.	3.06	6		9.06	
Minimum Charge	Job		150		150	

Components	Unit	Material	Labor	Equip.	Total	Specification
Hardware						
Doorknob						
Demolish	Ea.		11.15		11.15	Includes material and labor to install residential passage lockset, keyless bored type.
Install	Ea.	17.15	18.80		35.95	
Demolish and Install	Ea.	17.15	29.95		47.10	
Remove and Install	Ea.		15.04		15.04	
Clean	Ea.	.28	6.80		7.08	
Minimum Charge	Job		150		150	

Components	Unit	Material	Labor	Equip.	Total	Specification
Doorknob w / Lock						
Demolish	Ea.		11.15		11.15	Includes material and labor to install privacy lockset.
Install	Ea.	19.65	18.80		38.45	
Demolish and Install	Ea.	19.65	29.95		49.60	
Remove and Install	Ea.		15.04		15.04	
Clean	Ea.	.28	6.80		7.08	
Minimum Charge	Job		150		150	
Lever Handle						
Demolish	Ea.		11.15		11.15	Includes material and labor to install residential passage lockset, keyless bored type with lever handle.
Install	Ea.	27	30		57	
Demolish and Install	Ea.	27	41.15		68.15	
Remove and Install	Ea.		24.06		24.06	
Clean	Ea.	.28	6.80		7.08	
Minimum Charge	Job		150		150	
Deadbolt						
Demolish	Ea.		10.45		10.45	Includes material and labor to install a deadbolt.
Install	Ea.	36.50	21.50		58	
Demolish and Install	Ea.	36.50	31.95		68.45	
Remove and Install	Ea.		17.19		17.19	
Clean	Ea.	.28	6.80		7.08	
Minimum Charge	Job		150		150	
Grip Handle Entry						
Demolish	Ea.		9.55		9.55	Includes material and labor to install residential keyed entry lock set with grip type handle.
Install	Ea.	117	33.50		150.50	
Demolish and Install	Ea.	117	43.05		160.05	
Remove and Install	Ea.		26.74		26.74	
Clean	Ea.	.28	6.80		7.08	
Minimum Charge	Job		150		150	
Panic Device						
Demolish	Ea.		24		24	Includes material and labor to install touch bar, low profile, exit only.
Install	Ea.	320	50		370	
Demolish and Install	Ea.	320	74		394	
Remove and Install	Ea.		40.11		40.11	
Clean	Ea.	.13	12.10		12.23	
Minimum Charge	Job		150		150	
Closer						
Demolish	Ea.		6.25		6.25	Includes material and labor to install pneumatic heavy duty closer for exterior type doors.
Install	Ea.	152	50		202	
Demolish and Install	Ea.	152	56.25		208.25	
Remove and Install	Ea.		40.11		40.11	
Clean	Ea.	.13	9.10		9.23	
Minimum Charge	Job		150		150	
Peep-hole						
Demolish	Ea.		5.40		5.40	Includes material and labor to install peep-hole for door.
Install	Ea.	8.25	9.40		17.65	
Demolish and Install	Ea.	8.25	14.80		23.05	
Remove and Install	Ea.		7.52		7.52	
Clean	Ea.	.02	2.41		2.43	
Minimum Charge	Job		150		150	
Exit Sign						
Demolish	Ea.		16.65		16.65	Includes material and labor to install a wall mounted interior electric exit sign.
Install	Ea.	43	41		84	
Demolish and Install	Ea.	43	57.65		100.65	
Remove and Install	Ea.		32.92		32.92	
Clean	Ea.	.02	4.90		4.92	
Minimum Charge	Job		150		150	

Components	Unit	Material	Labor	Equip.	Total	Specification
Kickplate						
Demolish	Ea.		9.30		9.30	Includes material and labor to install aluminum kickplate, 10" x 28".
Install	Ea.	16.45	20		36.45	
Demolish and Install	Ea.	16.45	29.30		45.75	
Remove and Install	Ea.		16.04		16.04	
Clean	Ea.	.28	6.80		7.08	
Minimum Charge	Job		150		150	
Threshold						
Install	L.F.	2.49	3.01		5.50	Includes labor and material to install oak threshold.
Minimum Charge	Job		75		75	
Weatherstripping						
Install	L.F.	.26	.75		1.01	Includes labor and material to install rubber weatherstripping for door.
Minimum Charge	Job		75		75	
Metal Jamb						
6' 8" High						
Demolish	Ea.		16.70		16.70	Cost includes material and labor to install 18 gauge hollow metal door frame to fit a 4-1/2" jamb 6' 8" high and up to 3' 6" opening.
Install	Ea.	77	18.80		95.80	
Demolish and Install	Ea.	77	35.50		112.50	
Remove and Install	Ea.		18.80		18.80	
Clean	Ea.	.57	4.54		5.11	
Paint	Ea.	1.03	5.30		6.33	
Minimum Charge	Job		150		150	
7' High						
Demolish	Ea.		16.70		16.70	Cost includes material and labor to install 18 gauge hollow metal door frame to fit a 4-1/2" jamb 7' high and up to 3' 6" opening.
Install	Ea.	84	20		104	
Demolish and Install	Ea.	84	36.70		120.70	
Remove and Install	Ea.		20.05		20.05	
Clean	Ea.	.57	4.54		5.11	
Paint	Ea.	1.03	5.30		6.33	
Minimum Charge	Job		150		150	
8' High						
Demolish	Ea.		16.70		16.70	Cost includes material and labor to install 18 gauge hollow metal door frame to fit a 4-1/2" jamb 8' high and up to 3' 6" opening.
Install	Ea.	100	25		125	
Demolish and Install	Ea.	100	41.70		141.70	
Remove and Install	Ea.		25.07		25.07	
Clean	Ea.	.80	5.45		6.25	
Paint	Ea.	1.16	5.85		7.01	
Minimum Charge	Job		150		150	
9' High						
Demolish	Ea.		16.70		16.70	Cost includes material and labor to install 18 gauge hollow metal door frame to fit a 4-1/2" jamb 9' high and up to 3' 6" opening.
Install	Ea.	110	30		140	
Demolish and Install	Ea.	110	46.70		156.70	
Remove and Install	Ea.		30.08		30.08	
Clean	Ea.	.80	5.45		6.25	
Paint	Ea.	1.16	5.85		7.01	
Minimum Charge	Job		150		150	
Wood Jamb						
Demolish	Ea.		21		21	Includes material and labor to install flat pine jamb with square cut heads and rabbeted sides for 6' 8" high and 3-9/16" door including trim sets for both sides.
Install	Ea.	77	8.15		85.15	
Demolish and Install	Ea.	77	29.15		106.15	
Clean	Ea.	.09	5.45		5.54	
Paint	Ea.	5.55	17.60		23.15	
Minimum Charge	Job		75		75	
Stain						
Install	Ea.	3.93	22		25.93	Includes labor and material to stain single door and trim on both sides.
Minimum Charge	Job		132		132	

Components	Unit	Material	Labor	Equip.	Total	Specification
Shave & Refit						
Install	Ea.		25		25	Includes labor to shave and rework door to fit opening at the job site.
Minimum Charge	Job		75		75	

Sliding Patio Door	Unit	Material	Labor	Equip.	Total	Specification
Aluminum						
6′ Single Pane						
Demolish	Ea.		42		42	Cost includes material and labor to install 6′ x 6′ 8″ sliding anodized aluminum patio door with single pane tempered safety glass, anodized frame, screen, weatherstripped.
Install	Ea.	400	150		550	
Demolish and Install	Ea.	400	192		592	
Remove and Install	Ea.		120.32		120.32	
Clean	Ea.	1.71	13.65		15.36	
Minimum Charge	Job		150		150	
8′ Single Pane						
Demolish	Ea.		42		42	Cost includes material and labor to install 8′ x 6′ 8″ sliding aluminum patio door with single pane tempered safety glass, anodized frame, screen, weatherstripped.
Install	Ea.	455	201		656	
Demolish and Install	Ea.	455	243		698	
Remove and Install	Ea.		160.43		160.43	
Clean	Ea.	1.71	13.65		15.36	
Minimum Charge	Job		150		150	
10′ Single Pane						
Demolish	Ea.		42		42	Cost includes material and labor to install 10′ x 6′ 8″ sliding aluminum patio door with single pane tempered safety glass, anodized frame, screen, weatherstripped.
Install	Ea.	595	300		895	
Demolish and Install	Ea.	595	342		937	
Remove and Install	Ea.		240.64		240.64	
Clean	Ea.	1.71	13.65		15.36	
Minimum Charge	Job		150		150	
6′ Insulated						
Demolish	Ea.		42		42	Cost includes material and labor to install 6′ x 6′ 8″ sliding aluminum patio door with double insulated tempered glass, baked-on enamel finish, screen, weatherstripped.
Install	Ea.	525	150		675	
Demolish and Install	Ea.	525	192		717	
Remove and Install	Ea.		120.32		120.32	
Clean	Ea.	1.71	13.65		15.36	
Minimum Charge	Job		150		150	
8′ Insulated						
Demolish	Ea.		42		42	Cost includes material and labor to install 8′ x 6′ 8″ sliding aluminum patio door with insulated tempered glass, baked-on enamel finish, screen, weatherstripped.
Install	Ea.	620	201		821	
Demolish and Install	Ea.	620	243		863	
Remove and Install	Ea.		160.43		160.43	
Clean	Ea.	1.71	13.65		15.36	
Minimum Charge	Job		150		150	
10′ Insulated						
Demolish	Ea.		42		42	Cost includes material and labor to install 10′ x 6′ 8″ sliding aluminum patio door with double insulated tempered glass, baked-on enamel finish, screen, weatherstripped.
Install	Ea.	695	300		995	
Demolish and Install	Ea.	695	342		1037	
Remove and Install	Ea.		240.64		240.64	
Clean	Ea.	1.71	13.65		15.36	
Minimum Charge	Job		150		150	

Sliding Patio Door	Unit	Material	Labor	Equip.	Total	Specification
Vinyl						
6' Single Pane						
Demolish	Ea.		42		42	Cost includes material and labor to install 6' x 6' 8" sliding vinyl patio door with single pane tempered safety glass, anodized frame, screen, weatherstripped.
Install	Ea.	380	150		530	
Demolish and Install	Ea.	380	192		572	
Remove and Install	Ea.		120.32		120.32	
Clean	Ea.	1.71	13.65		15.36	
Minimum Charge	Job		150		150	
8' Single Pane						
Demolish	Ea.		42		42	Cost includes material and labor to install 8' x 6' 8" sliding vinyl patio door with single pane tempered safety glass, anodized frame, screen, weatherstripped.
Install	Ea.	435	201		636	
Demolish and Install	Ea.	435	243		678	
Remove and Install	Ea.		160.43		160.43	
Clean	Ea.	1.71	13.65		15.36	
Minimum Charge	Job		150		150	
10' Single Pane						
Demolish	Ea.		42		42	Cost includes material and labor to install 10' x 6' 8" sliding vinyl patio door with single pane tempered safety glass, anodized frame, screen, weatherstripped.
Install	Ea.	565	300		865	
Demolish and Install	Ea.	565	342		907	
Remove and Install	Ea.		240.64		240.64	
Clean	Ea.	1.71	13.65		15.36	
Minimum Charge	Job		150		150	
6' Insulated						
Demolish	Ea.		42		42	Cost includes material and labor to install 6' x 6' 8" sliding vinyl patio door with double insulated tempered glass, anodized frame, screen, weatherstripped.
Install	Ea.	595	150		745	
Demolish and Install	Ea.	595	192		787	
Remove and Install	Ea.		120.32		120.32	
Clean	Ea.	1.71	13.65		15.36	
Minimum Charge	Job		150		150	
8' Insulated						
Demolish	Ea.		42		42	Cost includes material and labor to install 8' x 6' 8" sliding vinyl patio door with double insulated tempered glass, anodized frame, screen, weatherstripped.
Install	Ea.	755	201		956	
Demolish and Install	Ea.	755	243		998	
Remove and Install	Ea.		160.43		160.43	
Clean	Ea.	1.71	13.65		15.36	
Minimum Charge	Job		150		150	
10' Insulated						
Demolish	Ea.		42		42	Cost includes material and labor to install 10' x 6' 8" sliding vinyl patio door with double insulated tempered glass, anodized frame, screen, weatherstripped.
Install	Ea.	860	300		1160	
Demolish and Install	Ea.	860	342		1202	
Remove and Install	Ea.		240.64		240.64	
Clean	Ea.	1.71	13.65		15.36	
Minimum Charge	Job		150		150	
Wood Framed						
6' Premium Insulated						
Demolish	Ea.		42		42	Cost includes material and labor to install 6' x 6' 8" sliding wood patio door with pine frame, double insulated glass, screen, lock and dead bolt, weatherstripped. Add for casing and finishing.
Install	Ea.	1050	150		1200	
Demolish and Install	Ea.	1050	192		1242	
Remove and Install	Ea.		120.32		120.32	
Clean	Ea.	1.71	13.65		15.36	
Minimum Charge	Job		150		150	

Exterior Doors and Windows

Sliding Patio Door	Unit	Material	Labor	Equip.	Total	Specification
8' Premium Insulated						
Demolish	Ea.		42		42	Cost includes material and labor to install 8' x 6' 8" sliding wood patio door with pine frame, double insulated glass, screen, lock and dead bolt, weatherstripped. Add for casing and finishing.
Install	Ea.	1200	201		1401	
Demolish and Install	Ea.	1200	243		1443	
Remove and Install	Ea.		160.43		160.43	
Clean	Ea.	1.71	13.65		15.36	
Minimum Charge	Job		150		150	
9' Premium Insulated						
Demolish	Ea.		42		42	Cost includes material and labor to install 9' x 6' 8" sliding wood patio door with pine frame, double insulated glass, screen, lock and dead bolt, weatherstripped. Add for casing and finishing.
Install	Ea.	1200	201		1401	
Demolish and Install	Ea.	1200	243		1443	
Remove and Install	Ea.		160.43		160.43	
Clean	Ea.	1.71	13.65		15.36	
Minimum Charge	Job		150		150	
Stain						
Install	Ea.	5.65	29.50		35.15	Includes labor and material to stain exterior type door and trim on both sides.
Minimum Charge	Job		132		132	
Casing Trim						
Demolish	Ea.		6.30		6.30	Cost includes material and labor to install 11/16" x 2-1/2" pine ranch style door casing for one side of a double door opening.
Install	Opng.	18.70	60		78.70	
Demolish and Install	Opng.	18.70	66.30		85	
Remove and Install	Opng.		48.13		48.13	
Clean	Opng.	.45	4.36		4.81	
Paint	Ea.	3.06	6		9.06	
Minimum Charge	Job		150		150	
Sliding Screen						
Demolish	Ea.		5.95		5.95	Includes material and labor to install sliding screen door, 6' 8" x 3'.
Install	Ea.	97.50	15.05		112.55	
Demolish and Install	Ea.	97.50	21		118.50	
Remove and Install	Ea.		12.03		12.03	
Clean	Ea.	.57	4.54		5.11	
Minimum Charge	Job		150		150	

Screen Door	Unit	Material	Labor	Equip.	Total	Specification
Aluminum						
Demolish	Ea.		28		28	Includes material and labor to install aluminum frame screen door with fiberglass screen cloth, closer, hinges, latch.
Install	Ea.	103	43		146	
Demolish and Install	Ea.	103	71		174	
Remove and Install	Ea.		34.38		34.38	
Clean	Ea.	2.73	9.10		11.83	
Paint	Ea.	2.94	10.55		13.49	
Minimum Charge	Job		150		150	
Wood						
Demolish	Ea.		28		28	Includes material and labor to install wood screen door with aluminum cloth screen. Frame and hardware not included.
Install	Ea.	134	50		184	
Demolish and Install	Ea.	134	78		212	
Remove and Install	Ea.		40.11		40.11	
Clean	Ea.	2.73	9.10		11.83	
Paint	Ea.	2.94	10.55		13.49	
Minimum Charge	Job		150		150	
Shave & Refit						
Install	Ea.		25		25	Includes labor to shave and rework door to fit opening at the job site.
Minimum Charge	Job		150		150	

Storm Door	Unit	Material	Labor	Equip.	Total	Specification
Good Grade						
Demolish	Ea.		28		28	Includes material and labor to install aluminum frame storm door with fiberglass screen cloth, upper glass panel, closer, hinges, latch.
Install	Ea.	155	43		198	
Demolish and Install	Ea.	155	71		226	
Remove and Install	Ea.		34.38		34.38	
Clean	Ea.	2.73	9.10		11.83	
Paint	Ea.	2.94	10.55		13.49	
Minimum Charge	Job		150		150	
Better Grade						
Demolish	Ea.		28		28	Includes material and labor to install aluminum frame storm door with fiberglass screen cloth, upper and lower glass panel, closer, latch.
Install	Ea.	200	43		243	
Demolish and Install	Ea.	200	71		271	
Remove and Install	Ea.		34.38		34.38	
Clean	Ea.	2.73	9.10		11.83	
Paint	Ea.	2.94	10.55		13.49	
Minimum Charge	Job		150		150	
Premium Grade						
Demolish	Ea.		28		28	Includes material and labor to install aluminum frame storm door with fiberglass screen cloth, upper and lower glass panel, closer, latch.
Install	Ea.	250	43		293	
Demolish and Install	Ea.	250	71		321	
Remove and Install	Ea.		34.38		34.38	
Clean	Ea.	2.73	9.10		11.83	
Paint	Ea.	2.94	10.55		13.49	
Minimum Charge	Job		150		150	
Insulated						
Demolish	Ea.		28		28	Includes material and labor to install aluminum frame storm door with fiberglass screen cloth, upper and lower glass panel, closer, latch.
Install	Ea.	230	21.50		251.50	
Demolish and Install	Ea.	230	49.50		279.50	
Remove and Install	Ea.		17.19		17.19	
Clean	Ea.	2.73	9.10		11.83	
Paint	Ea.	2.94	10.55		13.49	
Minimum Charge	Job		150		150	

Garage Door	Unit	Material	Labor	Equip.	Total	Specification
9′ One Piece Metal						
Demolish	Ea.		70.50		70.50	Includes material and labor to install one piece steel garage door 9′ x 7′ primed including hardware.
Install	Ea.	335	75		410	
Demolish and Install	Ea.	335	145.50		480.50	
Remove and Install	Ea.		60.16		60.16	
Clean	Ea.	3.42	27.50		30.92	
Paint	Ea.	6.05	66		72.05	
Minimum Charge	Job		150		150	
16′ One Piece Metal						
Demolish	Ea.		83.50		83.50	Includes material and labor to install one piece steel garage door 16′ x 7′ primed including hardware.
Install	Ea.	530	100		630	
Demolish and Install	Ea.	530	183.50		713.50	
Remove and Install	Ea.		80.21		80.21	
Clean	Ea.	6.80	36.50		43.30	
Paint	Ea.	12.05	75.50		87.55	
Minimum Charge	Job		150		150	

Garage Door		Unit	Material	Labor	Equip.	Total	Specification
9' Metal Sectional							
	Demolish	Ea.		70.50		70.50	Includes material and labor to install sectional metal garage door 9' x 7' primed including hardware.
	Install	Ea.	305	114		419	
	Demolish and Install	Ea.	305	184.50		489.50	
	Remove and Install	Ea.		91.15		91.15	
	Clean	Ea.	3.50	31		34.50	
	Paint	Ea.	12.05	75.50		87.55	
	Minimum Charge	Job		150		150	
16' Metal Sectional							
	Demolish	Ea.		83.50		83.50	Includes material and labor to install sectional metal garage door 16' x 7' primed including hardware.
	Install	Ea.	600	201		801	
	Demolish and Install	Ea.	600	284.50		884.50	
	Remove and Install	Ea.		160.43		160.43	
	Clean	Ea.	6.80	36.50		43.30	
	Paint	Ea.	12.05	75.50		87.55	
	Minimum Charge	Job		150		150	
Metal / Steel Roll Up							
8' x 8' Steel							
	Demolish	Ea.		95.50		95.50	Cost includes material and labor to install 8' x 8' steel overhead roll-up, chain hoist operated service door.
	Install	Ea.	790	470		1260	
	Demolish and Install	Ea.	790	565.50		1355.50	
	Remove and Install	Ea.		374		374	
	Clean	Ea.	3.50	31		34.50	
	Paint	Ea.	12.85	88		100.85	
	Minimum Charge	Job		750		750	
10' x 10' Steel							
	Demolish	Ea.		111		111	Cost includes material and labor to install 10' x 10' steel overhead roll-up, chain hoist operated service door.
	Install	Ea.	1050	535		1585	
	Demolish and Install	Ea.	1050	646		1696	
	Remove and Install	Ea.		427.43		427.43	
	Clean	Ea.	5.45	43.50		48.95	
	Paint	Ea.	20	88		108	
	Minimum Charge	Job		750		750	
12' x 12' Steel							
	Demolish	Ea.		134		134	Cost includes material and labor to install 12' x 12' steel overhead roll-up, chain hoist operated service door.
	Install	Ea.	1350	625		1975	
	Demolish and Install	Ea.	1350	759		2109	
	Remove and Install	Ea.		498.67		498.67	
	Clean	Ea.	7.85	48.50		56.35	
	Paint	Ea.	29	106		135	
	Minimum Charge	Job		750		750	
14' x 14' Steel							
	Demolish	Ea.		167		167	Cost includes material and labor to install 14' x 14' steel overhead roll-up, chain hoist operated service door.
	Install	Ea.	1800	935		2735	
	Demolish and Install	Ea.	1800	1102		2902	
	Remove and Install	Ea.		748		748	
	Clean	Ea.	10.75	62.50		73.25	
	Paint	Ea.	33.50	132		165.50	
	Minimum Charge	Job		750		750	
18' x 18' Steel							
	Demolish	Ea.		186		186	Cost includes material and labor to install 18' x 18' steel overhead roll-up, chain hoist operated service door.
	Install	Opng.	2350	1250		3600	
	Demolish and Install	Opng.	2350	1436		3786	
	Remove and Install	Opng.		997.33		997.33	
	Clean	Ea.	10.75	36.50		47.25	
	Paint	Ea.	65	264		329	
	Minimum Charge	Job		750		750	
Automatic Motor							
	Install	Ea.	209	30		239	Includes labor and material to install electric opener motor, 1/3 HP.
	Minimum Charge	Job		150		150	

Garage Door	Unit	Material	Labor	Equip.	Total	Specification
9′ One Piece Wood						
Demolish	Ea.		70.50		70.50	Includes material and labor to install one piece wood garage door 9′ x 7′ primed including hardware.
Install	Ea.	370	75		445	
Demolish and Install	Ea.	370	145.50		515.50	
Remove and Install	Ea.		60.16		60.16	
Clean	Ea.	3.42	27.50		30.92	
Paint	Ea.	6.05	66		72.05	
Minimum Charge	Job		150		150	
9′ Wood Sectional						
Demolish	Ea.		70.50		70.50	Includes material and labor to install sectional wood garage door 9′ x 7′ unfinished including hardware.
Install	Ea.	485	75		560	
Demolish and Install	Ea.	485	145.50		630.50	
Remove and Install	Ea.		60.16		60.16	
Clean	Ea.	3.42	27.50		30.92	
Paint	Ea.	6.05	66		72.05	
Minimum Charge	Job		150		150	
16′ Wood Sectional						
Demolish	Ea.		83.50		83.50	Includes material and labor to install sectional fiberglass garage door 16′ x 7′ unfinished including hardware.
Install	Ea.	1050	100		1150	
Demolish and Install	Ea.	1050	183.50		1233.50	
Remove and Install	Ea.		80.21		80.21	
Clean	Ea.	6.80	36.50		43.30	
Paint	Ea.	12.05	75.50		87.55	
Minimum Charge	Job		150		150	
Electric Opener						
Demolish	Ea.		48.50		48.50	Includes material and labor to install electric opener, 1/3 HP economy.
Install	Ea.	249	56.50		305.50	
Demolish and Install	Ea.	249	105		354	
Remove and Install	Ea.		45.15		45.15	
Clean	Ea.	.13	12.10		12.23	
Minimum Charge	Job		150		150	
Springs						
Install	Pr.	35	75		110	Includes labor and material to replace springs for garage door.
Minimum Charge	Job		150		150	

Commercial Metal Door	Unit	Material	Labor	Equip.	Total	Specification
Pre-hung Steel Door						
2′ 8″ x 6′ 8″						
Demolish	Ea.		42		42	Cost includes material and labor to install 2′ 8″ x 6′ 8″ pre-hung 1-3/4″ 18 gauge steel door with steel frame, hinges, aluminum sill, weather-stripped.
Install	Ea.	315	18.80		333.80	
Demolish and Install	Ea.	315	60.80		375.80	
Remove and Install	Ea.		18.80		18.80	
Clean	Ea.	2.73	9.10		11.83	
Paint	Ea.	3.76	22		25.76	
Minimum Charge	Job		150		150	
3′ x 6′ 8″						
Demolish	Ea.		42		42	Cost includes material and labor to install 3′ x 6′ 8″ pre-hung 1-3/4″ 18 gauge steel door with steel frame, hinges, aluminum sill, weather-stripped.
Install	Ea.	310	20		330	
Demolish and Install	Ea.	310	62		372	
Remove and Install	Ea.		20.05		20.05	
Clean	Ea.	2.73	9.10		11.83	
Paint	Ea.	3.76	22		25.76	
Minimum Charge	Job		150		150	

Commercial Metal Door	Unit	Material	Labor	Equip.	Total	Specification
3′ 6″ x 6′ 8″						
Demolish	Ea.		42		42	Cost includes material and labor to install 3′ 6″ x 6′ 8″ pre-hung 1-3/4″ 18 gauge steel door with steel frame, hinges, aluminum sill, weather-stripped.
Install	Ea.	360	23		383	
Demolish and Install	Ea.	360	65		425	
Remove and Install	Ea.		23.14		23.14	
Clean	Ea.	2.73	9.10		11.83	
Paint	Ea.	3.76	22		25.76	
Minimum Charge	Job		150		150	
4′ x 6′ 8″						
Demolish	Ea.		42		42	Cost includes material and labor to install 4′ x 6′ 8″ pre-hung 1-3/4″ 18 gauge steel door with steel frame, hinges, aluminum sill, weather-stripped.
Install	Ea.	390	30		420	
Demolish and Install	Ea.	390	72		462	
Remove and Install	Ea.		30.08		30.08	
Clean	Ea.	2.73	9.10		11.83	
Paint	Ea.	3.76	22		25.76	
Minimum Charge	Job		150		150	
Metal Jamb						
6′ 8″ High						
Demolish	Ea.		16.70		16.70	Cost includes material and labor to install 18 gauge hollow metal door frame to fit a 4-1/2″ jamb 6′ 8″ high and up to 3′ 6″ opening.
Install	Ea.	77	18.80		95.80	
Demolish and Install	Ea.	77	35.50		112.50	
Remove and Install	Ea.		18.80		18.80	
Clean	Ea.	.57	4.54		5.11	
Paint	Ea.	1.03	5.30		6.33	
Minimum Charge	Job		150		150	
7′ High						
Demolish	Ea.		16.70		16.70	Cost includes material and labor to install 18 gauge hollow metal door frame to fit a 4-1/2″ jamb 7′ high and up to 3′ 6″ opening.
Install	Ea.	84	20		104	
Demolish and Install	Ea.	84	36.70		120.70	
Remove and Install	Ea.		20.05		20.05	
Clean	Ea.	.57	4.54		5.11	
Paint	Ea.	1.03	5.30		6.33	
Minimum Charge	Job		150		150	
8′ High						
Demolish	Ea.		16.70		16.70	Cost includes material and labor to install 18 gauge hollow metal door frame to fit a 4-1/2″ jamb 8′ high and up to 3′ 6″ opening.
Install	Ea.	100	25		125	
Demolish and Install	Ea.	100	41.70		141.70	
Remove and Install	Ea.		25.07		25.07	
Clean	Ea.	.80	5.45		6.25	
Paint	Ea.	1.16	5.85		7.01	
Minimum Charge	Job		150		150	
9′ High						
Demolish	Ea.		16.70		16.70	Cost includes material and labor to install 18 gauge hollow metal door frame to fit a 4-1/2″ jamb 9′ high and up to 3′ 6″ opening.
Install	Ea.	110	30		140	
Demolish and Install	Ea.	110	46.70		156.70	
Remove and Install	Ea.		30.08		30.08	
Clean	Ea.	.80	5.45		6.25	
Paint	Ea.	1.16	5.85		7.01	
Minimum Charge	Job		150		150	
Hardware						
Doorknob w / Lock						
Demolish	Ea.		11.15		11.15	Includes material and labor to install keyed entry lock set, bored type including knob trim, strike and strike box for standard commercial application.
Install	Ea.	100	25		125	
Demolish and Install	Ea.	100	36.15		136.15	
Remove and Install	Ea.		20.05		20.05	
Clean	Ea.	.28	6.80		7.08	
Minimum Charge	Job		150		150	

Commercial Metal Door	Unit	Material	Labor	Equip.	Total	Specification
Lever Handle						
Demolish	Ea.		11.15		11.15	Includes material and labor to install passage lockset, keyless bored type with lever handle, non-locking standard commercial latchset.
Install	Ea.	55	30		85	
Demolish and Install	Ea.	55	41.15		96.15	
Remove and Install	Ea.		24.06		24.06	
Clean	Ea.	.28	6.80		7.08	
Minimum Charge	Job		150		150	
Deadbolt						
Demolish	Ea.		10.45		10.45	Includes material and labor to install a deadbolt.
Install	Ea.	36.50	21.50		58	
Demolish and Install	Ea.	36.50	31.95		68.45	
Remove and Install	Ea.		17.19		17.19	
Clean	Ea.	.28	6.80		7.08	
Minimum Charge	Job		150		150	
Panic Device						
Demolish	Ea.		24		24	Includes material and labor to install touch bar, low profile, exit only.
Install	Ea.	320	50		370	
Demolish and Install	Ea.	320	74		394	
Remove and Install	Ea.		40.11		40.11	
Clean	Ea.	.13	12.10		12.23	
Minimum Charge	Job		150		150	
Closer						
Demolish	Ea.		6.25		6.25	Includes material and labor to install pneumatic heavy duty model for exterior type doors.
Install	Ea.	161	50		211	
Demolish and Install	Ea.	161	56.25		217.25	
Remove and Install	Ea.		40.11		40.11	
Clean	Ea.	.13	9.10		9.23	
Minimum Charge	Job		150		150	
Peep-Hole						
Demolish	Ea.		5.40		5.40	Includes material and labor to install peep-hole for door.
Install	Ea.	8.25	9.40		17.65	
Demolish and Install	Ea.	8.25	14.80		23.05	
Remove and Install	Ea.		7.52		7.52	
Clean	Ea.	.02	2.41		2.43	
Minimum Charge	Job		150		150	
Exit Sign						
Demolish	Ea.		16.65		16.65	Includes material and labor to install a wall mounted interior electric exit sign.
Install	Ea.	43	41		84	
Demolish and Install	Ea.	43	57.65		100.65	
Remove and Install	Ea.		32.92		32.92	
Clean	Ea.	.02	4.90		4.92	
Minimum Charge	Job		150		150	
Kickplate						
Demolish	Ea.		9.30		9.30	Includes material and labor to install stainless steel kickplate, 10″ x 36″.
Install	Ea.	43	20		63	
Demolish and Install	Ea.	43	29.30		72.30	
Remove and Install	Ea.		16.04		16.04	
Clean	Ea.	.28	6.80		7.08	
Minimum Charge	Job		150		150	

Aluminum Window	Unit	Material	Labor	Equip.	Total	Specification
Single Hung Two Light						
2′ x 2′						
Demolish	Ea.		15.15		15.15	Cost includes material and labor to install 2′ x 2′ single hung, two light, double glazed aluminum window including screen.
Install	Ea.	58	68		126	
Demolish and Install	Ea.	58	83.15		141.15	
Remove and Install	Ea.		54.40		54.40	
Clean	Ea.	.03	5.60		5.63	
Minimum Charge	Job		187		187	

Aluminum Window	Unit	Material	Labor	Equip.	Total	Specification
2' x 2' 6"						
Demolish	Ea.		15.15		15.15	Cost includes material and labor to install 2' x 2' 6" single hung, two light, double glazed aluminum window including screen.
Install	Ea.	62.50	68		130.50	
Demolish and Install	Ea.	62.50	83.15		145.65	
Remove and Install	Ea.		54.40		54.40	
Clean	Ea.	.03	5.60		5.63	
Minimum Charge	Job		187		187	
3' x 1' 6"						
Demolish	Ea.		15.15		15.15	Cost includes material and labor to install 3' x 1' 6" single hung, two light, double glazed aluminum window including screen.
Install	Ea.	59	75		134	
Demolish and Install	Ea.	59	90.15		149.15	
Remove and Install	Ea.		59.84		59.84	
Clean	Ea.	.02	4.90		4.92	
Minimum Charge	Job		187		187	
3' x 2'						
Demolish	Ea.		15.15		15.15	Cost includes material and labor to install 3' x 2' single hung, two light, double glazed aluminum window including screen.
Install	Ea.	71	75		146	
Demolish and Install	Ea.	71	90.15		161.15	
Remove and Install	Ea.		59.84		59.84	
Clean	Ea.	.02	4.90		4.92	
Minimum Charge	Job		187		187	
3' x 2' 6"						
Demolish	Ea.		15.15		15.15	Cost includes material and labor to install 3' x 2' 6" single hung, two light, double glazed aluminum window including screen.
Install	Ea.	74.50	75		149.50	
Demolish and Install	Ea.	74.50	90.15		164.65	
Remove and Install	Ea.		59.84		59.84	
Clean	Ea.	.03	5.60		5.63	
Minimum Charge	Job		187		187	
3' x 3'						
Demolish	Ea.		15.15		15.15	Cost includes material and labor to install 3' x 3' single hung, two light, double glazed aluminum window including screen.
Install	Ea.	83	75		158	
Demolish and Install	Ea.	83	90.15		173.15	
Remove and Install	Ea.		59.84		59.84	
Clean	Ea.	.03	5.60		5.63	
Minimum Charge	Job		187		187	
3' x 3' 6"						
Demolish	Ea.		15.15		15.15	Cost includes material and labor to install 3' x 3' 6" single hung, two light, double glazed aluminum window including screen.
Install	Ea.	87.50	75		162.50	
Demolish and Install	Ea.	87.50	90.15		177.65	
Remove and Install	Ea.		59.84		59.84	
Clean	Ea.	.04	6.55		6.59	
Minimum Charge	Job		187		187	
3' x 4'						
Demolish	Ea.		15.15		15.15	Cost includes material and labor to install 3' x 4' single hung, two light, double glazed aluminum window including screen.
Install	Ea.	95.50	75		170.50	
Demolish and Install	Ea.	95.50	90.15		185.65	
Remove and Install	Ea.		59.84		59.84	
Clean	Ea.	.04	6.55		6.59	
Minimum Charge	Job		187		187	
3' x 5'						
Demolish	Ea.		22		22	Cost includes material and labor to install 3' x 5' single hung, two light, double glazed aluminum window including screen.
Install	Ea.	106	75		181	
Demolish and Install	Ea.	106	97		203	
Remove and Install	Ea.		59.84		59.84	
Clean	Ea.	.06	7.85		7.91	
Minimum Charge	Job		187		187	

Aluminum Window	Unit	Material	Labor	Equip.	Total	Specification
3′ x 6′						
Demolish	Ea.		22		22	Cost includes material and labor to install 3′ x 6′ single hung, two light, double glazed aluminum window including screen.
Install	Ea.	114	83		197	
Demolish and Install	Ea.	114	105		219	
Remove and Install	Ea.		66.49		66.49	
Clean	Ea.	.07	9.20		9.27	
Minimum Charge	Job		187		187	
Double Hung Two Light						
2′ x 2′						
Demolish	Ea.		15.15		15.15	Cost includes material and labor to install 2′ x 2′ double hung, two light, double glazed aluminum window including screen.
Install	Ea.	99.50	68		167.50	
Demolish and Install	Ea.	99.50	83.15		182.65	
Remove and Install	Ea.		54.40		54.40	
Clean	Ea.	.02	4.90		4.92	
Minimum Charge	Job		187		187	
2′ x 2′ 6″						
Demolish	Ea.		15.15		15.15	Cost includes material and labor to install 2′ x 2′ 6″ double hung, two light, double glazed aluminum window including screen.
Install	Ea.	108	68		176	
Demolish and Install	Ea.	108	83.15		191.15	
Remove and Install	Ea.		54.40		54.40	
Clean	Ea.	.02	4.90		4.92	
Minimum Charge	Job		187		187	
3′ x 1′ 6″						
Demolish	Ea.		15.15		15.15	Cost includes material and labor to install 3′ x 1′ 6″ double hung, two light, double glazed aluminum window including screen.
Install	Ea.	256	75		331	
Demolish and Install	Ea.	256	90.15		346.15	
Remove and Install	Ea.		59.84		59.84	
Clean	Ea.	.02	4.90		4.92	
Minimum Charge	Job		187		187	
3′ x 2′						
Demolish	Ea.		15.15		15.15	Cost includes material and labor to install 3′ x 2′ double hung, two light, double glazed aluminum window including screen.
Install	Ea.	271	75		346	
Demolish and Install	Ea.	271	90.15		361.15	
Remove and Install	Ea.		59.84		59.84	
Clean	Ea.	.02	4.90		4.92	
Minimum Charge	Job		187		187	
3′ x 2′ 6″						
Demolish	Ea.		15.15		15.15	Cost includes material and labor to install 3′ x 2′ 6″ double hung, two light, double glazed aluminum window including screen.
Install	Ea.	264	75		339	
Demolish and Install	Ea.	264	90.15		354.15	
Remove and Install	Ea.		59.84		59.84	
Clean	Ea.	.03	5.60		5.63	
Minimum Charge	Job		187		187	
3′ x 3′						
Demolish	Ea.		15.15		15.15	Cost includes material and labor to install 3′ x 3′ double hung, two light, double glazed aluminum window including screen.
Install	Ea.	286	75		361	
Demolish and Install	Ea.	286	90.15		376.15	
Remove and Install	Ea.		59.84		59.84	
Clean	Ea.	.03	5.60		5.63	
Minimum Charge	Job		187		187	
3′ x 3′ 6″						
Demolish	Ea.		15.15		15.15	Cost includes material and labor to install 3′ x 3′ 6″ double hung, two light, double glazed aluminum window including screen.
Install	Ea.	315	75		390	
Demolish and Install	Ea.	315	90.15		405.15	
Remove and Install	Ea.		59.84		59.84	
Clean	Ea.	.04	6.55		6.59	
Minimum Charge	Job		187		187	

Aluminum Window	Unit	Material	Labor	Equip.	Total	Specification
3′ x 4′						
Demolish	Ea.		15.15		15.15	Cost includes material and labor to install 3′ x 4′ double hung, two light, double glazed aluminum window including screen.
Install	Ea.	325	75		400	
Demolish and Install	Ea.	325	90.15		415.15	
Remove and Install	Ea.		59.84		59.84	
Clean	Ea.	.04	6.55		6.59	
Minimum Charge	Job		187		187	
3′ x 5′						
Demolish	Ea.		22		22	Cost includes material and labor to install 3′ x 5′ double hung, two light, double glazed aluminum window including screen.
Install	Ea.	365	75		440	
Demolish and Install	Ea.	365	97		462	
Remove and Install	Ea.		59.84		59.84	
Clean	Ea.	.06	7.85		7.91	
Minimum Charge	Job		187		187	
3′ x 6′						
Demolish	Ea.		22		22	Cost includes material and labor to install 3′ x 6′ double hung, two light, double glazed aluminum window including screen.
Install	Ea.	480	83		563	
Demolish and Install	Ea.	480	105		585	
Remove and Install	Ea.		66.49		66.49	
Clean	Ea.	.07	9.20		9.27	
Minimum Charge	Job		187		187	
Sliding Sash						
1′ 6″ x 3′						
Demolish	Ea.		15.15		15.15	Cost includes material and labor to install 1′ 6″ x 3′ sliding sash, two light, double glazed aluminum window including screen.
Install	Ea.	55.50	68		123.50	
Demolish and Install	Ea.	55.50	83.15		138.65	
Remove and Install	Ea.		54.40		54.40	
Clean	Ea.	.02	4.90		4.92	
Minimum Charge	Job		187		187	
2′ x 2′						
Demolish	Ea.		15.15		15.15	Cost includes material and labor to install 2′ x 2′ sliding sash, two light, double glazed aluminum window including screen.
Install	Ea.	50.50	75		125.50	
Demolish and Install	Ea.	50.50	90.15		140.65	
Remove and Install	Ea.		59.84		59.84	
Clean	Ea.	.02	4.90		4.92	
Minimum Charge	Job		187		187	
3′ x 2′						
Demolish	Ea.		15.15		15.15	Cost includes material and labor to install 3′ x 2′ sliding sash, two light, double glazed aluminum window including screen.
Install	Ea.	64.50	75		139.50	
Demolish and Install	Ea.	64.50	90.15		154.65	
Remove and Install	Ea.		59.84		59.84	
Clean	Ea.	.02	4.90		4.92	
Minimum Charge	Job		187		187	
3′ x 2′ 6″						
Demolish	Ea.		15.15		15.15	Cost includes material and labor to install 3′ x 2′ 6″ sliding sash, two light, double glazed aluminum window including screen.
Install	Ea.	68	75		143	
Demolish and Install	Ea.	68	90.15		158.15	
Remove and Install	Ea.		59.84		59.84	
Clean	Ea.	.03	5.60		5.63	
Minimum Charge	Job		187		187	
3′ x 3′						
Demolish	Ea.		15.15		15.15	Cost includes material and labor to install 3′ x 3′ sliding sash, two light, double glazed aluminum window including screen.
Install	Ea.	69	75		144	
Demolish and Install	Ea.	69	90.15		159.15	
Remove and Install	Ea.		59.84		59.84	
Clean	Ea.	.03	5.60		5.63	
Minimum Charge	Job		187		187	

Aluminum Window		Unit	Material	Labor	Equip.	Total	Specification
3' x 3' 6"							
	Demolish	Ea.		15.15		15.15	Cost includes material and labor to install 3' x 3' 6" sliding sash, two light, double glazed aluminum window including screen.
	Install	Ea.	81.50	75		156.50	
	Demolish and Install	Ea.	81.50	90.15		171.65	
	Remove and Install	Ea.		59.84		59.84	
	Clean	Ea.	.04	6.55		6.59	
	Minimum Charge	Job		187		187	
3' x 4'							
	Demolish	Ea.		15.15		15.15	Cost includes material and labor to install 3' x 4' sliding sash, two light, double glazed aluminum window including screen.
	Install	Ea.	89	83		172	
	Demolish and Install	Ea.	89	98.15		187.15	
	Remove and Install	Ea.		66.49		66.49	
	Clean	Ea.	.04	6.55		6.59	
	Minimum Charge	Job		187		187	
3' x 5'							
	Demolish	Ea.		22		22	Cost includes material and labor to install 3' x 5' sliding sash, two light, double glazed aluminum window including screen.
	Install	Ea.	113	83		196	
	Demolish and Install	Ea.	113	105		218	
	Remove and Install	Ea.		66.49		66.49	
	Clean	Ea.	.06	7.85		7.91	
	Minimum Charge	Job		187		187	
4' x 5'							
	Demolish	Ea.		22		22	Cost includes material and labor to install 4' x 5' sliding sash, two light, double glazed aluminum window including screen.
	Install	Ea.	114	107		221	
	Demolish and Install	Ea.	114	129		243	
	Remove and Install	Ea.		85.49		85.49	
	Clean	Ea.	.09	11.20		11.29	
	Minimum Charge	Job		187		187	
5' x 5'							
	Demolish	Ea.		22		22	Cost includes material and labor to install 5' x 5' sliding sash, two light, double glazed aluminum window including screen.
	Install	Ea.	156	107		263	
	Demolish and Install	Ea.	156	129		285	
	Remove and Install	Ea.		85.49		85.49	
	Clean	Ea.	.09	11.20		11.29	
	Minimum Charge	Job		187		187	
Awning Sash							
26" x 26", 2 Light							
	Demolish	Ea.		15.15		15.15	Cost includes material and labor to install 26" x 26" two light awning, double glazed aluminum window including screen.
	Install	Ea.	244	83		327	
	Demolish and Install	Ea.	244	98.15		342.15	
	Remove and Install	Ea.		66.49		66.49	
	Clean	Ea.	.03	5.60		5.63	
	Minimum Charge	Job		187		187	
26" x 37", 2 Light							
	Demolish	Ea.		15.15		15.15	Cost includes material and labor to install 26" x 37" two light awning, double glazed aluminum window including screen.
	Install	Ea.	277	83		360	
	Demolish and Install	Ea.	277	98.15		375.15	
	Remove and Install	Ea.		66.49		66.49	
	Clean	Ea.	.03	5.60		5.63	
	Minimum Charge	Job		187		187	
38" x 26", 3 Light							
	Demolish	Ea.		15.15		15.15	Cost includes material and labor to install 38" x 26" three light awning, double glazed aluminum window including screen.
	Install	Ea.	310	83		393	
	Demolish and Install	Ea.	310	98.15		408.15	
	Remove and Install	Ea.		66.49		66.49	
	Clean	Ea.	.04	6.55		6.59	
	Minimum Charge	Job		187		187	

Aluminum Window	Unit	Material	Labor	Equip.	Total	Specification
38″ x 37″, 3 Light						
Demolish	Ea.		15.15		15.15	Cost includes material and labor to install 38″ x 37″ three light awning, double glazed aluminum window including screen.
Install	Ea.	320	93.50		413.50	
Demolish and Install	Ea.	320	108.65		428.65	
Remove and Install	Ea.		74.80		74.80	
Clean	Ea.	.04	6.55		6.59	
Minimum Charge	Job		187		187	
38″ x 53″, 3 Light						
Demolish	Ea.		22		22	Cost includes material and labor to install 38″ x 53″ three light awning, double glazed aluminum window including screen.
Install	Ea.	515	93.50		608.50	
Demolish and Install	Ea.	515	115.50		630.50	
Remove and Install	Ea.		74.80		74.80	
Clean	Ea.	.06	7.85		7.91	
Minimum Charge	Job		187		187	
50″ x 37″, 4 Light						
Demolish	Ea.		22		22	Cost includes material and labor to install 50″ x 37″ four light awning, double glazed aluminum window including screen.
Install	Ea.	525	93.50		618.50	
Demolish and Install	Ea.	525	115.50		640.50	
Remove and Install	Ea.		74.80		74.80	
Clean	Ea.	.06	7.85		7.91	
Minimum Charge	Job		187		187	
50″ x 53″, 4 Light						
Demolish	Ea.		22		22	Cost includes material and labor to install 50″ x 53″ four light awning, double glazed aluminum window including screen.
Install	Ea.	865	107		972	
Demolish and Install	Ea.	865	129		994	
Remove and Install	Ea.		85.49		85.49	
Clean	Ea.	.07	9.20		9.27	
Minimum Charge	Job		187		187	
63″ x 37″, 5 Light						
Demolish	Ea.		22		22	Cost includes material and labor to install 63″ x 37″ five light awning, double glazed aluminum window including screen.
Install	Ea.	675	107		782	
Demolish and Install	Ea.	675	129		804	
Remove and Install	Ea.		85.49		85.49	
Clean	Ea.	.07	9.20		9.27	
Minimum Charge	Job		187		187	
63″ x 53″, 5 Light						
Demolish	Ea.		22		22	Cost includes material and labor to install 63″ x 53″ five light awning, double glazed aluminum window including screen.
Install	Ea.	930	107		1037	
Demolish and Install	Ea.	930	129		1059	
Remove and Install	Ea.		85.49		85.49	
Clean	Ea.	.09	11.20		11.29	
Minimum Charge	Job		187		187	
Mullion Bars						
Install	Ea.	8.90	30		38.90	Includes labor and material to install mullions for sliding aluminum window.
Minimum Charge	Job		187		187	
Glass Louvered						
2′ x 2′						
Demolish	Ea.		15.15		15.15	Cost includes material and labor to install 2′ x 2′ aluminum window including 4″ glass louvers, hardware, screen.
Install	Ea.	45.50	68		113.50	
Demolish and Install	Ea.	45.50	83.15		128.65	
Remove and Install	Ea.		54.40		54.40	
Clean	Ea.	.02	4.90		4.92	
Minimum Charge	Job		187		187	

Aluminum Window	Unit	Material	Labor	Equip.	Total	Specification
2′ x 2′ 6″						
Demolish	Ea.		15.15		15.15	Cost includes material and labor to install 2′ x 2′ 6″ aluminum window including 4″ glass louvers, hardware, screen.
Install	Ea.	55.50	68		123.50	
Demolish and Install	Ea.	55.50	83.15		138.65	
Remove and Install	Ea.		54.40		54.40	
Clean	Ea.	.02	4.90		4.92	
Minimum Charge	Job		187		187	
3′ x 1′ 6″						
Demolish	Ea.		15.15		15.15	Cost includes material and labor to install 3′ x 1′ 6″ aluminum window including 4″ glass louvers, hardware, screen.
Install	Ea.	58.50	75		133.50	
Demolish and Install	Ea.	58.50	90.15		148.65	
Remove and Install	Ea.		59.84		59.84	
Clean	Ea.	.02	4.90		4.92	
Minimum Charge	Job		187		187	
3′ x 2′						
Demolish	Ea.		15.15		15.15	Cost includes material and labor to install 3′ x 2′ aluminum window including 4″ glass louvers, hardware, screen.
Install	Ea.	67	75		142	
Demolish and Install	Ea.	67	90.15		157.15	
Remove and Install	Ea.		59.84		59.84	
Clean	Ea.	.02	4.90		4.92	
Minimum Charge	Job		187		187	
3′ x 2′ 6″						
Demolish	Ea.		15.15		15.15	Cost includes material and labor to install 3′ x 2′ 6″ aluminum window including 4″ glass louvers, hardware, screen.
Install	Ea.	75	75		150	
Demolish and Install	Ea.	75	90.15		165.15	
Remove and Install	Ea.		59.84		59.84	
Clean	Ea.	.03	5.60		5.63	
Minimum Charge	Job		187		187	
3′ x 3′						
Demolish	Ea.		15.15		15.15	Cost includes material and labor to install 3′ x 3′ aluminum window including 4″ glass louvers, hardware, screen.
Install	Ea.	85.50	83		168.50	
Demolish and Install	Ea.	85.50	98.15		183.65	
Remove and Install	Ea.		66.49		66.49	
Clean	Ea.	.03	5.60		5.63	
Minimum Charge	Job		187		187	
3′ x 3′ 6″						
Demolish	Ea.		15.15		15.15	Cost includes material and labor to install 3′ x 3′ 6″ aluminum window including 4″ glass louvers, hardware, screen.
Install	Ea.	97	83		180	
Demolish and Install	Ea.	97	98.15		195.15	
Remove and Install	Ea.		66.49		66.49	
Clean	Ea.	.04	6.55		6.59	
Minimum Charge	Job		187		187	
3′ x 4′						
Demolish	Ea.		15.15		15.15	Cost includes material and labor to install 3′ x 4′ aluminum window including 4″ glass louvers, hardware, screen.
Install	Ea.	108	83		191	
Demolish and Install	Ea.	108	98.15		206.15	
Remove and Install	Ea.		66.49		66.49	
Clean	Ea.	.04	6.55		6.59	
Minimum Charge	Job		187		187	
3′ x 5′						
Demolish	Ea.		22		22	Cost includes material and labor to install 3′ x 5′ aluminum window with 4″ glass louvers, hardware, screen.
Install	Ea.	138	93.50		231.50	
Demolish and Install	Ea.	138	115.50		253.50	
Remove and Install	Ea.		74.80		74.80	
Clean	Ea.	.06	7.85		7.91	
Minimum Charge	Job		187		187	

Aluminum Window	Unit	Material	Labor	Equip.	Total	Specification
3' x 6'						
Demolish	Ea.		22		22	Cost includes material and labor to install 3' x 6' aluminum window with 4" glass louvers, hardware, screen.
Install	Ea.	160	93.50		253.50	
Demolish and Install	Ea.	160	115.50		275.50	
Remove and Install	Ea.		74.80		74.80	
Clean	Ea.	.07	9.20		9.27	
Minimum Charge	Job		187		187	
Bay						
4' 8" x 6' 4"						
Demolish	Ea.		16.15		16.15	Cost includes material and labor to install 4' 8" x 6' 4" angle bay double hung unit, pine frame with aluminum cladding, picture center sash, hardware, screens.
Install	Ea.	1025	75		1100	
Demolish and Install	Ea.	1025	91.15		1116.15	
Remove and Install	Ea.		60.16		60.16	
Clean	Ea.	.12	13.05		13.17	
Minimum Charge	Job		187		187	
4' 8" x 7' 2"						
Demolish	Ea.		16.15		16.15	Cost includes material and labor to install 4' 8" x 7' 2" angle bay double hung unit, pine frame with aluminum cladding, picture center sash, hardware, screens.
Install	Ea.	1050	75		1125	
Demolish and Install	Ea.	1050	91.15		1141.15	
Remove and Install	Ea.		60.16		60.16	
Clean	Ea.	.12	13.05		13.17	
Minimum Charge	Job		187		187	
4' 8" x 7' 10"						
Demolish	Ea.		16.15		16.15	Cost includes material and labor to install 4' 8" x 7' 10" angle bay double hung unit, pine frame with aluminum cladding, picture center sash, hardware, screens.
Install	Ea.	1050	75		1125	
Demolish and Install	Ea.	1050	91.15		1141.15	
Remove and Install	Ea.		60.16		60.16	
Clean	Ea.	.12	13.05		13.17	
Minimum Charge	Job		187		187	
4' 8" x 8' 6"						
Demolish	Ea.		18.65		18.65	Cost includes material and labor to install 4' 8" x 8' 6" angle bay double hung unit, pine frame with aluminum cladding, picture center sash, hardware, screens.
Install	Ea.	1200	86		1286	
Demolish and Install	Ea.	1200	104.65		1304.65	
Remove and Install	Ea.		68.75		68.75	
Clean	Ea.	.02	17.40		17.42	
Minimum Charge	Job		187		187	
4' 8" x 8' 10"						
Demolish	Ea.		18.65		18.65	Cost includes material and labor to install 4' 8" x 8' 10" angle bay double hung unit, pine frame with aluminum cladding, picture center sash, hardware, screens.
Install	Ea.	1275	86		1361	
Demolish and Install	Ea.	1275	104.65		1379.65	
Remove and Install	Ea.		68.75		68.75	
Clean	Ea.	.02	17.40		17.42	
Minimum Charge	Job		187		187	
4' 8" x 9' 2"						
Demolish	Ea.		18.65		18.65	Cost includes material and labor to install 4' 8" x 9' 2" angle bay double hung unit, pine frame with aluminum cladding, picture center sash, hardware, screens.
Install	Ea.	1325	150		1475	
Demolish and Install	Ea.	1325	168.65		1493.65	
Remove and Install	Ea.		120.32		120.32	
Clean	Ea.	.02	17.40		17.42	
Minimum Charge	Job		187		187	
4' 8" x 9' 10"						
Demolish	Ea.		18.65		18.65	Cost includes material and labor to install 4' 8" x 9' 10" angle bay double hung unit, pine frame with aluminum cladding, picture center sash, hardware, screens.
Install	Ea.	1375	150		1525	
Demolish and Install	Ea.	1375	168.65		1543.65	
Remove and Install	Ea.		120.32		120.32	
Clean	Ea.	.02	17.40		17.42	
Minimum Charge	Job		187		187	

Aluminum Window	Unit	Material	Labor	Equip.	Total	Specification
Casing						
Demolish	Ea.		6.95		6.95	Cost includes material and labor to install 11/16" x 2-1/2" pine ranch style window casing.
Install	Opng.	22	23		45	
Demolish and Install	Opng.	22	29.95		51.95	
Remove and Install	Opng.		18.51		18.51	
Clean	Opng.	.39	2.18		2.57	
Paint	Ea.	3.06	5.30		8.36	
Minimum Charge	Job		187		187	
Storm Window						
2' x 3' 6"						
Demolish	Ea.		8.95		8.95	Includes material and labor to install residential aluminum storm window with mill finish, 2' x 3' 5" high.
Install	Ea.	71.50	20		91.50	
Demolish and Install	Ea.	71.50	28.95		100.45	
Remove and Install	Ea.		16.04		16.04	
Clean	Ea.	.02	4.90		4.92	
Minimum Charge	Job		187		187	
2' 6" x 5'						
Demolish	Ea.		8.95		8.95	Includes material and labor to install residential aluminum storm window with mill finish, 2' 6" x 5' high.
Install	Ea.	80	21.50		101.50	
Demolish and Install	Ea.	80	30.45		110.45	
Remove and Install	Ea.		17.19		17.19	
Clean	Ea.	.04	6.55		6.59	
Minimum Charge	Job		187		187	
4' x 6'						
Demolish	Ea.		17.30		17.30	Includes material and labor to install residential aluminum storm window, 4' x 6' high.
Install	Ea.	95	24		119	
Demolish and Install	Ea.	95	41.30		136.30	
Remove and Install	Ea.		19.25		19.25	
Clean	Ea.	.08	9.80		9.88	
Minimum Charge	Job		187		187	
Hardware						
Weatherstripping						
Install	L.F.	.25	2.51		2.76	Includes labor and material to install vinyl V strip weatherstripping for double-hung window.
Minimum Charge	Job		100		100	
Trim Set						
Demolish	Ea.		6.95		6.95	Cost includes material and labor to install 11/16" x 2-1/2" pine ranch style window casing.
Install	Opng.	22	23		45	
Demolish and Install	Opng.	22	29.95		51.95	
Remove and Install	Opng.		18.51		18.51	
Clean	Opng.	.39	2.18		2.57	
Paint	Ea.	3.06	5.30		8.36	
Minimum Charge	Job		150		150	

Vinyl Window	Unit	Material	Labor	Equip.	Total	Specification
Double Hung						
2' x 2' 6"						
Demolish	Ea.		11		11	Cost includes material and labor to install 2' x 2' 6" double hung, two light, vinyl window with screen.
Install	Ea.	163	40		203	
Demolish and Install	Ea.	163	51		214	
Remove and Install	Ea.		32.09		32.09	
Clean	Ea.	.02	4.90		4.92	
Minimum Charge	Job		300		300	

Vinyl Window

Vinyl Window	Unit	Material	Labor	Equip.	Total	Specification
2' x 3' 6"						
Demolish	Ea.		11		11	Cost includes material and labor to install 2' x 3' 6" double hung, two light, vinyl window with screen.
Install	Ea.	167	43		210	
Demolish and Install	Ea.	167	54		221	
Remove and Install	Ea.		34.38		34.38	
Clean	Ea.	.03	5.60		5.63	
Minimum Charge	Job		300		300	
2' 6" x 4' 6"						
Demolish	Ea.		11		11	Cost includes material and labor to install 2' 6" x 4' 6" double hung, two light, vinyl window with screen.
Install	Ea.	202	46.50		248.50	
Demolish and Install	Ea.	202	57.50		259.50	
Remove and Install	Ea.		37.02		37.02	
Clean	Ea.	.04	6.55		6.59	
Minimum Charge	Job		300		300	
3' x 4'						
Demolish	Ea.		11		11	Includes material and labor to install medium size double hung, 3' x 4', two light, vinyl window with screen.
Install	Ea.	213	60		273	
Demolish and Install	Ea.	213	71		284	
Remove and Install	Ea.		48.13		48.13	
Clean	Ea.	.04	6.55		6.59	
Minimum Charge	Job		300		300	
3' x 4' 6"						
Demolish	Ea.		13.45		13.45	Cost includes material and labor to install 3' x 4' 6" double hung, two light, vinyl window with screen.
Install	Ea.	231	67		298	
Demolish and Install	Ea.	231	80.45		311.45	
Remove and Install	Ea.		53.48		53.48	
Clean	Ea.	.06	7.85		7.91	
Minimum Charge	Job		300		300	
4' x 4' 6"						
Demolish	Ea.		13.45		13.45	Cost includes material and labor to install 4' x 4' 6" double hung, two light, vinyl window with screen.
Install	Ea.	250	75		325	
Demolish and Install	Ea.	250	88.45		338.45	
Remove and Install	Ea.		60.16		60.16	
Clean	Ea.	.07	9.20		9.27	
Minimum Charge	Job		300		300	
4' x 6'						
Demolish	Ea.		13.45		13.45	Includes material and labor to install large double hung, 4' x 6', two light, vinyl window with screen.
Install	Ea.	294	86		380	
Demolish and Install	Ea.	294	99.45		393.45	
Remove and Install	Ea.		68.75		68.75	
Clean	Ea.	.09	11.20		11.29	
Minimum Charge	Job		300		300	
Trim Set						
Demolish	Ea.		6.95		6.95	Cost includes material and labor to install 11/16" x 2-1/2" pine ranch style window casing.
Install	Opng.	22	23		45	
Demolish and Install	Opng.	22	29.95		51.95	
Remove and Install	Opng.		18.51		18.51	
Clean	Opng.	.39	2.18		2.57	
Paint	Ea.	3.06	5.30		8.36	
Minimum Charge	Job		187		187	

Wood Window

Wood Window	Unit	Material	Labor	Equip.	Total	Specification
2' 2" x 3' 4"						
Demolish	Ea.		11		11	Cost includes material and labor to install 2' 2" x 3' 4" double hung, two light, double glazed wood window with screen.
Install	Ea.	176	40		216	
Demolish and Install	Ea.	176	51		227	
Remove and Install	Ea.		32.09		32.09	
Clean	Ea.	.03	5.60		5.63	
Paint	Ea.	1.90	11		12.90	
Minimum Charge	Job		150		150	

Wood Window		Unit	Material	Labor	Equip.	Total	Specification
2' 2" x 4' 4"							
	Demolish	Ea.		11		11	Cost includes material and labor to install 2' 2" x 4' 4" double hung, two light, double glazed wood window with screen.
	Install	Ea.	198	43		241	
	Demolish and Install	Ea.	198	54		252	
	Remove and Install	Ea.		34.38		34.38	
	Clean	Ea.	.03	5.60		5.63	
	Paint	Ea.	1.90	11		12.90	
	Minimum Charge	Job		150		150	
2' 6" x 3' 4"							
	Demolish	Ea.		11		11	Cost includes material and labor to install 2' 6" x 3' 4" double hung, two light, double glazed wood window with screen.
	Install	Ea.	185	46.50		231.50	
	Demolish and Install	Ea.	185	57.50		242.50	
	Remove and Install	Ea.		37.02		37.02	
	Clean	Ea.	.04	6.55		6.59	
	Paint	Ea.	2.44	13.20		15.64	
	Minimum Charge	Job		150		150	
2' 6" x 4'							
	Demolish	Ea.		11		11	Cost includes material and labor to install 2' 6" x 4' double hung, two light, double glazed wood window with screen.
	Install	Ea.	200	50		250	
	Demolish and Install	Ea.	200	61		261	
	Remove and Install	Ea.		40.11		40.11	
	Clean	Ea.	.04	6.55		6.59	
	Paint	Ea.	2.44	13.20		15.64	
	Minimum Charge	Job		150		150	
2' 6" x 4' 8"							
	Demolish	Ea.		11		11	Cost includes material and labor to install 2' 6" x 4' 8" double hung, two light, double glazed wood window with screen.
	Install	Ea.	220	50		270	
	Demolish and Install	Ea.	220	61		281	
	Remove and Install	Ea.		40.11		40.11	
	Clean	Ea.	.04	6.55		6.59	
	Paint	Ea.	2.44	13.20		15.64	
	Minimum Charge	Job		150		150	
2' 10" x 3' 4"							
	Demolish	Ea.		11		11	Cost includes material and labor to install 2' 10" x 3' 4" double hung, two light, double glazed wood window with screen.
	Install	Ea.	195	60		255	
	Demolish and Install	Ea.	195	71		266	
	Remove and Install	Ea.		48.13		48.13	
	Clean	Ea.	.04	6.55		6.59	
	Paint	Ea.	2.44	13.20		15.64	
	Minimum Charge	Job		150		150	
2' 10" x 4'							
	Demolish	Ea.		11		11	Cost includes material and labor to install 2' 10" x 4' double hung, two light, double glazed wood window with screen.
	Install	Ea.	215	60		275	
	Demolish and Install	Ea.	215	71		286	
	Remove and Install	Ea.		48.13		48.13	
	Clean	Ea.	.04	6.55		6.59	
	Paint	Ea.	2.44	13.20		15.64	
	Minimum Charge	Job		150		150	
3' 7" x 3' 4"							
	Demolish	Ea.		11		11	Cost includes material and labor to install 3' 7" x 3' 4" double hung, two light, double glazed wood window with screen.
	Install	Ea.	220	67		287	
	Demolish and Install	Ea.	220	78		298	
	Remove and Install	Ea.		53.48		53.48	
	Clean	Ea.	.04	6.55		6.59	
	Paint	Ea.	2.44	13.20		15.64	
	Minimum Charge	Job		150		150	

Wood Window		Unit	Material	Labor	Equip.	Total	Specification
3′ 7″ x 5′ 4″							
	Demolish	Ea.		13.45		13.45	Cost includes material and labor to install 3′ 7″ x 5′ 4″ double hung, two light, double glazed wood window with screen.
	Install	Ea.	283	67		350	
	Demolish and Install	Ea.	283	80.45		363.45	
	Remove and Install	Ea.		53.48		53.48	
	Clean	Ea.	.09	11.20		11.29	
	Paint	Ea.	4.07	22		26.07	
	Minimum Charge	Job		150		150	
4′ 3″ x 5′ 4″							
	Demolish	Ea.		13.45		13.45	Cost includes material and labor to install 4′ 3″ x 5′ 4″ double hung, two light, double glazed wood window with screen.
	Install	Ea.	445	75		520	
	Demolish and Install	Ea.	445	88.45		533.45	
	Remove and Install	Ea.		60.16		60.16	
	Clean	Ea.	.09	11.20		11.29	
	Paint	Ea.	2.62	26.50		29.12	
	Minimum Charge	Job		150		150	
Casement							
2′ x 3′ 4″							
	Demolish	Ea.		11		11	Cost includes material and labor to install 2′ x 3′ 4 ″ casement window with 3/4″ insulated glass, screens, weatherstripping, hardware.
	Install	Ea.	185	30		215	
	Demolish and Install	Ea.	185	41		226	
	Remove and Install	Ea.		24.06		24.06	
	Clean	Ea.	.03	5.60		5.63	
	Paint	Ea.	1.90	11		12.90	
	Minimum Charge	Job		300		300	
2′ x 4′							
	Demolish	Ea.		11		11	Cost includes material and labor to install 2′ x 4′ casement window with 3/4″ insulated glass, screens, weatherstripping, hardware.
	Install	Ea.	207	33.50		240.50	
	Demolish and Install	Ea.	207	44.50		251.50	
	Remove and Install	Ea.		26.74		26.74	
	Clean	Ea.	.03	5.60		5.63	
	Paint	Ea.	1.90	11		12.90	
	Minimum Charge	Job		300		300	
2′ x 5′							
	Demolish	Ea.		11		11	Cost includes material and labor to install 2′ x 5′ casement window with 3/4″ insulated glass, screens, weatherstripping, hardware.
	Install	Ea.	232	35.50		267.50	
	Demolish and Install	Ea.	232	46.50		278.50	
	Remove and Install	Ea.		28.31		28.31	
	Clean	Ea.	.04	6.55		6.59	
	Paint	Ea.	2.44	13.20		15.64	
	Minimum Charge	Job		300		300	
2′ x 6′							
	Demolish	Ea.		11		11	Cost includes material and labor to install 2′ x 6′ casement window with 3/4″ insulated glass, screens, weatherstripping, hardware.
	Install	Ea.	262	37.50		299.50	
	Demolish and Install	Ea.	262	48.50		310.50	
	Remove and Install	Ea.		30.08		30.08	
	Clean	Ea.	.04	6.55		6.59	
	Paint	Ea.	2.44	13.20		15.64	
	Minimum Charge	Job		300		300	
4′ x 3′ 4″							
	Demolish	Ea.		13.45		13.45	Cost includes material and labor to install 4′ x 3′ 4 ″ casement windows with 3/4″ insulated glass, screens, weatherstripping, hardware.
	Install	Ea.	535	40		575	
	Demolish and Install	Ea.	535	53.45		588.45	
	Remove and Install	Ea.		32.09		32.09	
	Clean	Ea.	.04	6.55		6.59	
	Paint	Ea.	1.31	13.20		14.51	
	Minimum Charge	Job		300		300	

Wood Window	Unit	Material	Labor	Equip.	Total	Specification
4′ x 4′						
Demolish	Ea.		13.45		13.45	Cost includes material and labor to install 4′ x 4′ casement windows with 3/4″ insulated glass, screens, weatherstripping, hardware.
Install	Ea.	590	40		630	
Demolish and Install	Ea.	590	53.45		643.45	
Remove and Install	Ea.		32.09		32.09	
Clean	Ea.	.07	9.20		9.27	
Paint	Ea.	1.55	16.50		18.05	
Minimum Charge	Job		300		300	
4′ x 5′						
Demolish	Ea.		13.45		13.45	Cost includes material and labor to install 4′ x 5′ casement windows with 3/4″ insulated glass, screens, weatherstripping, hardware.
Install	Ea.	660	43		703	
Demolish and Install	Ea.	660	56.45		716.45	
Remove and Install	Ea.		34.38		34.38	
Clean	Ea.	.09	11.20		11.29	
Paint	Ea.	2.02	22		24.02	
Minimum Charge	Job		300		300	
4′ x 6′						
Demolish	Ea.		13.45		13.45	Cost includes material and labor to install 4′ x 6′ casement windows with 3/4″ insulated glass, screens, weatherstripping, hardware.
Install	Ea.	750	50		800	
Demolish and Install	Ea.	750	63.45		813.45	
Remove and Install	Ea.		40.11		40.11	
Clean	Ea.	.09	11.20		11.29	
Paint	Ea.	2.38	26.50		28.88	
Minimum Charge	Job		300		300	
Casement Bow						
4′ 8″ x 8′ 1″						
Demolish	Ea.		18.65		18.65	Includes material and labor to install casement bow window unit with 4 lights, pine frame, 1/2″ insulated glass, weatherstripping, hardware, screens.
Install	Ea.	1400	113		1513	
Demolish and Install	Ea.	1400	131.65		1531.65	
Remove and Install	Ea.		90.24		90.24	
Clean	Ea.	.12	13.05		13.17	
Paint	Ea.	4.40	53		57.40	
Minimum Charge	Job		300		300	
4′ 8″ x 10′						
Demolish	Ea.		18.65		18.65	Includes material and labor to install casement bow window unit with 5 lights, pine frame, 1/2″ insulated glass, weatherstripping, hardware, screens.
Install	Ea.	1525	150		1675	
Demolish and Install	Ea.	1525	168.65		1693.65	
Remove and Install	Ea.		120.32		120.32	
Clean	Ea.	.02	17.40		17.42	
Paint	Ea.	5.45	66		71.45	
Minimum Charge	Job		300		300	
5′ 4″ x 8′ 1″						
Demolish	Ea.		18.65		18.65	Includes material and labor to install casement bow window unit with 4 lights, pine frame, 1/2″ insulated glass, weatherstripping, hardware, screens.
Install	Ea.	1575	113		1688	
Demolish and Install	Ea.	1575	131.65		1706.65	
Remove and Install	Ea.		90.24		90.24	
Clean	Ea.	.02	17.40		17.42	
Paint	Ea.	4.99	53		57.99	
Minimum Charge	Job		300		300	
5′ 4″ x 10′						
Demolish	Ea.		18.65		18.65	Includes material and labor to install casement bow window unit with 5 lights, pine frame, 1/2″ insulated glass, weatherstripping, hardware, screens.
Install	Ea.	1725	150		1875	
Demolish and Install	Ea.	1725	168.65		1893.65	
Remove and Install	Ea.		120.32		120.32	
Clean	Ea.	.02	17.40		17.42	
Paint	Ea.	6.35	66		72.35	
Minimum Charge	Job		300		300	

Wood Window		Unit	Material	Labor	Equip.	Total	Specification
6' x 8' 1"							
	Demolish	Ea.		18.65		18.65	Includes material and labor to install casement bow window unit with 5 lights, pine frame, 1/2" insulated glass, weatherstripping, hardware, screens.
	Install	Ea.	1525	150		1675	
	Demolish and Install	Ea.	1525	168.65		1693.65	
	Remove and Install	Ea.		120.32		120.32	
	Clean	Ea.	.02	17.40		17.42	
	Paint	Ea.	5.70	66		71.70	
	Minimum Charge	Job		300		300	
6' x 10'							
	Demolish	Ea.		18.65		18.65	Includes material and labor to install casement bow window unit with 5 lights, pine frame, 1/2" insulated glass, weatherstripping, hardware, screens.
	Install	Ea.	1800	150		1950	
	Demolish and Install	Ea.	1800	168.65		1968.65	
	Remove and Install	Ea.		120.32		120.32	
	Clean	Ea.	.02	17.40		17.42	
	Paint	Ea.	7.15	88		95.15	
	Minimum Charge	Job		300		300	
Casing							
	Demolish	Ea.		6.95		6.95	Cost includes material and labor to install 11/16" x 2-1/2" pine ranch style window casing.
	Install	Opng.	22	23		45	
	Demolish and Install	Opng.	22	29.95		51.95	
	Remove and Install	Opng.		18.51		18.51	
	Clean	Opng.	.39	2.18		2.57	
	Paint	Ea.	3.06	5.30		8.36	
	Minimum Charge	Job		187		187	
Trim Set							
	Demolish	Ea.		5.30		5.30	Cost includes material and labor to install 11/16" x 2-1/2" pine ranch style window casing.
	Install	Opng.	22	23		45	
	Demolish and Install	Opng.	22	28.30		50.30	
	Remove and Install	Opng.		18.51		18.51	
	Clean	Opng.	.39	2.18		2.57	
	Paint	Ea.	3.06	5.30		8.36	
	Minimum Charge	Job		150		150	
Wood Picture Window							
3' 6" x 4'							
	Demolish	Ea.		13.45		13.45	Includes material and labor to install a wood framed picture window including exterior trim.
	Install	Ea.	300	50		350	
	Demolish and Install	Ea.	300	63.45		363.45	
	Clean	Ea.	.09	11.20		11.29	
	Paint	Ea.	1.67	20.50		22.17	
	Minimum Charge	Job		300		300	
4' x 4' 6"							
	Demolish	Ea.		13.45		13.45	Includes material and labor to install a wood framed picture window including exterior trim.
	Install	Ea.	325	54.50		379.50	
	Demolish and Install	Ea.	325	67.95		392.95	
	Clean	Ea.	.09	11.20		11.29	
	Paint	Ea.	1.67	20.50		22.17	
	Minimum Charge	Job		300		300	
5' x 4'							
	Demolish	Ea.		13.45		13.45	Includes material and labor to install a wood framed picture window including exterior trim.
	Install	Ea.	405	54.50		459.50	
	Demolish and Install	Ea.	405	67.95		472.95	
	Clean	Ea.	.09	11.20		11.29	
	Paint	Ea.	3.56	33		36.56	
	Minimum Charge	Job		300		300	

Wood Window	Unit	Material	Labor	Equip.	Total	Specification
6' x 4' 6"						
Demolish	Ea.		13.45		13.45	Includes material and labor to install a wood framed picture window including exterior trim.
Install	Ea.	525	60		585	
Demolish and Install	Ea.	525	73.45		598.45	
Clean	Ea.	.09	11.20		11.29	
Paint	Ea.	3.56	33		36.56	
Minimum Charge	Job		300		300	

Window Screens	Unit	Material	Labor	Equip.	Total	Specification
Small						
Demolish	Ea.		2.52		2.52	Includes material and labor to install small window screen, 4 S.F.
Install	Ea.	12.45	9.10		21.55	
Demolish and Install	Ea.	12.45	11.62		24.07	
Remove and Install	Ea.		7.29		7.29	
Clean	Ea.	.28	4.54		4.82	
Minimum Charge	Job		75		75	
Standard Size						
Demolish	Ea.		2.52		2.52	Includes material and labor to install medium window screen, 12 S.F.
Install	Ea.	25.50	9.10		34.60	
Demolish and Install	Ea.	25.50	11.62		37.12	
Remove and Install	Ea.		7.29		7.29	
Clean	Ea.	.28	4.54		4.82	
Minimum Charge	Job		75		75	
Large						
Demolish	Ea.		2.52		2.52	Includes material and labor to install large window screen, 16 S.F.
Install	Ea.	33.50	9.10		42.60	
Demolish and Install	Ea.	33.50	11.62		45.12	
Remove and Install	Ea.		7.29		7.29	
Clean	Ea.	.28	4.54		4.82	
Minimum Charge	Job		75		75	

Window Glass	Unit	Material	Labor	Equip.	Total	Specification
Plate, Clear, 1/4"						
Demolish	S.F.		1.21		1.21	Includes material and labor to install clear glass.
Install	S.F.	6.20	4.85		11.05	
Demolish and Install	S.F.	6.20	6.06		12.26	
Remove and Install	S.F.		3.88		3.88	
Clean	S.F.	.02	.13		.15	
Minimum Charge	Job		146		146	
Insulated						
Demolish	S.F.		1.21		1.21	Includes material and labor to install insulated glass.
Install	S.F.	10.20	8.30		18.50	
Demolish and Install	S.F.	10.20	9.51		19.71	
Remove and Install	S.F.		6.66		6.66	
Clean	S.F.	.02	.13		.15	
Minimum Charge	Job		146		146	
Tinted						
Demolish	S.F.		1.21		1.21	Includes material and labor to install tinted glass.
Install	S.F.	6.25	4.48		10.73	
Demolish and Install	S.F.	6.25	5.69		11.94	
Remove and Install	S.F.		3.58		3.58	
Clean	S.F.	.02	.13		.15	
Minimum Charge	Job		146		146	

Window Glass		Unit	Material	Labor	Equip.	Total	Specification
Obscure / Florentine							
	Demolish	S.F.		1.21		1.21	Includes material and labor to install obscure glass.
	Install	S.F.	7.10	4.16		11.26	
	Demolish and Install	S.F.	7.10	5.37		12.47	
	Remove and Install	S.F.		3.33		3.33	
	Clean	S.F.	.02	.13		.15	
	Minimum Charge	Job		146		146	
Laminated Safety							
	Demolish	S.F.		1.21		1.21	Includes material and labor to install clear laminated glass.
	Install	S.F.	8.45	6.45		14.90	
	Demolish and Install	S.F.	8.45	7.66		16.11	
	Remove and Install	S.F.		5.18		5.18	
	Clean	S.F.	.02	.13		.15	
	Minimum Charge	Job		146		146	
Wired							
	Demolish	S.F.		1.21		1.21	Includes material and labor to install rough obscure wire glass.
	Install	S.F.	11.75	4.31		16.06	
	Demolish and Install	S.F.	11.75	5.52		17.27	
	Remove and Install	S.F.		3.45		3.45	
	Clean	S.F.	.02	.13		.15	
	Minimum Charge	Job		146		146	
Tempered							
	Demolish	S.F.		1.21		1.21	Includes material and labor to install clear glass.
	Install	S.F.	6.20	4.85		11.05	
	Demolish and Install	S.F.	6.20	6.06		12.26	
	Remove and Install	S.F.		3.88		3.88	
	Clean	S.F.	.02	.13		.15	
	Minimum Charge	Job		146		146	
Stained							
	Demolish	S.F.		1.21		1.21	Includes material and labor to install stained glass.
	Install	S.F.	48	4.85		52.85	
	Demolish and Install	S.F.	48	6.06		54.06	
	Remove and Install	S.F.		3.88		3.88	
	Clean	S.F.	.02	.13		.15	
	Minimum Charge	Job		146		146	
Etched							
	Demolish	S.F.		1.21		1.21	Includes material and labor to install etched glass.
	Install	S.F.	39	4.85		43.85	
	Demolish and Install	S.F.	39	6.06		45.06	
	Remove and Install	S.F.		3.88		3.88	
	Clean	S.F.	.02	.13		.15	
	Minimum Charge	Job		146		146	
Wall Mirror							
	Demolish	S.F.		1.21		1.21	Includes material and labor to install distortion-free float glass mirror, 1/4" thick, cut and polished edges.
	Install	S.F.	6.75	4.66		11.41	
	Demolish and Install	S.F.	6.75	5.87		12.62	
	Remove and Install	S.F.		3.73		3.73	
	Clean	S.F.	.02	.13		.15	
	Minimum Charge	Job		146		146	
Bullet Resistant							
	Demolish	S.F.		1.21		1.21	Includes material and labor to install bullet resistant clear laminated glass.
	Install	S.F.	82	58		140	
	Demolish and Install	S.F.	82	59.21		141.21	
	Remove and Install	S.F.		46.59		46.59	
	Clean	S.F.	.02	.13		.15	
	Minimum Charge	Job		146		146	

Storefront System	Unit	Material	Labor	Equip.	Total	Specification
Clear Glass						
Stub Wall to 8'						
Demolish	S.F.		1.89		1.89	Includes material and labor to install commercial grade 4-1/2" section, anodized aluminum with clear glass in upper section and safety glass in lower section.
Install	S.F.	16.15	8		24.15	
Demolish and Install	S.F.	16.15	9.89		26.04	
Remove and Install	S.F.		6.40		6.40	
Clean	S.F.		.26		.26	
Minimum Charge	Job		187		187	
Floor to 10'						
Demolish	S.F.		1.89		1.89	Includes material and labor to install commercial grade 4-1/2" section, anodized aluminum with clear glass in upper section and safety glass in lower section.
Install	S.F.	17.70	8		25.70	
Demolish and Install	S.F.	17.70	9.89		27.59	
Remove and Install	S.F.		6.40		6.40	
Clean	S.F.		.26		.26	
Minimum Charge	Job		187		187	
Polished Plate Glass						
Stub Wall to 9'						
Demolish	S.F.		1.89		1.89	Cost includes material and labor to install 4' x 9' polished plate glass window built on 8" stub wall.
Install	S.F.	12.25	8		20.25	
Demolish and Install	S.F.	12.25	9.89		22.14	
Remove and Install	S.F.		6.40		6.40	
Clean	S.F.		.26		.26	
Minimum Charge	Job		187		187	
Stub Wall to 13'						
Demolish	S.F.		1.89		1.89	Cost includes material and labor to install 4' x 13' polished plate glass window built on 8" stub wall.
Install	S.F.	12.80	8		20.80	
Demolish and Install	S.F.	12.80	9.89		22.69	
Remove and Install	S.F.		6.40		6.40	
Clean	S.F.		.26		.26	
Minimum Charge	Job		187		187	
Floor to 9'						
Demolish	S.F.		1.89		1.89	Cost includes material and labor to install 4' x 9' polished plate glass window built on floor.
Install	S.F.	13.30	8		21.30	
Demolish and Install	S.F.	13.30	9.89		23.19	
Remove and Install	S.F.		6.40		6.40	
Clean	S.F.		.26		.26	
Minimum Charge	Job		187		187	
Floor to 13'						
Demolish	S.F.		1.89		1.89	Cost includes material and labor to install 4' x 13' polished plate glass window built on floor.
Install	S.F.	14.30	8		22.30	
Demolish and Install	S.F.	14.30	9.89		24.19	
Remove and Install	S.F.		6.40		6.40	
Clean	S.F.		.26		.26	
Minimum Charge	Job		187		187	
Clear Plate Glass						
Floor to 9'						
Demolish	S.F.		1.89		1.89	Cost includes material and labor to install 4' x 9' clear plate glass window built on floor.
Install	S.F.	19.20	8		27.20	
Demolish and Install	S.F.	19.20	9.89		29.09	
Remove and Install	S.F.		6.40		6.40	
Clean	S.F.		.26		.26	
Minimum Charge	Job		187		187	

Storefront System	Unit	Material	Labor	Equip.	Total	Specification
Floor to 11′						
Demolish	S.F.		1.89		1.89	Cost includes material and labor to install 4′ x 11′ clear plate glass window built on floor.
Install	S.F.	18.65	8		26.65	
Demolish and Install	S.F.	18.65	9.89		28.54	
Remove and Install	S.F.		6.40		6.40	
Clean	S.F.		.26		.26	
Minimum Charge	Job		187		187	
Floor to 13′						
Demolish	S.F.		1.89		1.89	Cost includes material and labor to install 4′ x 13′ clear plate glass window built on floor.
Install	S.F.	18.80	8		26.80	
Demolish and Install	S.F.	18.80	9.89		28.69	
Remove and Install	S.F.		6.40		6.40	
Clean	S.F.		.26		.26	
Minimum Charge	Job		187		187	
Decorative Type						
Demolish	S.F.		1.89		1.89	Includes material and labor to install decorative storefront system.
Install	S.F.	14.65	5.80		20.45	
Demolish and Install	S.F.	14.65	7.69		22.34	
Remove and Install	S.F.		4.66		4.66	
Clean	S.F.		.26		.26	
Minimum Charge	Job		146		146	
Decorative Curved						
Demolish	S.F.		1.89		1.89	Includes material and labor to install decorative curved storefront system.
Install	S.F.	27.50	11.65		39.15	
Demolish and Install	S.F.	27.50	13.54		41.04	
Remove and Install	S.F.		9.32		9.32	
Clean	S.F.		.26		.26	
Minimum Charge	Job		146		146	
Entrance						
Narrow Stile 3′ x 7″						
Demolish	Ea.		16.70		16.70	Cost includes material and labor to install 3′ x 7′ narrow stile door, center pivot concealed closer.
Install	Ea.	700	201		901	
Demolish and Install	Ea.	700	217.70		917.70	
Remove and Install	Ea.		160.43		160.43	
Clean	Ea.	3.42	27.50		30.92	
Minimum Charge	Job		187		187	
Tempered Glass						
Demolish	Ea.		16.70		16.70	Cost includes material and labor to install 3′ x 7′ tempered glass door, center pivot concealed closer,.
Install	Ea.	2550	405		2955	
Demolish and Install	Ea.	2550	421.70		2971.70	
Remove and Install	Ea.		323.46		323.46	
Clean	Ea.	3.42	27.50		30.92	
Minimum Charge	Job		187		187	
Heavy Section						
Demolish	Ea.		16.70		16.70	Includes material and labor to install wide stile door with 3′ 6″ x 7′ opening.
Install	Ea.	790	375		1165	
Demolish and Install	Ea.	790	391.70		1181.70	
Remove and Install	Ea.		299.20		299.20	
Clean	Ea.	.68	9.10		9.78	
Minimum Charge	Job		580		580	

Curtain Wall System	Unit	Material	Labor	Equip.	Total	Specification
Regular Weight						
Bronze Anodized						
Demolish	S.F.		1.89		1.89	Includes material and labor to install structural curtain wall system with bronze anodized frame of heavy weight, float glass.
Install	S.F.	30.50	8		38.50	
Demolish and Install	S.F.	30.50	9.89		40.39	
Clean	S.F.		.26		.26	
Minimum Charge	Job		150		150	
Black Anodized						
Demolish	S.F.		1.89		1.89	Includes material and labor to install structural curtain wall system with black anodized frame of regular weight, float glass.
Install	S.F.	32.50	8		40.50	
Demolish and Install	S.F.	32.50	9.89		42.39	
Clean	S.F.		.26		.26	
Minimum Charge	Job		150		150	
Heavy Weight						
Bronze Anodized						
Demolish	S.F.		1.89		1.89	Includes material and labor to install structural curtain wall system with bronze anodized frame of heavy weight, float glass.
Install	S.F.	30.50	8		38.50	
Demolish and Install	S.F.	30.50	9.89		40.39	
Clean	S.F.		.26		.26	
Minimum Charge	Job		150		150	
Black Anodized						
Demolish	S.F.		1.89		1.89	Includes material and labor to install structural curtain wall system with black anodized frame of heavy weight, float glass.
Install	S.F.	31	8		39	
Demolish and Install	S.F.	31	9.89		40.89	
Clean	S.F.		.26		.26	
Minimum Charge	Job		150		150	
Glass						
1" Insulated						
Install	S.F.	12.50	7.75		20.25	Cost includes labor and material to install 1" insulated glass.
Minimum Charge	Job		146		146	
Heat Absorbing						
Install	S.F.	7.25	7.75		15	Includes labor an material to install heat absorbing glass.
Minimum Charge	Job		146		146	
Low Transmission						
Install	S.F.	4.77	7.75		12.52	Includes labor an material to install low transmission glass.
Minimum Charge	Job		146		146	

Skylight	Unit	Material	Labor	Equip.	Total	Specification
Dome						
22" x 22"						
Demolish	Ea.		59		59	Cost includes material and labor to install 22" x 22" fixed dome skylight.
Install	Ea.	46	91		137	
Demolish and Install	Ea.	46	150		196	
Remove and Install	Ea.		72.92		72.92	
Clean	Ea.	.31	7.80		8.11	
Minimum Charge	Job		545		545	
22" x 46"						
Demolish	Ea.		59		59	Cost includes material and labor to install 22" x 46" fixed dome skylight.
Install	Ea.	104	109		213	
Demolish and Install	Ea.	104	168		272	
Remove and Install	Ea.		87.50		87.50	
Clean	Ea.		12.10		12.10	
Minimum Charge	Job		545		545	

Skylight	Unit	Material	Labor	Equip.	Total	Specification
30″ x 30″						
Demolish	Ea.		59		59	Cost includes material and labor to install 30″ x 30″ fixed dome skylight.
Install	Ea.	87	91		178	
Demolish and Install	Ea.	87	150		237	
Remove and Install	Ea.		72.92		72.92	
Clean	Ea.		12.10		12.10	
Minimum Charge	Job		545		545	
30″ x 46″						
Demolish	Ea.		59		59	Cost includes material and labor to install 30″ x 46″ fixed dome skylight.
Install	Ea.	151	109		260	
Demolish and Install	Ea.	151	168		319	
Remove and Install	Ea.		87.50		87.50	
Clean	Ea.		12.10		12.10	
Minimum Charge	Job		545		545	
Fixed						
22″ x 27″						
Demolish	Ea.		59		59	Includes material and labor to install double glazed skylight, 22″ x 27″, fixed.
Install	Ea.	221	91		312	
Demolish and Install	Ea.	221	150		371	
Remove and Install	Ea.		72.92		72.92	
Clean	Ea.	.31	7.80		8.11	
Minimum Charge	Job		545		545	
22″ x 46″						
Demolish	Ea.		59		59	Includes material and labor to install double glazed skylight, 22″ x 46″, fixed.
Install	Ea.	277	109		386	
Demolish and Install	Ea.	277	168		445	
Remove and Install	Ea.		87.50		87.50	
Clean	Ea.		12.10		12.10	
Minimum Charge	Job		545		545	
44″ x 46″						
Demolish	Ea.		59		59	Includes material and labor to install double glazed skylight, 44″ x 46″, fixed.
Install	Ea.	395	109		504	
Demolish and Install	Ea.	395	168		563	
Remove and Install	Ea.		87.50		87.50	
Clean	Ea.		13.65		13.65	
Minimum Charge	Job		545		545	
Operable						
22″ x 27″						
Demolish	Ea.		59		59	Includes material and labor to install double glazed skylight, 22″ x 27″, operable.
Install	Ea.	330	91		421	
Demolish and Install	Ea.	330	150		480	
Remove and Install	Ea.		72.92		72.92	
Clean	Ea.	.31	7.80		8.11	
Minimum Charge	Job		545		545	
22″ x 46″						
Demolish	Ea.		59		59	Includes material and labor to install double glazed skylight, 22″ x 46″, operable.
Install	Ea.	385	109		494	
Demolish and Install	Ea.	385	168		553	
Remove and Install	Ea.		87.50		87.50	
Clean	Ea.		12.10		12.10	
Minimum Charge	Job		545		545	
44″ x 46″						
Demolish	Ea.		59		59	Includes material and labor to install double glazed skylight, 44″ x 46″, operable.
Install	Ea.	525	109		634	
Demolish and Install	Ea.	525	168		693	
Remove and Install	Ea.		87.50		87.50	
Clean	Ea.		13.65		13.65	
Minimum Charge	Job		545		545	

Industrial Steel Window

Industrial Steel Window	Unit	Material	Labor	Equip.	Total	Specification
100% Fixed						
Demolish	S.F.		1.55		1.55	Includes material and labor to install commercial grade steel windows, including glazing.
Install	S.F.	26	3.74		29.74	
Demolish and Install	S.F.	26	5.29		31.29	
Remove and Install	S.F.		2.99		2.99	
Clean	Ea.	.12	.82		.94	
Paint	S.F.	.11	.59		.70	
Minimum Charge	Job		187		187	
50% Vented						
Demolish	S.F.		1.55		1.55	Includes material and labor to install commercial grade steel windows, including glazing.
Install	S.F.	37.50	3.74		41.24	
Demolish and Install	S.F.	37.50	5.29		42.79	
Remove and Install	S.F.		2.99		2.99	
Clean	Ea.	.12	.82		.94	
Paint	S.F.	.11	.59		.70	
Minimum Charge	Job		187		187	

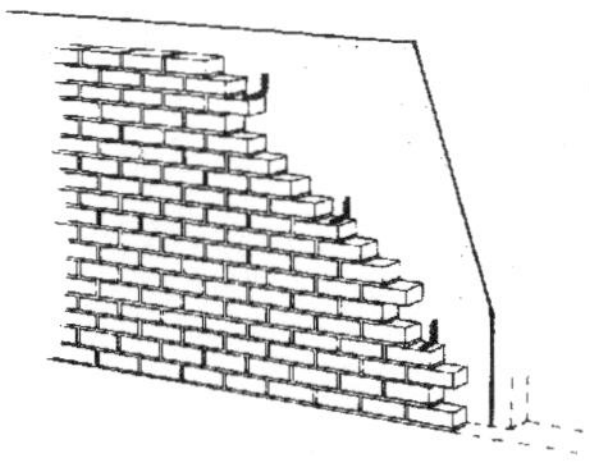

Brick Veneer

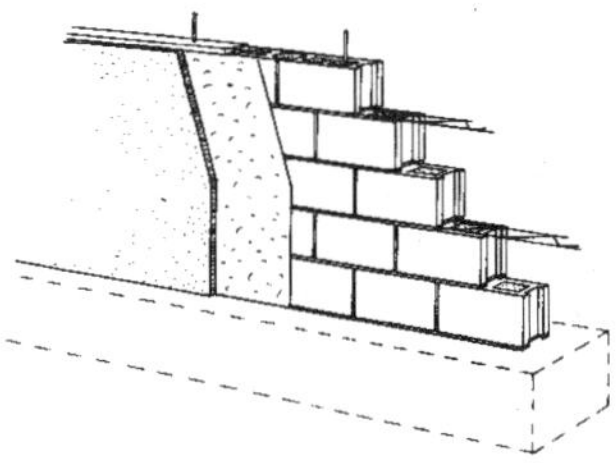

Concrete Masonry Units

Masonry Wall	Unit	Material	Labor	Equip.	Total	Specification
Unreinforced CMU						
4″ x 8″ x 16″						
Demolish	S.F.		1.35		1.35	Includes material and labor to install normal weight 4″ x 8″ x 16″ block.
Install	S.F.	1.06	2.99		4.05	
Demolish and Install	S.F.	1.06	4.34		5.40	
Clean	S.F.	.03	.34		.37	
Paint	S.F.	.12	.60		.72	
Minimum Charge	Job		152		152	
6″ x 8″ x 16″						
Demolish	S.F.		1.42		1.42	Includes material and labor to install normal weight 6″ x 8″ x 16″ block.
Install	S.F.	1.54	3.12		4.66	
Demolish and Install	S.F.	1.54	4.54		6.08	
Clean	S.F.	.03	.34		.37	
Paint	S.F.	.12	.60		.72	
Minimum Charge	Job		152		152	
8″ x 8″ x 16″						
Demolish	S.F.		1.61		1.61	Includes material and labor to install normal weight 8″ x 8″ x 16″ block.
Install	S.F.	1.66	3.43		5.09	
Demolish and Install	S.F.	1.66	5.04		6.70	
Clean	S.F.	.03	.34		.37	
Paint	S.F.	.12	.60		.72	
Minimum Charge	Job		152		152	
12″ x 8″ x 16″						
Demolish	S.F.		1.61		1.61	Includes material and labor to install normal weight 12″ x 8″ x 16″ block.
Install	S.F.	2.42	5.20		7.62	
Demolish and Install	S.F.	2.42	6.81		9.23	
Clean	S.F.	.03	.34		.37	
Paint	S.F.	.12	.60		.72	
Minimum Charge	Job		152		152	
Reinforced CMU						
4″ x 8″ x 16″						
Demolish	S.F.		1.35	.39	1.74	Includes material and labor to install normal weight 4″ x 8″ x 16″ block, mortar and 1/2″ re-bar installed vertically every 24″ O.C.
Install	S.F.	1.14	3.05		4.19	
Demolish and Install	S.F.	1.14	4.40	.39	5.93	
Clean	S.F.	.03	.34		.37	
Paint	S.F.	.12	.60		.72	
Minimum Charge	Job		267		267	
6″ x 8″ x 16″						
Demolish	S.F.		1.42	.41	1.83	Includes material and labor to install normal weight 6″ x 8″ x 16″ block, mortar and 1/2″ re-bar installed vertically every 24″ O.C.
Install	S.F.	1.63	3.19		4.82	
Demolish and Install	S.F.	1.63	4.61	.41	6.65	
Clean	S.F.	.03	.34		.37	
Paint	S.F.	.12	.60		.72	
Minimum Charge	Job		267		267	

Masonry Wall		Unit	Material	Labor	Equip.	Total	Specification
8" x 8" x 16"							
	Demolish	S.F.		1.50	.43	1.93	Includes material and labor to install normal weight 8" x 8" x 16" block, mortar and 1/2" re-bar installed vertically every 24" O.C.
	Install	S.F.	1.75	3.48		5.23	
	Demolish and Install	S.F.	1.75	4.98	.43	7.16	
	Clean	S.F.	.03	.34		.37	
	Paint	S.F.	.12	.60		.72	
	Minimum Charge	Job		267		267	
12" x 8" x 16"							
	Demolish	S.F.		1.54	.45	1.99	Includes material and labor to install normal weight 12" x 8" x 16" block, mortar and 1/2" re-bar installed vertically every 24" O.C.
	Install	S.F.	2.52	5.35		7.87	
	Demolish and Install	S.F.	2.52	6.89	.45	9.86	
	Clean	S.F.	.03	.34		.37	
	Paint	S.F.	.12	.60		.72	
	Minimum Charge	Job		267		267	
Structural Brick							
4" Single Row							
	Demolish	S.F.		1.94	.56	2.50	Cost includes material and labor to install 4" red brick wall, running bond with 3/8" concave joints, mortar and ties.
	Install	S.F.	2.99	6.40		9.39	
	Demolish and Install	S.F.	2.99	8.34	.56	11.89	
	Clean	S.F.	.03	.29		.32	
	Paint	S.F.	.12	.60		.72	
	Minimum Charge	Job		267		267	
8" Double Row							
	Demolish	S.F.		1.94	.56	2.50	Cost includes material and labor to install 8" red brick wall, running bond, with 3/8" concave joints, mortar and wall ties.
	Install	S.F.	4.70	10.15		14.85	
	Demolish and Install	S.F.	4.70	12.09	.56	17.35	
	Clean	S.F.	.03	.29		.32	
	Paint	S.F.	.12	.60		.72	
	Minimum Charge	Job		267		267	
12" Triple Row							
	Demolish	S.F.		2.69	.78	3.47	Cost includes material and labor to install 12" red brick wall, running bond, with 3/8" concave joints, mortar and wall ties.
	Install	S.F.	7.10	14.45		21.55	
	Demolish and Install	S.F.	7.10	17.14	.78	25.02	
	Clean	S.F.	.03	.29		.32	
	Paint	S.F.	.12	.60		.72	
	Minimum Charge	Job		267		267	
Brick Veneer							
4" Red Common							
	Demolish	S.F.		1.94	.56	2.50	Cost includes material and labor to install 4" common red face brick, running bond with 3/8" concave joints, mortar and wall ties.
	Install	S.F.	2.23	5.95		8.18	
	Demolish and Install	S.F.	2.23	7.89	.56	10.68	
	Clean	S.F.	.03	.29		.32	
	Paint	S.F.	.12	.60		.72	
	Minimum Charge	Job		152		152	
4" Used							
	Demolish	S.F.		1.94	.56	2.50	Cost includes material and labor to install 4" red brick wall, running bond with 3/8" concave joints, mortar and ties.
	Install	S.F.	2.99	6.40		9.39	
	Demolish and Install	S.F.	2.99	8.34	.56	11.89	
	Clean	S.F.	.03	.29		.32	
	Paint	S.F.	.12	.60		.72	
	Minimum Charge	Job		152		152	

Masonry Wall	Unit	Material	Labor	Equip.	Total	Specification
6″ Jumbo						
Demolish	S.F.		1.94	.56	2.50	Cost includes material and labor to install 6″ red jumbo brick veneer, running bond with 3/8″ concave joints, mortar and ties.
Install	S.F.	3.97	3.16		7.13	
Demolish and Install	S.F.	3.97	5.10	.56	9.63	
Clean	S.F.	.03	.29		.32	
Paint	S.F.	.12	.60		.72	
Minimum Charge	Job		152		152	
Norman (11-1/2″)						
Demolish	S.F.		1.94	.56	2.50	Cost includes material and labor to install 4″ norman red brick wall, running bond with 3/8″ concave joints, mortar and ties.
Install	S.F.	3.94	4.29		8.23	
Demolish and Install	S.F.	3.94	6.23	.56	10.73	
Clean	S.F.	.03	.29		.32	
Paint	S.F.	.12	.60		.72	
Minimum Charge	Job		152		152	
Glazed						
Demolish	S.F.		1.94	.56	2.50	Includes material, labor and equipment to install glazed brick.
Install	S.F.	7.70	6.55		14.25	
Demolish and Install	S.F.	7.70	8.49	.56	16.75	
Clean	S.F.	.03	.29		.32	
Paint	S.F.	.12	.60		.72	
Minimum Charge	Job		152		152	
Slumpstone						
6″ x 4″						
Demolish	S.F.		1.35		1.35	Cost includes material and labor to install 6″ x 4″ x 16″ concrete slump block.
Install	S.F.	4.41	3.34		7.75	
Demolish and Install	S.F.	4.41	4.69		9.10	
Clean	S.F.	.03	.34		.37	
Paint	S.F.	.12	.60		.72	
Minimum Charge	Job		152		152	
6″ x 6″						
Demolish	S.F.		1.42		1.42	Cost includes material and labor to install 6″ x 6″ x 16″ concrete slump block.
Install	S.F.	4.08	3.45		7.53	
Demolish and Install	S.F.	4.08	4.87		8.95	
Clean	S.F.	.03	.34		.37	
Paint	S.F.	.12	.60		.72	
Minimum Charge	Job		152		152	
8″ x 6″						
Demolish	S.F.		1.42		1.42	Cost includes material and labor to install 8″ x 6″ x 16″ concrete slump block.
Install	S.F.	5.35	3.57		8.92	
Demolish and Install	S.F.	5.35	4.99		10.34	
Clean	S.F.	.03	.34		.37	
Paint	S.F.	.12	.60		.72	
Minimum Charge	Job		152		152	
12″ x 4″						
Demolish	S.F.		1.35		1.35	Cost includes material and labor to install 12″ x 4″ x 16″ concrete slump block.
Install	S.F.	9.40	4.11		13.51	
Demolish and Install	S.F.	9.40	5.46		14.86	
Clean	S.F.	.03	.34		.37	
Paint	S.F.	.12	.60		.72	
Minimum Charge	Job		152		152	
12″ x 6″						
Demolish	S.F.		1.42		1.42	Cost includes material and labor to install 12″ x 6″ x 16″ concrete slump block.
Install	S.F.	8.55	4.46		13.01	
Demolish and Install	S.F.	8.55	5.88		14.43	
Clean	S.F.	.03	.34		.37	
Paint	S.F.	.12	.60		.72	
Minimum Charge	Job		152		152	

Masonry Wall	Unit	Material	Labor	Equip.	Total	Specification
Stucco						
Three Coat						
Demolish	S.Y.		1.67		1.67	Includes material and labor to install stucco on exterior masonry walls.
Install	S.Y.	3.58	6.40	.48	10.46	
Demolish and Install	S.Y.	3.58	8.07	.48	12.13	
Clean	S.Y.		3.03		3.03	
Paint	S.F.	.12	.43		.55	
Minimum Charge	Job		136		136	
Three Coat w / Wire Mesh						
Demolish	S.Y.		3.03		3.03	Includes material and labor to install stucco on wire mesh over wood framing.
Install	S.Y.	4.38	24.50	1.54	30.42	
Demolish and Install	S.Y.	4.38	27.53	1.54	33.45	
Clean	S.Y.		3.03		3.03	
Paint	S.F.	.12	.43		.55	
Minimum Charge	Job		136		136	
Fill Cracks						
Install	S.F.	.40	2.18		2.58	Includes labor and material to fill-in hairline cracks up to 1/8″ with filler, including sand and fill per S.F.
Minimum Charge	Job		136		136	
Furring						
1″ x 2″						
Demolish	L.F.		.10		.10	Cost includes material and labor to install 1″ x 2″ furring on masonry wall.
Install	L.F.	.22	.61		.83	
Demolish and Install	L.F.	.22	.71		.93	
Minimum Charge	Job		150		150	
2″ x 2″						
Demolish	L.F.		.11		.11	Cost includes material and labor to install 2″ x 2″ furring on masonry wall.
Install	L.F.	.21	.67		.88	
Demolish and Install	L.F.	.21	.78		.99	
Minimum Charge	Job		150		150	
2″ x 4″						
Demolish	L.F.		.12		.12	Cost includes material and labor to install 2″ x 4″ furring on masonry wall.
Install	L.F.	.36	.76		1.12	
Demolish and Install	L.F.	.36	.88		1.24	
Remove and Install	L.F.		.61		.61	
Minimum Charge	Job		150		150	
Maintenance / Cleaning						
Repoint						
Install	S.F.	.30	3.80		4.10	Includes labor and material to cut and repoint brick, hard mortar, running bond.
Minimum Charge	Hr.		38		38	
Acid Etch						
Install	S.F.	.02	.54		.56	Includes labor and material to acid wash smooth brick.
Minimum Charge	Job		560	92	652	
Pressure Wash						
Clean	S.F.		.74	.12	.86	Includes labor and material to high pressure water wash masonry, average.
Minimum Charge	Job		560	92	652	
Waterproof						
Install	S.F.	.18	.56		.74	Includes labor and material to install masory waterproofing.
Minimum Charge	Job		141		141	

Masonry

Chimney

Chimney	Unit	Material	Labor	Equip.	Total	Specification
Demo & Haul Exterior						
Demolish	V.L.F.		22	6.35	28.35	Includes minimum labor and equipment to remove masonry and haul debris to truck or dumpster.
Minimum Charge	Job		121	35	156	
Demo & Haul Interior						
Demolish	V.L.F.		27.50	7.90	35.40	Includes minimum labor and equipment to remove masonry and haul debris to truck or dumpster.
Minimum Charge	Job		121	35	156	
Install New (w / flue)						
Brick						
Install	V.L.F.	26	33.50		59.50	Includes material and labor to install brick chimney with flue from ground level up. Foundation not included.
Minimum Charge	Job		152		152	
Natural Stone						
Install	L.F.	58	53.50		111.50	Includes labor and material to install natural stone chimney.
Minimum Charge	Job		152		152	
Install From Roof Up						
Brick						
Install	V.L.F.	26	33.50		59.50	Includes material and labor to install brick chimney with flue from ground level up. Foundation not included.
Minimum Charge	Job		152		152	
Natural Stone						
Install	L.F.	58	44.50		102.50	Includes labor and material to install natural stone chimney.
Minimum Charge	Job		152		152	
Install Base Pad						
Install	Ea.	95	166	1.49	262.49	Includes labor and material to replace spread footing, 5′ square x 16″ deep.
Minimum Charge	Job		695	36	731	
Metal Cap						
Install	Ea.	39.50	19		58.50	Includes labor and materials to install bird and squirrel screens in a chimney flue.
Minimum Charge	Job		150		150	

Fireplace

Fireplace	Unit	Material	Labor	Equip.	Total	Specification
Common Brick						
30″						
Demolish	Ea.		865	251	1116	Includes material and labor to install a complete unit which includes common brick, foundation, damper, flue lining stack to 15′ high and a 30″ fire box.
Install	Ea.	1350	2675		4025	
Demolish and Install	Ea.	1350	3540	251	5141	
Minimum Charge	Job		152		152	
36″						
Demolish	Ea.		865	251	1116	Includes material and labor to install a complete unit which includes common brick, foundation, damper, flue lining stack to 15′ high and a 36″ fire box.
Install	Ea.	1425	2975		4400	
Demolish and Install	Ea.	1425	3840	251	5516	
Minimum Charge	Job		152		152	
42″						
Demolish	Ea.		865	251	1116	Includes material and labor to install a complete unit which includes common brick, foundation, damper, flue lining stack to 15′ high and a 42″ fire box.
Install	Ea.	1550	3350		4900	
Demolish and Install	Ea.	1550	4215	251	6016	
Minimum Charge	Job		152		152	

Fireplace	Unit	Material	Labor	Equip.	Total	Specification
48″						
Demolish	Ea.		865	251	1116	Includes material and labor to install a complete unit which includes common brick, foundation, damper, flue lining stack to 15′ high and a 48″ fire box.
Install	Ea.	1650	3800		5450	
Demolish and Install	Ea.	1650	4665	251	6566	
Minimum Charge	Job		152		152	
Raised Hearth						
Brick Hearth						
Demolish	S.F.		3.78	1.10	4.88	Includes material and labor to install a brick fireplace hearth based on 30″ x 29″ opening.
Install	S.F.	12.80	22.50		35.30	
Demolish and Install	S.F.	12.80	26.28	1.10	40.18	
Clean	S.F.	.03	.29		.32	
Minimum Charge	Job		152		152	
Marble Hearth / Facing						
Demolish	Ea.		24	7.05	31.05	Includes material and labor to install marble fireplace hearth.
Install	Ea.	585	90.50		675.50	
Demolish and Install	Ea.	585	114.50	7.05	706.55	
Remove and Install	Ea.		72.45		72.45	
Clean	Ea.	21.50	4.36		25.86	
Minimum Charge	Job		152		152	
Breast (face) Brick						
Demolish	S.F.		3.78	1.10	4.88	Includes material and labor to install standard red brick, running bond.
Install	S.F.	3.04	6.25		9.29	
Demolish and Install	S.F.	3.04	10.03	1.10	14.17	
Clean	S.F.		.74	.12	.86	
Minimum Charge	Hr.		38		38	
Fire Box						
Demolish	Ea.		194	56.50	250.50	Includes material and labor to install fireplace box only (110 brick).
Install	Ea.	138	267		405	
Demolish and Install	Ea.	138	461	56.50	655.50	
Clean	Ea.	.10	9.80		9.90	
Minimum Charge	Hr.		38		38	
Natural Stone						
Demolish	S.F.		3.78	1.10	4.88	Includes material and labor to install high priced stone, split or rock face.
Install	S.F.	12.65	11.45		24.10	
Demolish and Install	S.F.	12.65	15.23	1.10	28.98	
Clean	S.F.	.06	.67		.73	
Minimum Charge	Job		267		267	
Slumpstone Veneer						
Demolish	S.F.		1.42		1.42	Cost includes material and labor to install 8″ x 6″ x 16″ concrete slump block.
Install	S.F.	5.35	3.57		8.92	
Demolish and Install	S.F.	5.35	4.99		10.34	
Clean	S.F.	.03	.34		.37	
Paint	S.F.	.12	.60		.72	
Minimum Charge	Job		152		152	
Prefabricated						
Firebox						
Demolish	Ea.		34.50		34.50	Includes material and labor to install up to 43″ radiant heat, zero clearance, prefabricated fireplace. Flues, doors or blowers not included.
Install	Ea.	1375	273		1648	
Demolish and Install	Ea.	1375	307.50		1682.50	
Remove and Install	Ea.		218.76		218.76	
Clean	Ea.	1.36	13.65		15.01	
Minimum Charge	Job		150		150	

Fireplace	Unit	Material	Labor	Equip.	Total	Specification
Flue						
Demolish	V.L.F.		4.04		4.04	Includes material and labor to install prefabricated chimney vent, 8" exposed metal flue.
Install	V.L.F.	10.60	11.40		22	
Demolish and Install	V.L.F.	10.60	15.44		26.04	
Remove and Install	V.L.F.		9.10		9.10	
Clean	L.F.	.55	2.73		3.28	
Minimum Charge	Job		296		296	
Blower						
Demolish	Ea.		3.23		3.23	Includes material and labor to install blower for metal fireplace flue.
Install	Ea.	99.50	20.50		120	
Demolish and Install	Ea.	99.50	23.73		123.23	
Remove and Install	Ea.		16.44		16.44	
Clean	Ea.	2.73	27.50		30.23	
Minimum Charge	Job		296		296	
Accessories						
Wood Mantle						
Demolish	L.F.		4.04		4.04	Includes material and labor to install wood fireplace mantel.
Install	L.F.	5.85	8.60		14.45	
Demolish and Install	L.F.	5.85	12.64		18.49	
Remove and Install	L.F.		8.59		8.59	
Clean	Ea.	.02	.60		.62	
Paint	L.F.	.07	7.15		7.22	
Minimum Charge	Job		150		150	
Colonial Style						
Demolish	Ea.		37		37	Includes material and labor to install wood fireplace mantel.
Install	Opng.	835	150		985	
Demolish and Install	Opng.	835	187		1022	
Remove and Install	Opng.		150.40		150.40	
Clean	Ea.		.44		.44	
Paint	Ea.	11.30	99		110.30	
Minimum Charge	Job		150		150	
Modern Design						
Demolish	Ea.		37		37	Includes material and labor to install wood fireplace mantel.
Install	Opng.	173	60		233	
Demolish and Install	Opng.	173	97		270	
Remove and Install	Opng.		60.16		60.16	
Clean	Ea.		.44		.44	
Paint	Ea.	11.30	99		110.30	
Minimum Charge	Job		150		150	
Glass Door Unit						
Demolish	Ea.		5.05		5.05	Includes material and labor to install a fireplace glass door unit.
Install	Ea.	124	6.35		130.35	
Demolish and Install	Ea.	124	11.40		135.40	
Remove and Install	Ea.		6.33		6.33	
Clean	Ea.	.33	8.40		8.73	
Minimum Charge	Job		150		150	
Fire Screen						
Demolish	Ea.		2.52		2.52	Includes material and labor to install a fireplace fire screen.
Install	Ea.	38	3.17		41.17	
Demolish and Install	Ea.	38	5.69		43.69	
Remove and Install	Ea.		3.17		3.17	
Clean	Ea.	.17	7.25		7.42	
Minimum Charge	Job		150		150	

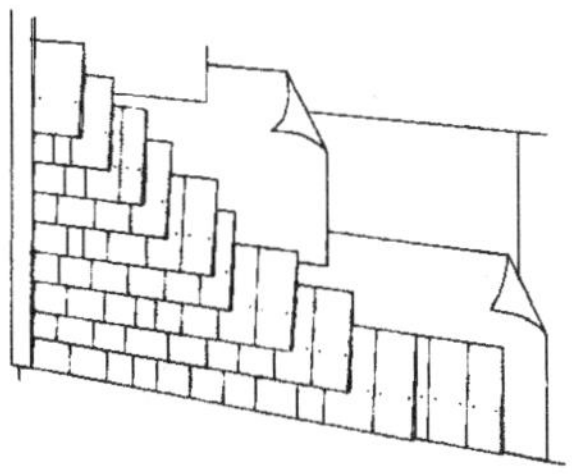

Shingles

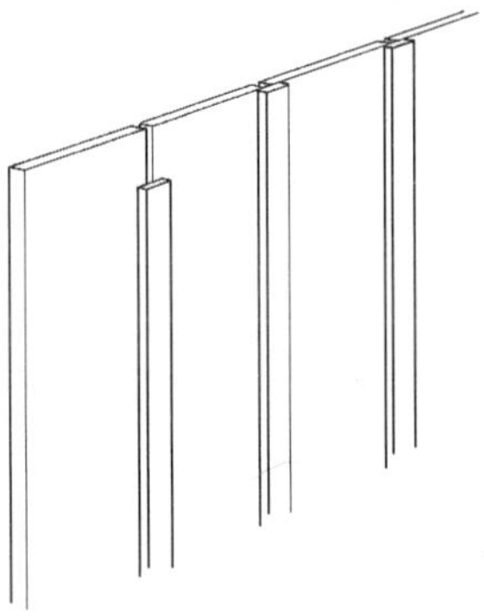

Board & Batten

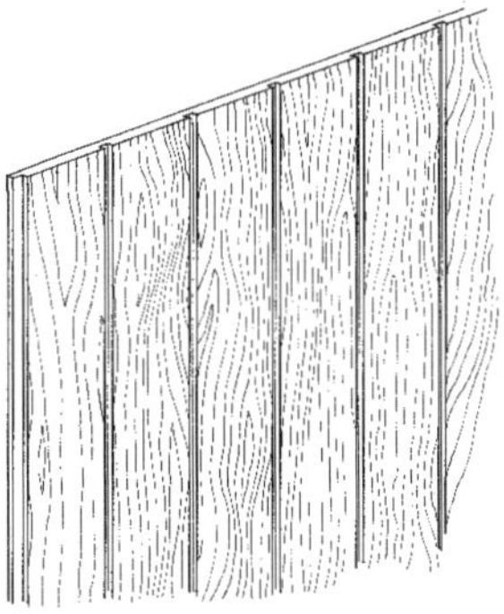

Texture 1-11

Exterior Siding	Unit	Material	Labor	Equip.	Total	Specification
Trim						
1" x 2"						
Demolish	L.F.		.20		.20	Cost includes material and labor to install 1" x 2" pine trim.
Install	L.F.	.33	.91		1.24	
Demolish and Install	L.F.	.33	1.11		1.44	
Clean	L.F.		.15		.15	
Paint	L.F.	.02	.41		.43	
Minimum Charge	Job		150		150	
1" x 3"						
Demolish	L.F.		.20		.20	Cost includes material and labor to install 1" x 3" pine trim.
Install	L.F.	.21	1		1.21	
Demolish and Install	L.F.	.21	1.20		1.41	
Clean	L.F.	.01	.15		.16	
Paint	L.F.	.02	.41		.43	
Minimum Charge	Job		150		150	
1" x 4"						
Demolish	L.F.		.20		.20	Cost includes material and labor to install 1" x 4" pine trim.
Install	L.F.	.67	1.20		1.87	
Demolish and Install	L.F.	.67	1.40		2.07	
Clean	L.F.	.01	.15		.16	
Paint	L.F.	.02	.41		.43	
Minimum Charge	Job		150		150	
1" x 6"						
Demolish	L.F.		.20		.20	Cost includes material and labor to install 1" x 6" pine trim.
Install	L.F.	1.18	1.20		2.38	
Demolish and Install	L.F.	1.18	1.40		2.58	
Clean	L.F.	.01	.15		.16	
Paint	L.F.	.07	.41		.48	
Minimum Charge	Job		150		150	
1" x 8"						
Demolish	L.F.		.22		.22	Cost includes material and labor to install 1" x 8" pine trim.
Install	L.F.	1.45	1.50		2.95	
Demolish and Install	L.F.	1.45	1.72		3.17	
Clean	L.F.	.02	.16		.18	
Paint	L.F.	.07	.41		.48	
Minimum Charge	Job		150		150	
Aluminum						
8" Non-Insulated						
Demolish	S.F.		.55		.55	Includes material and labor to install horizontal, colored clapboard, 8" to 10" wide, plain.
Install	S.F.	1.35	1.19		2.54	
Demolish and Install	S.F.	1.35	1.74		3.09	
Remove and Install	S.F.		1.19		1.19	
Clean	S.F.		.22		.22	
Minimum Charge	Job		150		150	

Exterior Siding	Unit	Material	Labor	Equip.	Total	Specification
8″ Insulated						
Demolish	S.F.		.55		.55	Includes material and labor to install horizontal, colored insulated clapboard, 8″ wide, plain.
Install	S.F.	1.53	1.19		2.72	
Demolish and Install	S.F.	1.53	1.74		3.27	
Remove and Install	S.F.		1.19		1.19	
Clean	S.F.		.22		.22	
Minimum Charge	Job		150		150	
12″ Non-Insulated						
Demolish	S.F.		.55		.55	Includes material and labor to install horizontal, colored clapboard, plain.
Install	S.F.	1.42	1.03		2.45	
Demolish and Install	S.F.	1.42	1.58		3	
Remove and Install	S.F.		1.03		1.03	
Clean	S.F.		.22		.22	
Minimum Charge	Job		150		150	
12″ Insulated						
Demolish	S.F.		.55		.55	Includes material and labor to install horizontal, colored insulated clapboard, plain.
Install	S.F.	1.67	1.03		2.70	
Demolish and Install	S.F.	1.67	1.58		3.25	
Remove and Install	S.F.		1.03		1.03	
Clean	S.F.		.22		.22	
Minimum Charge	Job		150		150	
Clapboard 8″ to 10″						
Demolish	S.F.		.55		.55	Includes material and labor to install horizontal, colored clapboard, 8″ to 10″ wide, plain.
Install	S.F.	1.35	1.19		2.54	
Demolish and Install	S.F.	1.35	1.74		3.09	
Remove and Install	S.F.		1.19		1.19	
Clean	S.F.		.22		.22	
Minimum Charge	Job		150		150	
Corner Strips						
Demolish	L.F.		.29		.29	Includes material and labor to install corners for horizontal lapped aluminum siding.
Install	V.L.F.	1.87	1.19		3.06	
Demolish and Install	V.L.F.	1.87	1.48		3.35	
Remove and Install	V.L.F.		1.19		1.19	
Clean	L.F.		.15		.15	
Minimum Charge	Job		150		150	
Vinyl						
Non-Insulated						
Demolish	S.F.		.43		.43	Cost includes material and labor to install 10″ non-insulated solid vinyl siding.
Install	S.F.	.76	1.18		1.94	
Demolish and Install	S.F.	.76	1.61		2.37	
Remove and Install	S.F.		.94		.94	
Clean	S.F.		.22		.22	
Minimum Charge	Job		150		150	
Insulated						
Demolish	S.F.		.43		.43	Includes material and labor to install insulated solid vinyl siding panels.
Install	S.F.	.84	1.18		2.02	
Demolish and Install	S.F.	.84	1.61		2.45	
Remove and Install	S.F.		.94		.94	
Clean	S.F.		.22		.22	
Minimum Charge	Job		150		150	
Corner Strips						
Demolish	L.F.		.27		.27	Includes material and labor to install corner posts for vinyl siding panels.
Install	L.F.	1.27	1.47		2.74	
Demolish and Install	L.F.	1.27	1.74		3.01	
Remove and Install	L.F.		1.17		1.17	
Clean	L.F.		.15		.15	
Minimum Charge	Job		150		150	

Exterior Siding	Unit	Material	Labor	Equip.	Total	Specification
Hardboard						
4' x 8' Panel						
Demolish	S.F.		.45		.45	Cost includes material and labor to install 4' x 8' masonite siding 7/16" thick.
Install	S.F.	1.02	.75		1.77	
Demolish and Install	S.F.	1.02	1.20		2.22	
Clean	S.F.		.22		.22	
Paint	S.F.	.23	.89		1.12	
Minimum Charge	Job		150		150	
Lap						
Demolish	S.F.		.45		.45	Cost includes material and labor to install 8" or 12" masonite lap siding 7/16" thick with wood grain finish.
Install	S.F.	.93	.78		1.71	
Demolish and Install	S.F.	.93	1.23		2.16	
Clean	S.F.		.22		.22	
Paint	S.F.	.23	.89		1.12	
Minimum Charge	Job		150		150	
Plywood Panel						
Douglas Fir (T-1-11)						
Demolish	S.F.		.33		.33	Includes material and labor to install Douglas fir 4' x 8' plywood exterior siding plain or patterns.
Install	S.F.	.95	.91		1.86	
Demolish and Install	S.F.	.95	1.24		2.19	
Remove and Install	S.F.		.91		.91	
Clean	S.F.		.22		.22	
Paint	S.F.	.23	.89		1.12	
Minimum Charge	Job		150		150	
Redwood						
Demolish	S.F.		.33		.33	Includes material and labor to install redwood 4' x 8' plywood exterior siding plain or patterns.
Install	S.F.	1.98	.91		2.89	
Demolish and Install	S.F.	1.98	1.24		3.22	
Remove and Install	S.F.		.91		.91	
Clean	S.F.		.22		.22	
Paint	S.F.	.23	.89		1.12	
Minimum Charge	Job		150		150	
Cedar						
Demolish	S.F.		.33		.33	Cost includes material and labor to install 4' x 8' cedar plywood exterior siding plain or patterns.
Install	S.F.	2.35	.91		3.26	
Demolish and Install	S.F.	2.35	1.24		3.59	
Remove and Install	S.F.		.91		.91	
Clean	S.F.		.22		.22	
Paint	S.F.	.23	.89		1.12	
Minimum Charge	Job		150		150	
Tongue and Groove						
Fir						
Demolish	S.F.		.61		.61	Includes material and labor to install fir tongue and groove exterior siding 1" x 8".
Install	S.F.	1.39	.80		2.19	
Demolish and Install	S.F.	1.39	1.41		2.80	
Remove and Install	S.F.		.64		.64	
Clean	S.F.		.22		.22	
Paint	S.F.	.23	.89		1.12	
Minimum Charge	Job		150		150	
Pine						
Demolish	S.F.		.61		.61	Includes material and labor to install pine tongue and groove exterior siding 1" x 8".
Install	S.F.	.67	1.64		2.31	
Demolish and Install	S.F.	.67	2.25		2.92	
Remove and Install	S.F.		1.64		1.64	
Clean	S.F.		.22		.22	
Paint	S.F.	.23	.89		1.12	
Minimum Charge	Job		150		150	

Exterior Siding

	Unit	Material	Labor	Equip.	Total	Specification
Cedar						
Demolish	S.F.		.61		.61	Includes material and labor to install cedar tongue and groove exterior siding 1″ x 8″.
Install	S.F.	3.04	.80		3.84	
Demolish and Install	S.F.	3.04	1.41		4.45	
Remove and Install	S.F.		.64		.64	
Clean	S.F.		.22		.22	
Paint	S.F.	.23	.89		1.12	
Minimum Charge	Job		150		150	
Redwood						
Demolish	S.F.		.61		.61	Includes material and labor to install redwood tongue and groove exterior siding.
Install	S.F.	2.93	2.05		4.98	
Demolish and Install	S.F.	2.93	2.66		5.59	
Remove and Install	S.F.		2.05		2.05	
Clean	S.F.		.22		.22	
Paint	S.F.	.23	.89		1.12	
Minimum Charge	Job		150		150	
Board and Batten						
Fir						
Demolish	S.F.		.61		.61	Cost includes material and labor to install 1″ x 8″ fir boards with 1″ x 3″ battens.
Install	S.F.	1.49	.75		2.24	
Demolish and Install	S.F.	1.49	1.36		2.85	
Remove and Install	S.F.		.60		.60	
Clean	S.F.		.22		.22	
Paint	S.F.	.23	.89		1.12	
Minimum Charge	Job		150		150	
Pine						
Demolish	S.F.		.61		.61	Cost includes material and labor to install 1″ x 10″ pine boards with 1″ x 3″ battens.
Install	S.F.	.77	.75		1.52	
Demolish and Install	S.F.	.77	1.36		2.13	
Remove and Install	S.F.		.60		.60	
Clean	S.F.		.22		.22	
Paint	S.F.	.23	.89		1.12	
Minimum Charge	Job		150		150	
Cedar						
Demolish	S.F.		.61		.61	Cost includes material and labor to install 1″ x 10″ cedar boards with 1″ x 3″ battens.
Install	S.F.	2.18	.75		2.93	
Demolish and Install	S.F.	2.18	1.36		3.54	
Remove and Install	S.F.		.60		.60	
Clean	S.F.		.22		.22	
Paint	S.F.	.23	.89		1.12	
Minimum Charge	Job		150		150	
Redwood						
Demolish	S.F.		.61		.61	Cost includes material and labor to install 1″ x 10″ redwood boards with 1″ x 3″ battens.
Install	S.F.	3.95	.75		4.70	
Demolish and Install	S.F.	3.95	1.36		5.31	
Remove and Install	S.F.		.60		.60	
Clean	S.F.		.22		.22	
Paint	S.F.	.23	.89		1.12	
Minimum Charge	Job		150		150	
Wood Shingle						
Demolish	Sq.		56.50		56.50	Includes material and labor to install wood shingles/shakes.
Install	Sq.	130	134		264	
Demolish and Install	Sq.	130	190.50		320.50	
Clean	Sq.		23		23	
Paint	Sq.	.15	79		79.15	
Minimum Charge	Job		150		150	

Exterior Siding	Unit	Material	Labor	Equip.	Total	Specification
Synthetic Stucco System						
Demolish	S.F.		2.02		2.02	Cost includes material and labor to install 2 coats of adhesive mixture, glass fiber mesh, 1" insulation board, and textured finish coat.
Install	S.F.	2.86	5.85	.44	9.15	
Demolish and Install	S.F.	2.86	7.87	.44	11.17	
Clean	S.F.		.20		.20	
Paint	S.F.	.12	.43		.55	
Minimum Charge	Job		640	48.50	688.50	

Front-end Loader

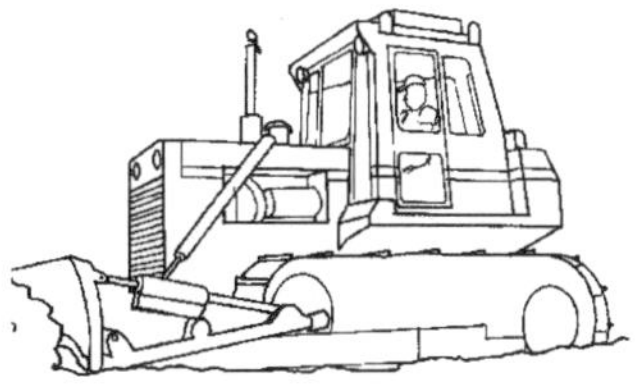

Bulldozer

Grading	Unit	Material	Labor	Equip.	Total	Specification
Hand (fine)						
Install	S.F.		.44		.44	Includes labor and material to install fine grade for slab on grade, hand grade.
Minimum Charge	Job		109		109	
Machine (fine)						
Install	S.Y.		.38	.37	.75	Includes labor and material to install fine grade for roadway, base or leveling course.
Minimum Charge	Job		258	251	509	

Asphalt Paving	Unit	Material	Labor	Equip.	Total	Specification
Seal Coat						
Install	S.F.	.06	.08		.14	Includes labor and material to install sealcoat for a small area.
Minimum Charge	Job		109		109	
Wearing Course						
Install	S.F.	.07	.23	.18	.48	Cost includes labor and material to install 1" thick wearing course.
Minimum Charge	Job		109		109	

Masonry Pavers	Unit	Material	Labor	Equip.	Total	Specification
Natural Concrete						
Sand Base						
Demolish	S.F.		2.42		2.42	Includes material and labor to install concrete pavers on sand base.
Install	S.F.	2	5.05		7.05	
Demolish and Install	S.F.	2	7.47		9.47	
Remove and Install	S.F.		4.05		4.05	
Clean	S.F.		.87		.87	
Minimum Charge	Job		152		152	
Mortar Base						
Demolish	S.F.		4	1.34	5.34	Includes material and labor to install concrete pavers on mortar base.
Install	S.F.	2.57	4.22		6.79	
Demolish and Install	S.F.	2.57	8.22	1.34	12.13	
Remove and Install	S.F.		3.38		3.38	
Clean	S.F.		.79		.79	
Minimum Charge	Job		152		152	

Masonry Pavers	Unit	Material	Labor	Equip.	Total	Specification
Adobe						
Sand Base						
Demolish	S.F.		2.42		2.42	Includes material and labor to install adobe pavers on sand base.
Install	S.F.	1.58	3.38		4.96	
Demolish and Install	S.F.	1.58	5.80		7.38	
Remove and Install	S.F.		2.70		2.70	
Clean	S.F.		.87		.87	
Minimum Charge	Job		152		152	
Mortar Base						
Demolish	S.F.		4	1.34	5.34	Includes material and labor to install adobe pavers on mortar base.
Install	S.F.	2.19	4.22		6.41	
Demolish and Install	S.F.	2.19	8.22	1.34	11.75	
Remove and Install	S.F.		3.38		3.38	
Clean	S.F.		.79		.79	
Minimum Charge	Job		152		152	
Standard Brick						
Sand Base						
Demolish	S.F.		2.42		2.42	Includes material and labor to install standard brick pavers on sand base.
Install	S.F.	2.44	5.35		7.79	
Demolish and Install	S.F.	2.44	7.77		10.21	
Remove and Install	S.F.		4.28		4.28	
Clean	S.F.		.87		.87	
Minimum Charge	Job		152		152	
Mortar Base						
Demolish	S.F.		4	1.34	5.34	Includes material and labor to install standard brick pavers on mortar base.
Install	S.F.	4.66	7.60		12.26	
Demolish and Install	S.F.	4.66	11.60	1.34	17.60	
Remove and Install	S.F.		6.08		6.08	
Clean	S.F.		.79		.79	
Minimum Charge	Job		152		152	
Deluxe Brick						
Sand Base						
Demolish	S.F.		2.42		2.42	Includes material and labor to install deluxe brick pavers on sand base.
Install	S.F.	4.59	4.22		8.81	
Demolish and Install	S.F.	4.59	6.64		11.23	
Remove and Install	S.F.		3.38		3.38	
Clean	S.F.		.87		.87	
Minimum Charge	Job		152		152	
Mortar Base						
Demolish	S.F.		4	1.34	5.34	Includes material and labor to install deluxe brick pavers on mortar base.
Install	S.F.	4.93	7.60		12.53	
Demolish and Install	S.F.	4.93	11.60	1.34	17.87	
Remove and Install	S.F.		6.08		6.08	
Clean	S.F.		.79		.79	
Minimum Charge	Job		152		152	

Fencing	Unit	Material	Labor	Equip.	Total	Specification
Block w / Footing						
4″						
Demolish	S.F.		1.35		1.35	Cost includes material and labor to install 4″ non-reinforced concrete block.
Install	S.F.	1.01	3.19		4.20	
Demolish and Install	S.F.	1.01	4.54		5.55	
Clean	S.F.	.13	.93	.15	1.21	
Paint	S.F.	.15	.35		.50	
Minimum Charge	Job		152		152	

Fencing	Unit	Material	Labor	Equip.	Total	Specification
4" Reinforced						
Demolish	S.F.		1.35	.39	1.74	Cost includes material and labor to install 4" reinforced concrete block.
Install	S.F.	1.10	3.23		4.33	
Demolish and Install	S.F.	1.10	4.58	.39	6.07	
Clean	S.F.	.13	.93	.15	1.21	
Paint	S.F.	.15	.35		.50	
Minimum Charge	Job		152		152	
6"						
Demolish	S.F.		1.42		1.42	Cost includes material and labor to install 6" non-reinforced concrete block.
Install	S.F.	1.50	3.43		4.93	
Demolish and Install	S.F.	1.50	4.85		6.35	
Clean	S.F.	.13	.93	.15	1.21	
Paint	S.F.	.15	.35		.50	
Minimum Charge	Job		152		152	
6" Reinforced						
Demolish	S.F.		1.42	.41	1.83	Cost includes material and labor to install 6" reinforced concrete block.
Install	S.F.	1.58	3.48		5.06	
Demolish and Install	S.F.	1.58	4.90	.41	6.89	
Clean	S.F.	.13	.93	.15	1.21	
Paint	S.F.	.15	.35		.50	
Minimum Charge	Job		152		152	
Cap for Block Wall						
Demolish	L.F.		1.51	.44	1.95	Includes labor and equipment to install precast concrete coping.
Install	L.F.	19.15	7.15		26.30	
Demolish and Install	L.F.	19.15	8.66	.44	28.25	
Clean	L.F.	.01	2.27		2.28	
Paint	S.F.	.15	.35		.50	
Minimum Charge	Job		152		152	
Block Pilaster						
Demolish	Ea.		104	30	134	Includes material and labor to install block pilaster for fence.
Install	Ea.	56.50	107		163.50	
Demolish and Install	Ea.	56.50	211	30	297.50	
Remove and Install	Ea.		106.98		106.98	
Paint	Ea.	3.76	11.75		15.51	
Minimum Charge	Job		152		152	
Block w / o Footing						
4"						
Demolish	S.F.		1.35		1.35	Cost includes material and labor to install 4" non-reinforced hollow lightweight block.
Install	S.F.	1.21	2.43		3.64	
Demolish and Install	S.F.	1.21	3.78		4.99	
Clean	S.F.	.13	.93	.15	1.21	
Paint	S.F.	.15	.35		.50	
Minimum Charge	Job		152		152	
4" Reinforced						
Demolish	S.F.		1.35	.39	1.74	Cost includes material and labor to install 4" reinforced hollow lightweight block.
Install	S.F.	1.36	3.16		4.52	
Demolish and Install	S.F.	1.36	4.51	.39	6.26	
Clean	S.F.	.13	.93	.15	1.21	
Paint	S.F.	.15	.35		.50	
Minimum Charge	Job		152		152	
6"						
Demolish	S.F.		1.42		1.42	Cost includes material and labor to install 6" non-reinforced hollow lightweight block.
Install	S.F.	1.64	2.67		4.31	
Demolish and Install	S.F.	1.64	4.09		5.73	
Clean	S.F.	.13	.93	.15	1.21	
Paint	S.F.	.15	.35		.50	
Minimum Charge	Job		152		152	

Fencing	Unit	Material	Labor	Equip.	Total	Specification
6" Reinforced						
Demolish	S.F.		1.42	.41	1.83	Cost includes material and labor to install 6" reinforced hollow lightweight block.
Install	S.F.	1.71	3.39		5.10	
Demolish and Install	S.F.	1.71	4.81	.41	6.93	
Clean	S.F.	.13	.93	.15	1.21	
Paint	S.F.	.15	.35		.50	
Minimum Charge	Job		152		152	
Cap for Block Wall						
Demolish	L.F.		1.51	.44	1.95	Includes labor and equipment to install precast concrete coping.
Install	L.F.	19.15	7.15		26.30	
Demolish and Install	L.F.	19.15	8.66	.44	28.25	
Clean	L.F.	.01	2.27		2.28	
Paint	S.F.	.15	.35		.50	
Minimum Charge	Job		152		152	
Block Pilaster						
Demolish	Ea.		104	30	134	Includes material and labor to install block pilaster for fence.
Install	Ea.	56.50	107		163.50	
Demolish and Install	Ea.	56.50	211	30	297.50	
Remove and Install	Ea.		106.98		106.98	
Paint	Ea.	3.76	11.75		15.51	
Minimum Charge	Job		152		152	
Chain Link						
3' High w / o Top Rail						
Demolish	L.F.		1.88	.63	2.51	Includes material and labor to install residential galvanized fence 3' high with 1-3/8" steel post 10' O.C. set without concrete.
Install	L.F.	2.39	1.30		3.69	
Demolish and Install	L.F.	2.39	3.18	.63	6.20	
Remove and Install	L.F.		1.04		1.04	
Clean	L.F.	.09	.79		.88	
Minimum Charge	Job		340		340	
4' High w / o Top Rail						
Demolish	L.F.		1.88	.63	2.51	Includes material and labor to install residential galvanized fence 4' high with 1-3/8" steel post 10' O.C. set without concrete.
Install	L.F.	2.53	1.60		4.13	
Demolish and Install	L.F.	2.53	3.48	.63	6.64	
Remove and Install	L.F.		1.28		1.28	
Clean	L.F.	.11	.87		.98	
Minimum Charge	Job		340		340	
5' High w / o Top Rail						
Demolish	L.F.		1.88	.63	2.51	Includes material and labor to install residential galvanized fence 5' high with 1-3/8" steel post 10' O.C. set without concrete.
Install	L.F.	2.97	2.10		5.07	
Demolish and Install	L.F.	2.97	3.98	.63	7.58	
Remove and Install	L.F.		1.68		1.68	
Clean	L.F.	.13	.97		1.10	
Minimum Charge	Job		340		340	
3' High w / Top Rail						
Demolish	L.F.		2	.67	2.67	Includes material and labor to install residential galvanized fence 3' high with 1-3/8" steel post 10' O.C. set without concrete.
Install	L.F.	2.94	1.36		4.30	
Demolish and Install	L.F.	2.94	3.36	.67	6.97	
Remove and Install	L.F.		1.09		1.09	
Clean	L.F.	.09	.81		.90	
Minimum Charge	Job		340		340	
4' High w / Top Rail						
Demolish	L.F.		2	.67	2.67	Includes material and labor to install residential galvanized fence 4' high with 1-5/8" steel post 10' O.C. set without concrete and 1-5/8" top rail with sleeves.
Install	L.F.	3.33	1.70		5.03	
Demolish and Install	L.F.	3.33	3.70	.67	7.70	
Remove and Install	L.F.		1.36		1.36	
Clean	L.F.	.11	.89		1	
Minimum Charge	Job		340		340	

Fencing	Unit	Material	Labor	Equip.	Total	Specification
5' High w / Top Rail						
Demolish	L.F.		2	.67	2.67	Includes material and labor to install residential galvanized fence 5' high with 1-5/8" steel post 10' O.C. set without concrete and 1-5/8" top rail with sleeves.
Install	L.F.	3.78	2.27		6.05	
Demolish and Install	L.F.	3.78	4.27	.67	8.72	
Remove and Install	L.F.		1.82		1.82	
Clean	L.F.	.13	.99		1.12	
Minimum Charge	Job		340		340	
6' High w / Top Rail						
Demolish	L.F.		2	.67	2.67	Includes material and labor to install residential galvanized fence 6' high with 2" steel post 10' O.C. set without concrete and 1-3/8" top rail with sleeves.
Install	L.F.	4.05	3.41		7.46	
Demolish and Install	L.F.	4.05	5.41	.67	10.13	
Remove and Install	L.F.		2.72		2.72	
Clean	L.F.	.17	1.12		1.29	
Minimum Charge	Job		340		340	
Barbed Wire						
Demolish	L.F.		1.13		1.13	Cost includes material and labor to install 3 strand galvanized barbed wire fence.
Install	L.F.	.26	.44	.10	.80	
Demolish and Install	L.F.	.26	1.57	.10	1.93	
Remove and Install	L.F.		.35	.08	.43	
Minimum Charge	Job		340	275	615	
6' Redwood						
Demolish	L.F.		7.55		7.55	Cost includes material and labor to install 6' high redwood fencing.
Install	L.F.	9.70	5.45		15.15	
Demolish and Install	L.F.	9.70	13		22.70	
Remove and Install	L.F.		4.36		4.36	
Clean	L.F.	.17	1.01		1.18	
Paint	L.F.	.61	2.78		3.39	
Minimum Charge	Job		340		340	
6' Red Cedar						
Demolish	L.F.		7.55		7.55	Cost includes material and labor to install 6' cedar fence.
Install	L.F.	14.75	5.45		20.20	
Demolish and Install	L.F.	14.75	13		27.75	
Remove and Install	L.F.		4.36		4.36	
Clean	L.F.	.17	1.01		1.18	
Paint	L.F.	.61	2.78		3.39	
Minimum Charge	Job		340		340	
6' Shadowbox						
Demolish	L.F.		7.55		7.55	Cost includes material and labor to install 6' pressure treated pine fence.
Install	L.F.	11.30	4.54		15.84	
Demolish and Install	L.F.	11.30	12.09		23.39	
Remove and Install	L.F.		3.63		3.63	
Clean	L.F.	.17	1.01		1.18	
Paint	L.F.	.61	2.78		3.39	
Minimum Charge	Job		340		340	
3' Picket						
Demolish	L.F.		7.55		7.55	Includes material, labor and equipment to install pressure treated pine picket fence.
Install	L.F.	4.27	4.87		9.14	
Demolish and Install	L.F.	4.27	12.42		16.69	
Remove and Install	L.F.		3.89		3.89	
Clean	L.F.	.09	.87		.96	
Paint	L.F.	.31	1.51		1.82	
Minimum Charge	Job		340		340	

Fencing	Unit	Material	Labor	Equip.	Total	Specification
Wrought Iron						
Demolish	L.F.		16.15		16.15	Cost includes material and labor to install 5′ high wrought iron fence with pickets at 4-1/2″ O.C.
Install	L.F.	16.70	9.10		25.80	
Demolish and Install	L.F.	16.70	25.25		41.95	
Remove and Install	L.F.		7.27		7.27	
Clean	L.F.		.87		.87	
Paint	L.F.	.56	3.11		3.67	
Minimum Charge	Job		340		340	
Post						
Dirt Based Wood						
Demolish	Ea.		18.90		18.90	Includes material and labor to install wood post set in earth.
Install	Ea.	10.40	85		95.40	
Demolish and Install	Ea.	10.40	103.90		114.30	
Remove and Install	Ea.		68.11		68.11	
Clean	Ea.		2.73		2.73	
Paint	Ea.	.67	1.06		1.73	
Minimum Charge	Job		340		340	
Concrete Based Wood						
Demolish	Ea.		26.50	8.90	35.40	Includes material and labor to install wood post set in concrete.
Install	Ea.	10.40	114		124.40	
Demolish and Install	Ea.	10.40	140.50	8.90	159.80	
Clean	Ea.		2.73		2.73	
Paint	Ea.	.67	1.06		1.73	
Minimum Charge	Job		340		340	
Block Pilaster						
Demolish	Ea.		104	30	134	Includes material and labor to install block pilaster for fence.
Install	Ea.	56.50	107		163.50	
Demolish and Install	Ea.	56.50	211	30	297.50	
Remove and Install	Ea.		106.98		106.98	
Paint	Ea.	3.76	11.75		15.51	
Minimum Charge	Job		152		152	
Gate						
Chain Link						
Demolish	Ea.		25		25	Cost includes material and labor to install 3′ wide chain link pass gate and hardware.
Install	Ea.	97	38		135	
Demolish and Install	Ea.	97	63		160	
Remove and Install	Ea.		30.27		30.27	
Clean	Ea.		3.03		3.03	
Minimum Charge	Job		340		340	
Redwood						
Demolish	Ea.		21.50		21.50	Includes material, labor and equipment to install redwood pass gate and hardware.
Install	Ea.	116	114		230	
Demolish and Install	Ea.	116	135.50		251.50	
Remove and Install	Ea.		90.82		90.82	
Clean	Ea.		3.03		3.03	
Paint	Ea.	1	5.30		6.30	
Minimum Charge	Job		340		340	
Red Cedar						
Demolish	Ea.		21.50		21.50	Cost includes material and labor to install 3′ wide red cedar pass gate and hardware.
Install	Ea.	66	38		104	
Demolish and Install	Ea.	66	59.50		125.50	
Remove and Install	Ea.		30.27		30.27	
Clean	Ea.		3.03		3.03	
Paint	Ea.	1	5.30		6.30	
Minimum Charge	Job		340		340	

Fencing	Unit	Material	Labor	Equip.	Total	Specification
Wrought Iron						
Demolish	Ea.		16.15		16.15	Includes material, labor and equipment to install wrought iron gate with hardware.
Install	Ea.	88	85		173	
Demolish and Install	Ea.	88	101.15		189.15	
Remove and Install	Ea.		68.11		68.11	
Clean	Ea.		3.03		3.03	
Paint	Ea.	1.11	5.30		6.41	
Minimum Charge	Job		340		340	

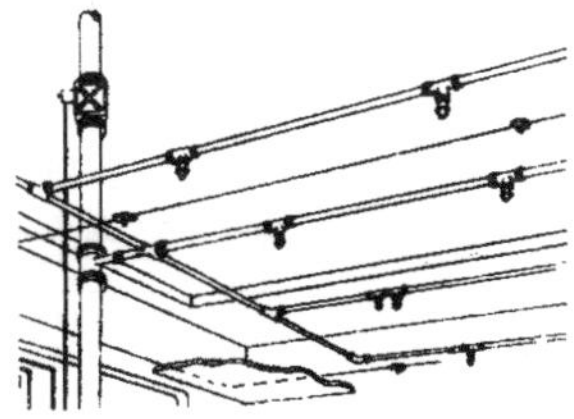

Fire-extinguishing System

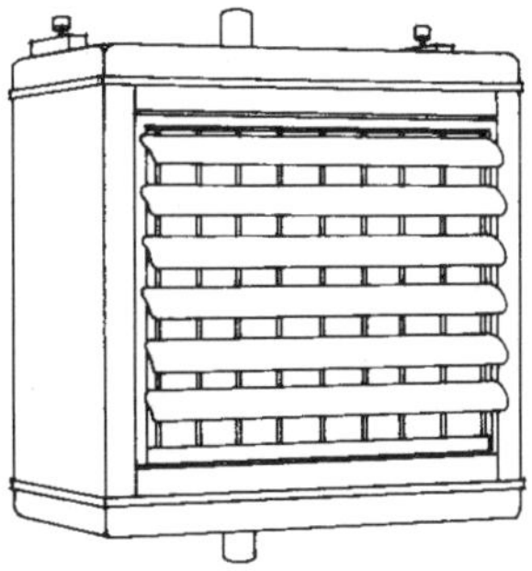

Space Heater

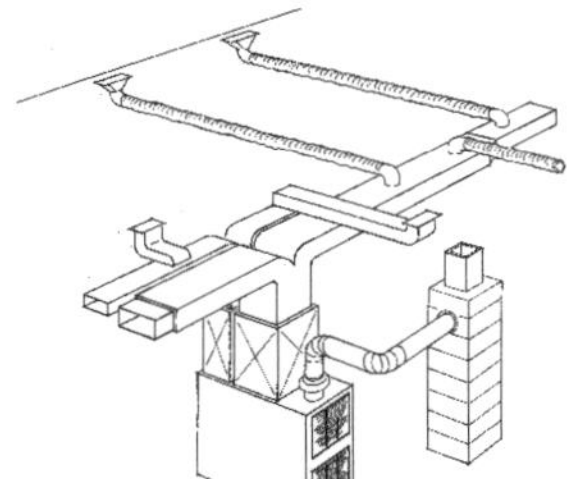

Hot-air Furnace

Plumbing	Unit	Material	Labor	Equip.	Total	Specification
Components						
3/4" Supply						
Demolish	L.F.		1.81		1.81	Includes material and labor to install type L copper tubing including couplings and hangers at 10' O.C.
Install	L.F.	1.95	4.28		6.23	
Demolish and Install	L.F.	1.95	6.09		8.04	
Remove and Install	L.F.		4.28		4.28	
Minimum Charge	Job		163		163	
1/2" Fixture Supply						
Install	Ea.	21	108		129	Includes labor and materials to install supply pipe only to a plumbing fixture.
2" Single Vent Stack						
Demolish	Ea.		30		30	Includes material and labor to install PVC vent stack pipe with couplings at 10' O.C.
Install	Ea.	11.85	163		174.85	
Demolish and Install	Ea.	11.85	193		204.85	
Remove and Install	Ea.		162.60		162.60	
Minimum Charge	Job		163		163	
4" Drain						
Demolish	L.F.		2.41		2.41	Includes material and labor to install DVW type PVC pipe with couplings at 10' O.C. and 3 strap hangers per 10' run.
Install	L.F.	6.25	5.30		11.55	
Demolish and Install	L.F.	6.25	7.71		13.96	
Remove and Install	L.F.		5.32		5.32	
Minimum Charge	Job		163		163	
1-1/2" Drain						
Demolish	L.F.		1.81		1.81	Includes material and labor to install DVW type PVC pipe with couplings at 10' O.C. and 3 strap hangers per 10' run.
Install	L.F.	1.65	3.46		5.11	
Demolish and Install	L.F.	1.65	5.27		6.92	
Remove and Install	L.F.		3.46		3.46	
Minimum Charge	Job		163		163	
Sewer / Septic						
Install	L.F.	2.43	3.06	1.02	6.51	Includes labor and material to install sewerage and drainage piping, 6" diameter.
Maintenance						
Flush Out / Unclog						
Clean	Job		163		163	Includes 4 hour minimum labor charge for a plumber.
Minimum Charge	Job		163		163	

Plumbing	Unit	Material	Labor	Equip.	Total	Specification
Vent Cap						
Large						
Demolish	Ea.		8.20		8.20	Includes labor and material to install a PVC plumbing pipe vent cap.
Install	Ea.	33	16.25		49.25	
Demolish and Install	Ea.	33	24.45		57.45	
Clean	Ea.	.39	4.95		5.34	
Paint	Ea.	.15	6		6.15	
Minimum Charge	Job		81.50		81.50	
Medium						
Demolish	Ea.		8.20		8.20	Includes material and labor to install a PVC plumbing pipe vent cap.
Install	Ea.	27.50	14.80		42.30	
Demolish and Install	Ea.	27.50	23		50.50	
Clean	Ea.	.39	4.95		5.34	
Paint	Ea.	.15	6		6.15	
Minimum Charge	Job		81.50		81.50	
Small						
Demolish	Ea.		8.20		8.20	Includes material and labor to install a PVC plumbing pipe vent cap.
Install	Ea.	20.50	13.55		34.05	
Demolish and Install	Ea.	20.50	21.75		42.25	
Clean	Ea.	.39	4.95		5.34	
Paint	Ea.	.15	6		6.15	
Minimum Charge	Job		81.50		81.50	

Plumbing	Unit	Material	Labor	Equip.	Total	Specification
Sump Pump						
1/4 HP						
Demolish	Ea.		32.50		32.50	Includes material and labor to install a submersible sump pump.
Install	Ea.	116	54		170	
Demolish and Install	Ea.	116	86.50		202.50	
Remove and Install	Ea.		43.36		43.36	
Clean	Ea.	.96	13.65		14.61	
Minimum Charge	Job		163		163	
1/3 HP						
Demolish	Ea.		32.50		32.50	Includes material and labor to install a submersible sump pump.
Install	Ea.	140	65		205	
Demolish and Install	Ea.	140	97.50		237.50	
Remove and Install	Ea.		52.03		52.03	
Clean	Ea.	.96	13.65		14.61	
Minimum Charge	Job		163		163	
1/2 HP						
Demolish	Ea.		32.50		32.50	Includes material and labor to install a submersible sump pump.
Install	Ea.	168	65		233	
Demolish and Install	Ea.	168	97.50		265.50	
Remove and Install	Ea.		52.03		52.03	
Clean	Ea.	.96	13.65		14.61	
Minimum Charge	Job		163		163	
Clean and Service						
Clean	Ea.		20.50		20.50	Includes labor to clean and service a sump pump.
Minimum Charge	Job		163		163	

Water Heater	Unit	Material	Labor	Equip.	Total	Specification
30 Gallon						
Demolish	Ea.		102		102	Includes material and labor to install a 30 gallon electric water heater.
Install	Ea.	297	148		445	
Demolish and Install	Ea.	297	250		547	
Remove and Install	Ea.		118.25		118.25	
Clean	Ea.	.96	13.65		14.61	
Minimum Charge	Job		163		163	
40 Gallon						
Demolish	Ea.		102		102	Includes material and labor to install a 40 gallon electric water heater.
Install	Ea.	320	163		483	
Demolish and Install	Ea.	320	265		585	
Remove and Install	Ea.		130.08		130.08	
Clean	Ea.	.96	13.65		14.61	
Minimum Charge	Job		163		163	
52 Gallon						
Demolish	Ea.		102		102	Includes material and labor to install a 52 gallon electric water heater.
Install	Ea.	370	163		533	
Demolish and Install	Ea.	370	265		635	
Remove and Install	Ea.		130.08		130.08	
Clean	Ea.	.96	13.65		14.61	
Minimum Charge	Job		163		163	
82 Gallon						
Demolish	Ea.		102		102	Includes material and labor to install an 82 gallon electric water heater.
Install	Ea.	560	203		763	
Demolish and Install	Ea.	560	305		865	
Remove and Install	Ea.		162.60		162.60	
Clean	Ea.	.96	13.65		14.61	
Minimum Charge	Job		163		163	
Gas-fired						
30 Gallon Gas-fired						
Demolish	Ea.		125		125	Includes material and labor to install a 30 gallon gas water heater.
Install	Ea.	350	163		513	
Demolish and Install	Ea.	350	288		638	
Remove and Install	Ea.		130.08		130.08	
Clean	Ea.	.96	13.65		14.61	
Minimum Charge	Job		163		163	
30 Gallon Gas-fired, Flue						
Install	Ea.	31.50	46.50		78	Includes labor and materials to install a flue for a water heater (up to 50 gallon size).
Minimum Charge	Job		163		163	
40 Gallon Gas-fired						
Demolish	Ea.		125		125	Includes material and labor to install a 40 gallon gas water heater.
Install	Ea.	360	171		531	
Demolish and Install	Ea.	360	296		656	
Remove and Install	Ea.		136.93		136.93	
Clean	Ea.	.96	13.65		14.61	
Minimum Charge	Job		163		163	
40 Gallon Gas-fired, Flue						
Install	Ea.	31.50	46.50		78	Includes labor and materials to install a flue for a water heater (up to 50 gallon size).
Minimum Charge	Job		163		163	
50 Gallon Gas-fired						
Demolish	Ea.		125		125	Includes material and labor to install a 50 gallon gas water heater.
Install	Ea.	430	181		611	
Demolish and Install	Ea.	430	306		736	
Remove and Install	Ea.		144.53		144.53	
Clean	Ea.	.96	13.65		14.61	
Minimum Charge	Job		163		163	

Water Heater	Unit	Material	Labor	Equip.	Total	Specification
50 Gallon Gas-fired, Flue						
Install	Ea.	31.50	46.50		78	Includes labor and materials to install a flue for a water heater (up to 50 gallon size).
Minimum Charge	Job		163		163	
Solar Hot Water System						
Water Heater						
Demolish	Ea.		81.50		81.50	Includes material and labor to install a solar hot water heat exchanger storage tank, 100 gallon.
Install	Ea.	770	217		987	
Demolish and Install	Ea.	770	298.50		1068.50	
Remove and Install	Ea.		173.44		173.44	
Minimum Charge	Job		164		164	
Solar Panel						
Demolish	S.F.		1.19		1.19	Includes material and labor to install solar panel.
Install	S.F.	18.75	2.14		20.89	
Demolish and Install	S.F.	18.75	3.33		22.08	
Remove and Install	S.F.		1.71		1.71	
Minimum Charge	Job		164		164	
Repair Plumbing						
Install	Job		164		164	Includes 4 hour minimum labor charge for a steamfitter or pipefitter.
Minimum Charge	Job		164		164	
Insulating Wrap						
Demolish	Ea.		3.76		3.76	Includes material and labor to install insulating wrap around a 100 gallon solar hot water heat exchanger storage tank.
Install	Ea.	71	10.15		81.15	
Demolish and Install	Ea.	71	13.91		84.91	
Remove and Install	Ea.		8.13		8.13	
Minimum Charge	Job		163		163	
Water Softener						
Demolish	Ea.		60		60	Includes material and labor to install a 14 GPM water softener.
Install	Ea.	1325	146		1471	
Demolish and Install	Ea.	1325	206		1531	
Remove and Install	Ea.		117.06		117.06	
Clean	Ea.	.96	13.65		14.61	
Minimum Charge	Ea.		163		163	

Ductwork	Unit	Material	Labor	Equip.	Total	Specification
Galvanized, Square or Rectangular						
6″						
Demolish	L.F.		1.52		1.52	Includes material and labor to install rectangular ductwork with a greatest dimension of 6″.
Install	L.F.	2.06	13.70		15.76	
Demolish and Install	L.F.	2.06	15.22		17.28	
Remove and Install	L.F.		10.96		10.96	
Clean	L.F.		1.75		1.75	
Minimum Charge	Job		164		164	
8″						
Demolish	L.F.		1.52		1.52	Includes material and labor to install rectangular ductwork with a greatest dimension of 8″.
Install	L.F.	2.74	13.70		16.44	
Demolish and Install	L.F.	2.74	15.22		17.96	
Remove and Install	L.F.		10.96		10.96	
Clean	L.F.		2.34		2.34	
Minimum Charge	Job		164		164	

Ductwork	Unit	Material	Labor	Equip.	Total	Specification
10"						
Demolish	L.F.		1.52		1.52	Includes material and labor to install rectangular ductwork with a greatest dimension of 10".
Install	L.F.	3.43	13.70		17.13	
Demolish and Install	L.F.	3.43	15.22		18.65	
Remove and Install	L.F.		10.96		10.96	
Clean	L.F.		2.92		2.92	
Minimum Charge	Job		164		164	
Galvanized, Spiral						
4"						
Demolish	L.F.		1.83		1.83	Cost includes material and labor to install 4" diameter spiral pre-formed steel galvanized ductwork.
Install	L.F.	.94	1.64		2.58	
Demolish and Install	L.F.	.94	3.47		4.41	
Remove and Install	L.F.		1.31		1.31	
Clean	L.F.		.92		.92	
Minimum Charge	Job		164		164	
6"						
Demolish	L.F.		1.83		1.83	Cost includes material and labor to install 6" diameter spiral pre-formed steel galvanized ductwork.
Install	L.F.	1.35	2.11		3.46	
Demolish and Install	L.F.	1.35	3.94		5.29	
Remove and Install	L.F.		1.69		1.69	
Clean	L.F.		1.38		1.38	
Minimum Charge	Job		164		164	
8"						
Demolish	L.F.		1.83		1.83	Cost includes material and labor to install 8" diameter spiral pre-formed steel galvanized ductwork.
Install	L.F.	1.78	2.96		4.74	
Demolish and Install	L.F.	1.78	4.79		6.57	
Remove and Install	L.F.		2.37		2.37	
Clean	L.F.		1.84		1.84	
Minimum Charge	Job		164		164	
10"						
Demolish	L.F.		2.28		2.28	Cost includes material and labor to install 10" diameter spiral pre-formed steel galvanized ductwork.
Install	L.F.	2.20	3.70		5.90	
Demolish and Install	L.F.	2.20	5.98		8.18	
Remove and Install	L.F.		2.96		2.96	
Clean	L.F.		2.30		2.30	
Minimum Charge	Job		164		164	
12"						
Demolish	L.F.		3.04		3.04	Cost includes material and labor to install 12" diameter spiral pre-formed steel galvanized ductwork.
Install	L.F.	2.71	4.93		7.64	
Demolish and Install	L.F.	2.71	7.97		10.68	
Remove and Install	L.F.		3.94		3.94	
Clean	L.F.		2.75		2.75	
Minimum Charge	Job		164		164	
16"						
Demolish	L.F.		6.10		6.10	Cost includes material and labor to install 16" diameter spiral pre-formed steel galvanized ductwork.
Install	L.F.	4.66	9.85		14.51	
Demolish and Install	L.F.	4.66	15.95		20.61	
Remove and Install	L.F.		7.89		7.89	
Clean	L.F.		3.67		3.67	
Minimum Charge	Job		164		164	
Flexible Fiberglass						
6" Non-insulated						
Demolish	L.F.		1.41		1.41	Cost includes material and labor to install 6" diameter non-insulated flexible fiberglass fabric on corrosion resistant metal ductwork.
Install	L.F.	1.57	2.11		3.68	
Demolish and Install	L.F.	1.57	3.52		5.09	
Remove and Install	L.F.		1.69		1.69	
Clean	L.F.		1.38		1.38	
Minimum Charge	Job		164		164	

Ductwork

Ductwork	Unit	Material	Labor	Equip.	Total	Specification
8″ Non-insulated						
Demolish	L.F.		2.03		2.03	Cost includes material and labor to install 8″ diameter non-insulated flexible fiberglass fabric on corrosion resistant metal ductwork.
Install	L.F.	2.15	2.96		5.11	
Demolish and Install	L.F.	2.15	4.99		7.14	
Remove and Install	L.F.		2.37		2.37	
Clean	L.F.		1.84		1.84	
Minimum Charge	Job		164		164	
10″ Non-insulated						
Demolish	L.F.		2.61		2.61	Cost includes material and labor to install 10″ diameter non-insulated flexible fiberglass fabric on corrosion resistant metal ductwork.
Install	L.F.	2.72	3.70		6.42	
Demolish and Install	L.F.	2.72	6.31		9.03	
Remove and Install	L.F.		2.96		2.96	
Clean	L.F.		2.30		2.30	
Minimum Charge	Job		164		164	
12″ Non-insulated						
Demolish	L.F.		3.65		3.65	Cost includes material and labor to install 12″ diameter non-insulated flexible fiberglass fabric on corrosion resistant metal ductwork.
Install	L.F.	3.28	4.93		8.21	
Demolish and Install	L.F.	3.28	8.58		11.86	
Remove and Install	L.F.		3.94		3.94	
Clean	L.F.		2.75		2.75	
Minimum Charge	Job		164		164	
4″ Insulated						
Demolish	L.F.		1.07		1.07	Cost includes material and labor to install 4″ diameter insulated flexible fiberglass fabric on corrosion resistant metal ductwork.
Install	L.F.	1.28	1.74		3.02	
Demolish and Install	L.F.	1.28	2.81		4.09	
Remove and Install	L.F.		1.39		1.39	
Clean	L.F.		.92		.92	
Minimum Charge	Job		164		164	
6″ Insulated						
Demolish	L.F.		1.41		1.41	Cost includes material and labor to install 6″ diameter insulated flexible fiberglass fabric on corrosion resistant metal ductwork.
Install	L.F.	1.62	2.28		3.90	
Demolish and Install	L.F.	1.62	3.69		5.31	
Remove and Install	L.F.		1.82		1.82	
Clean	L.F.		1.38		1.38	
Minimum Charge	Job		164		164	
8″ Insulated						
Demolish	L.F.		2.03		2.03	Cost includes material and labor to install 8″ diameter insulated flexible fiberglass fabric on corrosion resistant metal ductwork.
Install	L.F.	2.02	3.29		5.31	
Demolish and Install	L.F.	2.02	5.32		7.34	
Remove and Install	L.F.		2.63		2.63	
Clean	L.F.		1.84		1.84	
Minimum Charge	Job		164		164	
10″ Insulated						
Demolish	L.F.		2.61		2.61	Cost includes material and labor to install 10″ diameter insulated flexible fiberglass fabric on corrosion resistant metal ductwork.
Install	L.F.	2.49	4.23		6.72	
Demolish and Install	L.F.	2.49	6.84		9.33	
Remove and Install	L.F.		3.38		3.38	
Clean	L.F.		2.30		2.30	
Minimum Charge	Job		164		164	
12″ Insulated						
Demolish	L.F.		3.65		3.65	Cost includes material and labor to install 12″ diameter insulated flexible fiberglass fabric on corrosion resistant metal ductwork.
Install	L.F.	3.06	5.90		8.96	
Demolish and Install	L.F.	3.06	9.55		12.61	
Remove and Install	L.F.		4.73		4.73	
Clean	L.F.		2.75		2.75	
Minimum Charge	Job		164		164	

Heating	Unit	Material	Labor	Equip.	Total	Specification
Hot Water Baseboard						
Demolish	L.F.		5.60		5.60	Includes material and labor to install baseboard radiation.
Install	L.F.	5.90	9.80		15.70	
Demolish and Install	L.F.	5.90	15.40		21.30	
Remove and Install	L.F.		7.86		7.86	
Clean	L.F.	.96	3.41		4.37	
Minimum Charge	Job		163		163	
3'						
Demolish	Ea.		18.05		18.05	Includes material and labor to install fin tube baseboard radiation in 3' sections.
Install	Ea.	18.60	32.50		51.10	
Demolish and Install	Ea.	18.60	50.55		69.15	
Remove and Install	Ea.		26.19		26.19	
Clean	Ea.	.28	10.40		10.68	
Minimum Charge	Job		163		163	
4'						
Demolish	Ea.		24		24	Includes material and labor to install fin tube baseboard radiation in 4' sections.
Install	Ea.	25	37		62	
Demolish and Install	Ea.	25	61		86	
Remove and Install	Ea.		29.46		29.46	
Clean	Ea.	.55	13.65		14.20	
Minimum Charge	Job		163		163	
5'						
Demolish	Ea.		30		30	Includes material and labor to install fin tube baseboard radiation in 5' sections.
Install	Ea.	31	42		73	
Demolish and Install	Ea.	31	72		103	
Remove and Install	Ea.		33.67		33.67	
Clean	Ea.	.55	16.75		17.30	
Minimum Charge	Job		163		163	
6'						
Demolish	Ea.		36		36	Includes material and labor to install fin tube baseboard radiation in 6' sections.
Install	Ea.	37.50	49		86.50	
Demolish and Install	Ea.	37.50	85		122.50	
Remove and Install	Ea.		39.29		39.29	
Clean	Ea.	.55	19.80		20.35	
Minimum Charge	Job		163		163	
7'						
Demolish	Ea.		40		40	Includes material and labor to install fin tube baseboard radiation in 7' sections.
Install	Ea.	43.50	65		108.50	
Demolish and Install	Ea.	43.50	105		148.50	
Remove and Install	Ea.		52.03		52.03	
Clean	Ea.	.68	24		24.68	
Minimum Charge	Job		163		163	
9'						
Demolish	Ea.		60		60	Includes material and labor to install fin tube baseboard radiation in 9' sections.
Install	Ea.	56	93		149	
Demolish and Install	Ea.	56	153		209	
Remove and Install	Ea.		74.33		74.33	
Clean	Ea.	1.09	31		32.09	
Minimum Charge	Job		163		163	

Heating	Unit	Material	Labor	Equip.	Total	Specification
50 MBH						
Demolish	Ea.		29.50		29.50	Includes material and labor to install a cabinet space heater with an output of 50 MBH.
Install	Ea.	580	84		664	
Demolish and Install	Ea.	580	113.50		693.50	
Remove and Install	Ea.		67.35		67.35	
Clean	Ea.	.68	27.50		28.18	
Minimum Charge	Job		164		164	

Heating	Unit	Material	Labor	Equip.	Total	Specification
75 MBH						
Demolish	Ea.		29.50		29.50	Includes material and labor to install a cabinet space heater with an output of 75 MBH.
Install	Ea.	645	98		743	
Demolish and Install	Ea.	645	127.50		772.50	
Remove and Install	Ea.		78.57		78.57	
Clean	Ea.	.68	27.50		28.18	
Minimum Charge	Job		164		164	
125 MBH						
Demolish	Ea.		35.50		35.50	Includes material and labor to install a cabinet space heater with an output of 125 MBH.
Install	Ea.	855	118		973	
Demolish and Install	Ea.	855	153.50		1008.50	
Remove and Install	Ea.		94.28		94.28	
Clean	Ea.	.68	27.50		28.18	
Minimum Charge	Job		164		164	
175 MBH						
Demolish	Ea.		44		44	Includes material and labor to install a cabinet space heater with an output of 175 MBH.
Install	Ea.	1075	168		1243	
Demolish and Install	Ea.	1075	212		1287	
Remove and Install	Ea.		134.69		134.69	
Clean	Ea.	.68	27.50		28.18	
Minimum Charge	Job		164		164	

Furnace Unit	Unit	Material	Labor	Equip.	Total	Specification
Wall (gas fired)						
25 MBH w / Thermostat						
Demolish	Ea.		74		74	Includes material and labor to install a gas fired hot air furnace rated at 25 MBH.
Install	Ea.	395	118		513	
Demolish and Install	Ea.	395	192		587	
Remove and Install	Ea.		94.67		94.67	
Clean	Ea.	1.36	60.50		61.86	
Minimum Charge	Job		164		164	
35 MBH w / Thermostat						
Demolish	Ea.		74		74	Includes material and labor to install a gas fired hot air furnace rated at 35 MBH.
Install	Ea.	475	131		606	
Demolish and Install	Ea.	475	205		680	
Remove and Install	Ea.		105.19		105.19	
Clean	Ea.	1.36	60.50		61.86	
Minimum Charge	Job		164		164	
50 MBH w / Thermostat						
Demolish	Ea.		74		74	Includes material and labor to install a gas fired hot air furnace rated at 50 MBH.
Install	Ea.	775	148		923	
Demolish and Install	Ea.	775	222		997	
Remove and Install	Ea.		118.34		118.34	
Clean	Ea.	1.36	60.50		61.86	
Minimum Charge	Job		164		164	
60 MBH w / Thermostat						
Demolish	Ea.		74		74	Includes material and labor to install a gas fired hot air furnace rated at 60 MBH.
Install	Ea.	805	164		969	
Demolish and Install	Ea.	805	238		1043	
Remove and Install	Ea.		131.48		131.48	
Clean	Ea.	1.36	60.50		61.86	
Minimum Charge	Job		164		164	

Furnace Unit

Furnace Unit	Unit	Material	Labor	Equip.	Total	Specification
Floor (gas fired)						
30 MBH w / Thermostat						
Demolish	Ea.		74		74	Includes material and labor to install a gas fired hot air floor furnace rated at 30 MBH.
Install	Ea.	845	118		963	
Demolish and Install	Ea.	845	192		1037	
Remove and Install	Ea.		94.67		94.67	
Clean	Ea.	1.36	60.50		61.86	
Minimum Charge	Job		164		164	
50 MBH w / Thermostat						
Demolish	Ea.		74		74	Includes material and labor to install a gas fired hot air floor furnace rated at 50 MBH.
Install	Ea.	1075	148		1223	
Demolish and Install	Ea.	1075	222		1297	
Remove and Install	Ea.		118.34		118.34	
Clean	Ea.	1.36	60.50		61.86	
Minimum Charge	Job		164		164	
Gas Forced Air						
50 MBH						
Demolish	Ea.		74		74	Includes material and labor to install an up flow gas fired hot air furnace rated at 50 MBH.
Install	Ea.	755	148		903	
Demolish and Install	Ea.	755	222		977	
Remove and Install	Ea.		118.34		118.34	
Clean	Ea.	1.36	60.50		61.86	
Minimum Charge	Job		164		164	
80 MBH						
Demolish	Ea.		74		74	Includes material and labor to install an up flow gas fired hot air furnace rated at 80 MBH.
Install	Ea.	830	164		994	
Demolish and Install	Ea.	830	238		1068	
Remove and Install	Ea.		131.48		131.48	
Clean	Ea.	1.36	60.50		61.86	
Minimum Charge	Job		164		164	
100 MBH						
Demolish	Ea.		74		74	Includes material and labor to install an up flow gas fired hot air furnace rated at 100 MBH.
Install	Ea.	950	185		1135	
Demolish and Install	Ea.	950	259		1209	
Remove and Install	Ea.		147.92		147.92	
Clean	Ea.	1.36	60.50		61.86	
Minimum Charge	Job		164		164	
120 MBH						
Demolish	Ea.		74		74	Includes material and labor to install an up flow gas fired hot air furnace rated at 120 MBH.
Install	Ea.	1100	197		1297	
Demolish and Install	Ea.	1100	271		1371	
Remove and Install	Ea.		157.78		157.78	
Clean	Ea.	1.36	60.50		61.86	
Minimum Charge	Job		164		164	
Electric Forced Air						
31.4 MBH						
Demolish	Ea.		74		74	Includes material and labor to install an electric hot air furnace rated at 31.4 MBH.
Install	Ea.	765	168		933	
Demolish and Install	Ea.	765	242		1007	
Remove and Install	Ea.		134.44		134.44	
Clean	Ea.	1.36	60.50		61.86	
Minimum Charge	Job		164		164	

Furnace Unit	Unit	Material	Labor	Equip.	Total	Specification
47.1 MBH						
Demolish	Ea.		74		74	Includes material and labor to install an electric hot air furnace rated at 47.1 MBH.
Install	Ea.	860	176		1036	
Demolish and Install	Ea.	860	250		1110	
Remove and Install	Ea.		140.69		140.69	
Clean	Ea.	1.36	60.50		61.86	
Minimum Charge	Job		164		164	
62.9 MBH						
Demolish	Ea.		74		74	Includes material and labor to install an electric hot air furnace rated at 62.9 MBH.
Install	Ea.	1000	189		1189	
Demolish and Install	Ea.	1000	263		1263	
Remove and Install	Ea.		151.24		151.24	
Clean	Ea.	1.36	60.50		61.86	
Minimum Charge	Job		164		164	
78.5 MBH						
Demolish	Ea.		74		74	Includes material and labor to install an electric hot air furnace rated at 78.5 MBH.
Install	Ea.	1050	194		1244	
Demolish and Install	Ea.	1050	268		1318	
Remove and Install	Ea.		155.12		155.12	
Clean	Ea.	1.36	60.50		61.86	
Minimum Charge	Job		164		164	
110 MBH						
Demolish	Ea.		74		74	Includes material and labor to install an electric hot air furnace rated at 110 MBH.
Install	Ea.	1600	204		1804	
Demolish and Install	Ea.	1600	278		1878	
Remove and Install	Ea.		163.50		163.50	
Clean	Ea.	1.36	60.50		61.86	
Minimum Charge	Job		164		164	
A/C Junction Box						
Install	Ea.	47	82.50		129.50	Includes labor and materials install a heating system emergency shut-off switch. Cost includes up to 40 feet of non-metallic cable.
Minimum Charge	Job		164		164	

Boiler	Unit	Material	Labor	Equip.	Total	Specification
Gas Fired						
60 MBH						
Demolish	Ea.		315		315	Includes material and labor to install a gas fired cast iron boiler rated at 60 MBH.
Install	Ea.	1450	1075		2525	
Demolish and Install	Ea.	1450	1390		2840	
Remove and Install	Ea.		857.13		857.13	
Clean	Ea.	2.73	36.50		39.23	
Minimum Charge	Job		164		164	
Gas Fired						
125 MBH						
Demolish	Ea.		475		475	Includes material and labor to install a gas fired cast iron boiler rated at 125 MBH.
Install	Ea.	2550	1300		3850	
Demolish and Install	Ea.	2550	1775		4325	
Remove and Install	Ea.		1047.61		1047.61	
Clean	Ea.	2.73	36.50		39.23	
Minimum Charge	Job		164		164	

Boiler

Boiler	Unit	Material	Labor	Equip.	Total	Specification
Gas Fired						
400 MBH						
Demolish	Ea.		475		475	Includes material and labor to install a gas fired cast iron boiler rated at 400 MBH.
Install	Ea.	3400	1850		5250	
Demolish and Install	Ea.	3400	2325		5725	
Remove and Install	Ea.		1841.50		1841.50	
Clean	Ea.	2.73	36.50		39.23	
Minimum Charge	Job		164		164	

Heat Pump

Heat Pump	Unit	Material	Labor	Equip.	Total	Specification
2 Ton						
Demolish	Ea.		77		77	Includes material and labor to install a 2 ton heat pump.
Install	Ea.	1400	490		1890	
Demolish and Install	Ea.	1400	567		1967	
Remove and Install	Ea.		392.85		392.85	
Clean	Ea.	2.73	27.50		30.23	
Minimum Charge	Job		164		164	
3 Ton						
Demolish	Ea.		146		146	Includes material and labor to install a 3 ton heat pump.
Install	Ea.	1725	735		2460	
Demolish and Install	Ea.	1725	881		2606	
Remove and Install	Ea.		589.28		589.28	
Clean	Ea.	2.73	27.50		30.23	
Minimum Charge	Job		164		164	
4 Ton						
Demolish	Ea.		266		266	Includes material and labor to install a 4 ton heat pump.
Install	Ea.	2200	980		3180	
Demolish and Install	Ea.	2200	1246		3446	
Remove and Install	Ea.		785.71		785.71	
Clean	Ea.	2.73	27.50		30.23	
Minimum Charge	Job		164		164	
5 Ton						
Demolish	Ea.		390		390	Includes material and labor to install a 5 ton heat pump.
Install	Ea.	2525	1175		3700	
Demolish and Install	Ea.	2525	1565		4090	
Remove and Install	Ea.		942.85		942.85	
Clean	Ea.	2.73	27.50		30.23	
Minimum Charge	Job		164		164	
Thermostat						
Demolish	Ea.		9.10		9.10	Includes material and labor to install an electric thermostat, 2 set back.
Install	Ea.	122	41		163	
Demolish and Install	Ea.	122	50.10		172.10	
Remove and Install	Ea.		32.88		32.88	
Clean	Ea.	.06	6.80		6.86	
Minimum Charge	Job		164		164	
A/C Junction Box						
Install	Ea.	360	253		613	Includes labor and materials to install a heat pump circuit including up to 40 feet of non-metallic cable.
Minimum Charge	Job		165		165	

Heating

Heating	Unit	Material	Labor	Equip.	Total	Specification
Electric Baseboard						
Demolish	L.F.		4.57		4.57	Includes material and labor to install electric baseboard heating elements.
Install	L.F.	11.65	10.95		22.60	
Demolish and Install	L.F.	11.65	15.52		27.17	
Remove and Install	L.F.		8.78		8.78	
Clean	L.F.	.96	3.41		4.37	
Minimum Charge	Job		165		165	
2′ 6″						
Demolish	Ea.		15.25		15.25	Includes material and labor to install a 2′ 6″ long section of electric baseboard heating elements.
Install	Ea.	41.50	41		82.50	
Demolish and Install	Ea.	41.50	56.25		97.75	
Remove and Install	Ea.		32.92		32.92	
Clean	Ea.	.28	8.40		8.68	
Minimum Charge	Job		165		165	
3′						
Demolish	Ea.		16.65		16.65	Includes material and labor to install a 3′ long section of electric baseboard heating elements.
Install	Ea.	46	41		87	
Demolish and Install	Ea.	46	57.65		103.65	
Remove and Install	Ea.		32.92		32.92	
Clean	Ea.	.28	10.40		10.68	
Minimum Charge	Job		165		165	
4′						
Demolish	Ea.		18.30		18.30	Includes material and labor to install a 4′ long section of electric baseboard heating elements.
Install	Ea.	51.50	49		100.50	
Demolish and Install	Ea.	51.50	67.30		118.80	
Remove and Install	Ea.		39.31		39.31	
Clean	Ea.	.11	2.27		2.38	
Minimum Charge	Job		165		165	
5′						
Demolish	Ea.		20.50		20.50	Includes material and labor to install a 5′ long section of electric baseboard heating elements.
Install	Ea.	58.50	58		116.50	
Demolish and Install	Ea.	58.50	78.50		137	
Remove and Install	Ea.		46.20		46.20	
Clean	Ea.	.55	16.75		17.30	
Minimum Charge	Job		165		165	
6′						
Demolish	Ea.		23		23	Includes material and labor to install a 6′ long section of electric baseboard heating elements.
Install	Ea.	63.50	66		129.50	
Demolish and Install	Ea.	63.50	89		152.50	
Remove and Install	Ea.		52.67		52.67	
Clean	Ea.	.11	2.73		2.84	
Minimum Charge	Job		165		165	
8′						
Demolish	Ea.		30.50		30.50	Includes material and labor to install an 8′ long section of electric baseboard heating elements.
Install	Ea.	86.50	82.50		169	
Demolish and Install	Ea.	86.50	113		199.50	
Remove and Install	Ea.		65.84		65.84	
Clean	Ea.	.11	3.41		3.52	
Minimum Charge	Job		165		165	
10′						
Demolish	Ea.		36.50		36.50	Includes material and labor to install a 10′ long section of electric baseboard heating elements.
Install	Ea.	108	100		208	
Demolish and Install	Ea.	108	136.50		244.50	
Remove and Install	Ea.		79.81		79.81	
Clean	Ea.	1.09	34		35.09	
Minimum Charge	Job		165		165	

Rough Mechanical

Heating

	Unit	Material	Labor	Equip.	Total	Specification
Electric Radiant Wall Panel						
Demolish	Ea.		24.50		24.50	Includes material and labor to install an 18" x 23" electric radiant heat wall panel.
Install	Ea.	390	66		456	
Demolish and Install	Ea.	390	90.50		480.50	
Remove and Install	Ea.		52.67		52.67	
Clean	Ea.	.13	6.80		6.93	
Minimum Charge	Job		165		165	

Heating Control

	Unit	Material	Labor	Equip.	Total	Specification
Thermostat						
Demolish	Ea.		9.10		9.10	Includes material and labor to install an electric thermostat, 2 set back.
Install	Ea.	122	41		163	
Demolish and Install	Ea.	122	50.10		172.10	
Remove and Install	Ea.		32.88		32.88	
Clean	Ea.	.06	6.80		6.86	
Minimum Charge	Job		164		164	

Air Conditioning

	Unit	Material	Labor	Equip.	Total	Specification
2 Ton						
Demolish	Ea.		300		300	Includes material and labor to install a rooftop air conditioning unit, 2 ton. No hoisting is included.
Install	Ea.	1300	510		1810	
Demolish and Install	Ea.	1300	810		2110	
Remove and Install	Ea.		508.44		508.44	
Clean	Ea.	2.73	27.50		30.23	
Minimum Charge	Job		164		164	
3 Ton						
Demolish	Ea.		300		300	Includes material and labor to install a rooftop air conditioning unit, 3 ton. No hoisting is included.
Install	Ea.	1375	670		2045	
Demolish and Install	Ea.	1375	970		2345	
Remove and Install	Ea.		670.40		670.40	
Clean	Ea.	2.73	27.50		30.23	
Minimum Charge	Job		164		164	
4 Ton						
Demolish	Ea.		300		300	Includes material and labor to install a rooftop air conditioning unit, 4 ton. No hoisting is included.
Install	Ea.	1750	775		2525	
Demolish and Install	Ea.	1750	1075		2825	
Remove and Install	Ea.		777.41		777.41	
Clean	Ea.	2.73	27.50		30.23	
Minimum Charge	Job		164		164	
5 Ton						
Demolish	Ea.		300		300	Includes material and labor to install a rooftop air conditioning unit, 5 ton. No hoisting is included.
Install	Ea.	1975	840		2815	
Demolish and Install	Ea.	1975	1140		3115	
Remove and Install	Ea.		839.43		839.43	
Clean	Ea.	2.73	27.50		30.23	
Minimum Charge	Job		164		164	
Compressor						
2 Ton						
Demolish	Ea.		273		273	Includes material and labor to install a refrigeration compressor.
Install	Ea.	525	164		689	
Demolish and Install	Ea.	525	437		962	
Remove and Install	Ea.		130.88		130.88	
Clean	Ea.	2.73	27.50		30.23	
Minimum Charge	Job		164		164	

Air Conditioning

	Unit	Material	Labor	Equip.	Total	Specification
2.5 Ton						
Demolish	Ea.		273		273	Includes material and labor to install a refrigeration compressor.
Install	Ea.	400	177		577	
Demolish and Install	Ea.	400	450		850	
Remove and Install	Ea.		141.80		141.80	
Clean	Ea.	2.73	27.50		30.23	
Minimum Charge	Job		164		164	
3 Ton						
Demolish	Ea.		273		273	Includes material and labor to install a refrigeration compressor.
Install	Ea.	470	185		655	
Demolish and Install	Ea.	470	458		928	
Remove and Install	Ea.		147.97		147.97	
Clean	Ea.	2.73	27.50		30.23	
Minimum Charge	Job		164		164	
4 Ton						
Demolish	Ea.		273		273	Includes material and labor to install a refrigeration compressor.
Install	Ea.	665	149		814	
Demolish and Install	Ea.	665	422		1087	
Remove and Install	Ea.		118.98		118.98	
Clean	Ea.	2.73	27.50		30.23	
Minimum Charge	Job		164		164	
5 Ton						
Demolish	Ea.		273		273	Includes material and labor to install a refrigeration compressor.
Install	Ea.	820	203		1023	
Demolish and Install	Ea.	820	476		1296	
Remove and Install	Ea.		162.08		162.08	
Clean	Ea.	2.73	27.50		30.23	
Minimum Charge	Job		164		164	
Condensing Unit						
2 Ton Air-cooled						
Demolish	Ea.		78		78	Includes material and labor to install a 2 ton condensing unit.
Install	Ea.	920	281		1201	
Demolish and Install	Ea.	920	359		1279	
Remove and Install	Ea.		224.49		224.49	
Clean	Ea.	2.73	27.50		30.23	
Minimum Charge	Job		164		164	
2.5 Ton Air-cooled						
Demolish	Ea.		91		91	Includes material and labor to install a 2.5 ton condensing unit.
Install	Ea.	920	345		1265	
Demolish and Install	Ea.	920	436		1356	
Remove and Install	Ea.		277.31		277.31	
Clean	Ea.	2.73	27.50		30.23	
Minimum Charge	Job		164		164	
3 Ton Air-cooled						
Demolish	Ea.		109		109	Includes material and labor to install a 3 ton condensing unit.
Install	Ea.	1100	455		1555	
Demolish and Install	Ea.	1100	564		1664	
Remove and Install	Ea.		362.63		362.63	
Clean	Ea.	2.73	27.50		30.23	
Minimum Charge	Job		164		164	
4 Ton Air-cooled						
Demolish	Ea.		164		164	Includes material and labor to install a 4 ton condensing unit.
Install	Ea.	1400	655		2055	
Demolish and Install	Ea.	1400	819		2219	
Remove and Install	Ea.		523.80		523.80	
Clean	Ea.	2.73	27.50		30.23	
Minimum Charge	Job		164		164	

Air Conditioning	Unit	Material	Labor	Equip.	Total	Specification
5 Ton Air-cooled						
Demolish	Ea.		246		246	Includes material and labor to install a 5 ton condensing unit.
Install	Ea.	1700	980		2680	
Demolish and Install	Ea.	1700	1226		2926	
Remove and Install	Ea.		785.71		785.71	
Clean	Ea.	2.73	27.50		30.23	
Minimum Charge	Job		164		164	
Air Handler						
1200 CFM						
Demolish	Ea.		197		197	Includes material and labor to install a 1200 CFM capacity air handling unit.
Install	Ea.	1975	455		2430	
Demolish and Install	Ea.	1975	652		2627	
Remove and Install	Ea.		362.63		362.63	
Clean	Ea.	2.73	27.50		30.23	
Minimum Charge	Job		164		164	
2000 CFM						
Demolish	Ea.		197		197	Includes material and labor to install a 2000 CFM capacity air handling unit.
Install	Ea.	3300	535		3835	
Demolish and Install	Ea.	3300	732		4032	
Remove and Install	Ea.		428.57		428.57	
Clean	Ea.	2.73	27.50		30.23	
Minimum Charge	Job		164		164	
3000 CFM						
Demolish	Ea.		218		218	Includes material and labor to install a 3000 CFM capacity air handling unit.
Install	Ea.	4125	590		4715	
Demolish and Install	Ea.	4125	808		4933	
Remove and Install	Ea.		471.42		471.42	
Clean	Ea.	2.73	27.50		30.23	
Minimum Charge	Job		164		164	
Evaporative Cooler						
4,300 CFM, 1/3 HP						
Demolish	Ea.		80.50		80.50	Includes material and labor to install an evaporative cooler.
Install	Ea.	625	185		810	
Demolish and Install	Ea.	625	265.50		890.50	
Remove and Install	Ea.		147.92		147.92	
Clean	Ea.	1.09	27.50		28.59	
Minimum Charge	Job		164		164	
4,700 CFM, 1/2 HP						
Demolish	Ea.		80.50		80.50	Includes material and labor to install an evaporative cooler.
Install	Ea.	595	191		786	
Demolish and Install	Ea.	595	271.50		866.50	
Remove and Install	Ea.		152.69		152.69	
Clean	Ea.	1.09	27.50		28.59	
Minimum Charge	Job		164		164	
5,600 CFM, 3/4 HP						
Demolish	Ea.		80.50		80.50	Includes material and labor to install an evaporative cooler.
Install	Ea.	1025	197		1222	
Demolish and Install	Ea.	1025	277.50		1302.50	
Remove and Install	Ea.		157.78		157.78	
Clean	Ea.	1.09	27.50		28.59	
Minimum Charge	Job		164		164	
Master Cool Brand						
Demolish	Ea.		80.50		80.50	Includes material and labor to install an evaporative cooler.
Install	Ea.	625	204		829	
Demolish and Install	Ea.	625	284.50		909.50	
Remove and Install	Ea.		163.22		163.22	
Clean	Ea.	1.09	27.50		28.59	
Minimum Charge	Job		164		164	

Air Conditioning

Air Conditioning	Unit	Material	Labor	Equip.	Total	Specification
Motor						
Demolish	Ea.		45		45	Includes material and labor to install an evaporative cooler motor.
Install	Ea.	212	144		356	
Demolish and Install	Ea.	212	189		401	
Remove and Install	Ea.		115.07		115.07	
Clean	Ea.	1.09	27.50		28.59	
Minimum Charge	Job		164		164	
Water Pump						
Demolish	Ea.		45		45	Includes material and labor to install an evaporative cooler water pump.
Install	Ea.	39.50	82		121.50	
Demolish and Install	Ea.	39.50	127		166.50	
Remove and Install	Ea.		65.76		65.76	
Clean	Ea.	1.09	27.50		28.59	
Minimum Charge	Job		164		164	
Fan Belt						
Demolish	Ea.		9.05		9.05	Includes material and labor to install an evaporative cooler fan belt.
Install	Ea.	26.50	16.45		42.95	
Demolish and Install	Ea.	26.50	25.50		52	
Remove and Install	Ea.		13.15		13.15	
Minimum Charge	Job		164		164	
Pads (4 ea.)						
Demolish	Ea.		9.05		9.05	Includes material and labor to install pre-fabricated roof pads for an evaporative cooler.
Install	Ea.	29.50	41		70.50	
Demolish and Install	Ea.	29.50	50.05		79.55	
Remove and Install	Ea.		32.88		32.88	
Clean	Ea.	1.09	27.50		28.59	
Minimum Charge	Job		164		164	
Service call						
Install	Job		164		164	Includes 4 hour minimum labor charge for an sheetmetal worker.
Window Mounted						
Low Output BTU						
Install	Ea.	380	37.50		417.50	Includes material and labor to install a window air conditioning unit.
Remove and Install	Ea.		30.08		30.08	
Clean	Ea.	.41	6.80		7.21	
Minimum Charge	Job		164		164	
Mid-range Output BTU						
Demolish	Ea.		26.50		26.50	Includes material and labor to install a window air conditioning unit.
Install	Ea.	440	50		490	
Demolish and Install	Ea.	440	76.50		516.50	
Remove and Install	Ea.		40.11		40.11	
Clean	Ea.	.41	6.80		7.21	
Minimum Charge	Job		164		164	
High Output BTU						
Demolish	Ea.		33.50		33.50	Includes material and labor to install a window air conditioning unit.
Install	Ea.	515	50		565	
Demolish and Install	Ea.	515	83.50		598.50	
Remove and Install	Ea.		40.11		40.11	
Clean	Ea.	.41	6.80		7.21	
Minimum Charge	Job		164		164	
Hotel-type Hot / Cold						
Demolish	Ea.		55.50		55.50	Includes material and labor to install packaged air conditioning unit, 15000 BTU cooling, 13900 BTU heating.
Install	Ea.	1200	196		1396	
Demolish and Install	Ea.	1200	251.50		1451.50	
Remove and Install	Ea.		157.14		157.14	
Clean	Ea.	.55	13.65		14.20	
Minimum Charge	Job		164		164	

Air Conditioning	Unit	Material	Labor	Equip.	Total	Specification
Thermostat						
Demolish	Ea.		9.10		9.10	Includes material and labor to install an electric thermostat, 2 set back.
Install	Ea.	122	41		163	
Demolish and Install	Ea.	122	50.10		172.10	
Remove and Install	Ea.		32.88		32.88	
Clean	Ea.	.06	6.80		6.86	
Minimum Charge	Job		164		164	
Concrete Pad						
Demolish	S.F.		1.51		1.51	Includes material and labor to install air conditioner pad.
Install	S.F.	.95	3.43		4.38	
Demolish and Install	S.F.	.95	4.94		5.89	
Remove and Install	S.F.		2.74		2.74	
Clean	S.F.	.06	.44		.50	
Paint	S.F.	.12	.27		.39	
Minimum Charge	Job		164		164	
A/C Junction Box						
Install	Ea.	153	94		247	Includes labor and materials to install an air conditioning circuit including up to 40 feet of non-metallic cable.
Minimum Charge	Job		165		165	

Sprinkler System	Unit	Material	Labor	Equip.	Total	Specification
Exposed						
Wet Type						
Demolish	SF Flr.		.14		.14	Includes material and labor to install an exposed wet pipe fire sprinkler system.
Install	SF Flr.	.84	1.44		2.28	
Demolish and Install	SF Flr.	.84	1.58		2.42	
Remove and Install	SF Flr.		1.44		1.44	
Minimum Charge	Job		163		163	
Dry Type						
Demolish	SF Flr.		.14		.14	Includes material and labor to install an exposed dry pipe fire sprinkler system.
Install	SF Flr.	.96	1.44		2.40	
Demolish and Install	SF Flr.	.96	1.58		2.54	
Remove and Install	SF Flr.		1.44		1.44	
Minimum Charge	Job		163		163	
Concealed						
Wet Type						
Demolish	SF Flr.		.29		.29	Includes material and labor to install a concealed wet pipe fire sprinkler system.
Install	SF Flr.	.84	1.97		2.81	
Demolish and Install	SF Flr.	.84	2.26		3.10	
Remove and Install	SF Flr.		1.97		1.97	
Minimum Charge	Job		163		163	
Dry Type						
Demolish	SF Flr.		.29		.29	Includes material and labor to install a concealed dry pipe fire sprinkler system.
Install	SF Flr.	.96	1.97		2.93	
Demolish and Install	SF Flr.	.96	2.26		3.22	
Remove and Install	SF Flr.		1.97		1.97	
Minimum Charge	Job		163		163	

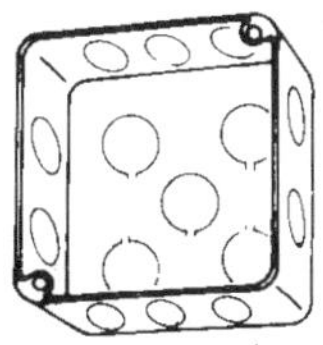

Junction Boxes

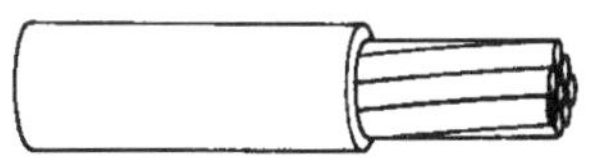

Cable

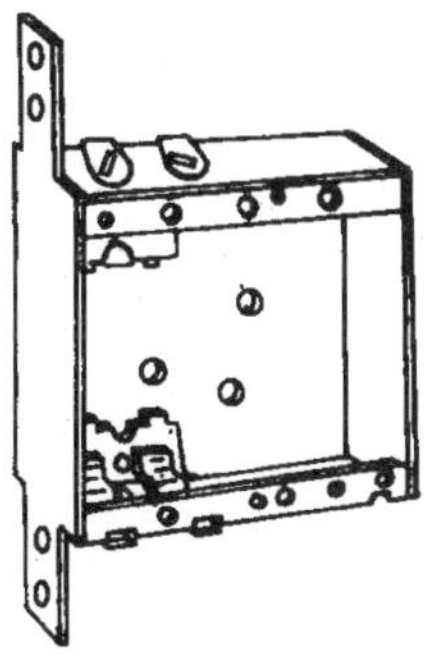

Switch Box

Electrical Per S.F.	Unit	Material	Labor	Equip.	Total	Specification
Residential						
Wiring, Lighting and Fixtures						
Install	S.F.	1.18	1.37		2.55	Includes labor and material to replace house wiring, lighting fixtures, and built-in appliances. Apply S.F. cost to floor area including garages but not basements.
Minimum Charge	Job		165		165	
Wiring Only						
Install	S.F.	.11	.54		.65	Includes labor and material to replace house wiring only. Apply S.F. cost to floor area including garages but not basements.

Main Service	Unit	Material	Labor	Equip.	Total	Specification
100 Amp.						
Demolish	Ea.		135		135	Cost includes material and labor to install 100 Amp. residential service panel including 12 breaker interior panel, 10 single pole breakers and hook-up. Wiring is not included.
Install	Ea.	450	277		727	
Demolish and Install	Ea.	450	412		862	
Remove and Install	Ea.		221.31		221.31	
Clean	Ea.	.28	54.50		54.78	
Minimum Charge	Job		165		165	
150 Amp.						
Demolish	Ea.		135		135	Cost includes material and labor to install 150 Amp. residential service, incl. 20 breaker exterior panel, main switch, 1 GFI and 14 single pole breakers, meter socket including hookup. Wiring is not included.
Install	Ea.	700	320		1020	
Demolish and Install	Ea.	700	455		1155	
Remove and Install	Ea.		255.69		255.69	
Clean	Ea.	.28	54.50		54.78	
Minimum Charge	Job		165		165	
200 Amp.						
Demolish	Ea.		135		135	Cost includes material and labor to install 200 Amp. residential service, incl. 20 breaker exterior panel, main switch, 1 GFI and 18 single pole breakers, meter socket including hookup. Wiring is not included.
Install	Ea.	920	365		1285	
Demolish and Install	Ea.	920	500		1420	
Remove and Install	Ea.		292.62		292.62	
Clean	Ea.	.28	54.50		54.78	
Minimum Charge	Job		165		165	

Main Service	Unit	Material	Labor	Equip.	Total	Specification
Masthead						
Demolish	Ea.		23		23	Includes material and labor to install galvanized steel conduit including masthead.
Install	Ea.	62.50	137		199.50	
Demolish and Install	Ea.	62.50	160		222.50	
Remove and Install	Ea.		109.73		109.73	
Minimum Charge	Job		165		165	
Allowance to Repair						
Install	Job		165		165	Minimum labor charge for residential wiring.
Minimum Charge	Job		165		165	

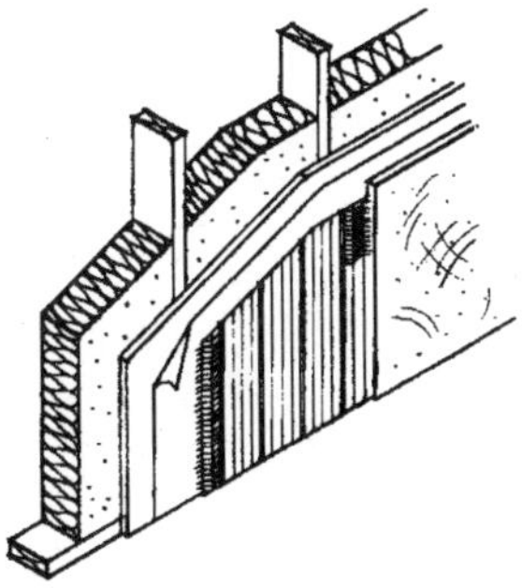

Wall Insulation

Floor Insulation		Unit	Material	Labor	Equip.	Total	Specification
Kraft Faced Batt							
3-1/2" (R-11)							
	Demolish	S.F.		.10		.10	Cost includes material and labor to install 3-1/2" (R-11) kraft paper or foil faced roll and batt fiberglass insulation, joists 16" O.C.
	Install	S.F.	.31	.43		.74	
	Demolish and Install	S.F.	.31	.53		.84	
	Minimum Charge	Job		150		150	
6" (R-19)							
	Demolish	S.F.		.10		.10	Cost includes material and labor to install 6" (R-19) kraft paper or foil faced roll and batt fiberglass insulation, joists 16" O.C.
	Install	S.F.	.42	.50		.92	
	Demolish and Install	S.F.	.42	.60		1.02	
	Minimum Charge	Job		150		150	
10" (R-30)							
	Demolish	S.F.		.10		.10	Cost includes material and labor to install 10" (R-30) kraft paper-faced roll and batt fiberglass insulation, joists 16" O.C.
	Install	S.F.	.75	.60		1.35	
	Demolish and Install	S.F.	.75	.70		1.45	
	Minimum Charge	Job		150		150	
12" (R-38)							
	Demolish	S.F.		.10		.10	Cost includes material and labor to install 12" (R-38) kraft paper-faced roll and batt fiberglass insulation, joists 16" O.C.
	Install	S.F.	.80	.63		1.43	
	Demolish and Install	S.F.	.80	.73		1.53	
	Minimum Charge	Job		150		150	
Foil Faced Batt							
3-1/2" (R-11)							
	Demolish	S.F.		.10		.10	Cost includes material and labor to install 3-1/2" (R-11) kraft paper or foil faced roll and batt fiberglass insulation, joists 16" O.C.
	Install	S.F.	.31	.43		.74	
	Demolish and Install	S.F.	.31	.53		.84	
	Minimum Charge	Job		150		150	
6" (R-19)							
	Demolish	S.F.		.10		.10	Cost includes material and labor to install 6" (R-19) kraft paper or foil faced roll and batt fiberglass insulation, joists 16" O.C.
	Install	S.F.	.42	.50		.92	
	Demolish and Install	S.F.	.42	.60		1.02	
	Minimum Charge	Job		150		150	
Unfaced Batt							
3-1/2" (R-11)							
	Demolish	S.F.		.10		.10	Cost includes material and labor to install 3-1/2" (R-11) unfaced roll and batt fiberglass insulation, joists 16" O.C.
	Install	S.F.	.23	.50		.73	
	Demolish and Install	S.F.	.23	.60		.83	
	Minimum Charge	Job		150		150	

Insulation

Floor Insulation

	Unit	Material	Labor	Equip.	Total	Specification
6" (R-19)						
Demolish	S.F.		.10		.10	Cost includes material and labor to install 6" (R-19) unfaced roll and batt fiberglass insulation, joists 16" O.C.
Install	S.F.	.37	.60		.97	
Demolish and Install	S.F.	.37	.70		1.07	
Minimum Charge	Job		150		150	
10" (R-30)						
Demolish	S.F.		.10		.10	Cost includes material and labor to install 10" (R-30) unfaced roll and batt fiberglass insulation, joists 16" O.C.
Install	S.F.	.75	.67		1.42	
Demolish and Install	S.F.	.75	.77		1.52	
Minimum Charge	Job		150		150	
12" (R-38)						
Demolish	S.F.		.10		.10	Cost includes material and labor to install 12" (R-38) unfaced roll and batt fiberglass insulation, joists 16" O.C.
Install	S.F.	.84	.71		1.55	
Demolish and Install	S.F.	.84	.81		1.65	
Minimum Charge	Job		150		150	
Rigid Board						
1/2"						
Demolish	S.F.		.10		.10	Includes material and labor to install rigid foam insulation board foil faced two sides.
Install	S.F.	.32	.38		.70	
Demolish and Install	S.F.	.32	.48		.80	
Minimum Charge	Job		150		150	
3/4"						
Demolish	S.F.		.10		.10	Includes material and labor to install rigid foam insulation board foil faced two sides.
Install	S.F.	.34	.38		.72	
Demolish and Install	S.F.	.34	.48		.82	
Minimum Charge	Job		150		150	
1-1/2"						
Demolish	S.F.		.10		.10	Includes material and labor to install rigid foam insulation board foil faced two sides.
Install	S.F.	.42	.40		.82	
Demolish and Install	S.F.	.42	.50		.92	
Minimum Charge	Job		150		150	
Support Wire						
Demolish	S.F.		.04		.04	Includes material and labor to install friction fit support wire.
Install	S.F.	.01	.08		.09	
Demolish and Install	S.F.	.01	.12		.13	
Minimum Charge	Job		150		150	
Vapor Barrier						
Demolish	S.F.		.05		.05	Cost includes material and labor to install 4 mil polyethylene vapor barrier 10' wide sheets with 6" overlaps.
Install	S.F.	.02	.08		.10	
Demolish and Install	S.F.	.02	.13		.15	
Minimum Charge	Job		150		150	

Wall Insulation

	Unit	Material	Labor	Equip.	Total	Specification
Kraft Faced Batt						
3-1/2" (R-11)						
Demolish	S.F.		.10		.10	Cost includes material and labor to install 3-1/2" (R-11) kraft paper-faced roll and batt fiberglass insulation, joist or studs 16" O.C.
Install	S.F.	.25	.19		.44	
Demolish and Install	S.F.	.25	.29		.54	
Minimum Charge	Job		150		150	
6" (R-19)						
Demolish	S.F.		.10		.10	Cost includes material and labor to install 6" (R-19) kraft paper-faced roll and batt fiberglass insulation, joist or studs 16" O.C.
Install	S.F.	.36	.22		.58	
Demolish and Install	S.F.	.36	.32		.68	
Minimum Charge	Job		150		150	

Wall Insulation		Unit	Material	Labor	Equip.	Total	Specification
10" (R-30)							
	Demolish	S.F.		.10		.10	Cost includes material and labor to install 10" (R-30) kraft paper-faced roll and batt fiberglass insulation, joist or studs 16" O.C.
	Install	S.F.	.66	.26		.92	
	Demolish and Install	S.F.	.66	.36		1.02	
	Minimum Charge	Job		150		150	
12" (R-38)							
	Demolish	S.F.		.10		.10	Cost includes material and labor to install 12" (R-38) kraft paper-faced roll and batt fiberglass insulation, joist or studs 16" O.C.
	Install	S.F.	.84	.30		1.14	
	Demolish and Install	S.F.	.84	.40		1.24	
	Minimum Charge	Job		150		150	
Foil Faced Batt							
3-1/2" (R-11)							
	Demolish	S.F.		.10		.10	Cost includes material and labor to install 3-1/2" (R-11) foil faced roll and batt fiberglass insulation, joist or studs 16" O.C.
	Install	S.F.	.37	.19		.56	
	Demolish and Install	S.F.	.37	.29		.66	
	Minimum Charge	Job		150		150	
6" (R-19)							
	Demolish	S.F.		.10		.10	Cost includes material and labor to install 6" (R-19) foil faced roll and batt fiberglass insulation, joist or studs 16" O.C.
	Install	S.F.	.45	.22		.67	
	Demolish and Install	S.F.	.45	.32		.77	
	Minimum Charge	Job		150		150	
Unfaced Batt							
3-1/2" (R-11)							
	Demolish	S.F.		.10		.10	Cost includes material and labor to install 3-1/2" (R-11) unfaced roll and batt fiberglass insulation, joist or studs 16" O.C.
	Install	S.F.	.23	.22		.45	
	Demolish and Install	S.F.	.23	.32		.55	
	Minimum Charge	Job		150		150	
6" (R-19)							
	Demolish	S.F.		.10		.10	Cost includes material and labor to install 6" (R-19) unfaced roll and batt fiberglass insulation, joist or studs 16" O.C.
	Install	S.F.	.37	.26		.63	
	Demolish and Install	S.F.	.37	.36		.73	
	Minimum Charge	Job		150		150	
10" (R-30)							
	Demolish	S.F.		.10		.10	Cost includes material and labor to install 10" (R-30) unfaced roll and batt fiberglass insulation, joist or studs 16" O.C.
	Install	S.F.	.66	.30		.96	
	Demolish and Install	S.F.	.66	.40		1.06	
	Minimum Charge	Job		150		150	
12" (R-38)							
	Demolish	S.F.		.10		.10	Cost includes material and labor to install 12" (R-38) unfaced roll and batt fiberglass insulation, joist or studs 16" O.C.
	Install	S.F.	.84	.30		1.14	
	Demolish and Install	S.F.	.84	.40		1.24	
	Minimum Charge	Job		150		150	
Blown							
R-11							
	Demolish	S.F.		.10		.10	Includes material and labor to install fiberglass blown-in insulation in wall with wood siding.
	Install	S.F.	.24	.99	.33	1.56	
	Demolish and Install	S.F.	.24	1.09	.33	1.66	
	Minimum Charge	Job		340	114	454	
R-19							
	Demolish	S.F.		.10		.10	Includes material and labor to install fiberglass blown-in insulation in wall with wood siding.
	Install	S.F.	.36	1.14	.38	1.88	
	Demolish and Install	S.F.	.36	1.24	.38	1.98	
	Minimum Charge	Job		340	114	454	

Insulation

Wall Insulation

Wall Insulation	Unit	Material	Labor	Equip.	Total	Specification
Rigid Board						
1/2"						
Demolish	S.F.		.12		.12	Includes material and labor to install rigid foam insulation board, foil faced, two sides.
Install	S.F.	.32	.38		.70	
Demolish and Install	S.F.	.32	.50		.82	
Minimum Charge	Job		150		150	
3/4"						
Demolish	S.F.		.12		.12	Includes material and labor to install rigid foam insulation board, foil faced, two sides.
Install	S.F.	.34	.38		.72	
Demolish and Install	S.F.	.34	.50		.84	
Minimum Charge	Job		150		150	
1-1/2"						
Demolish	S.F.		.12		.12	Includes material and labor to install rigid foam insulation board, foil faced, two sides.
Install	S.F.	.42	.41		.83	
Demolish and Install	S.F.	.42	.53		.95	
Minimum Charge	Job		150		150	
Vapor Barrier						
Demolish	S.F.		.03		.03	Cost includes material and labor to install 4 mil polyethylene vapor barrier 10' wide sheets with 6" overlaps.
Install	S.F.	.02	.08		.10	
Demolish and Install	S.F.	.02	.11		.13	
Minimum Charge	Job		150		150	

Ceiling Insulation

Ceiling Insulation	Unit	Material	Labor	Equip.	Total	Specification
Kraft Faced Batt						
3-1/2" (R-11)						
Demolish	S.F.		.10		.10	Cost includes material and labor to install 3-1/2" (R-11) kraft paper-faced roll and batt fiberglass insulation, joist or studs 16" O.C.
Install	S.F.	.25	.19		.44	
Demolish and Install	S.F.	.25	.29		.54	
Minimum Charge	Job		150		150	
6" (R-19)						
Demolish	S.F.		.10		.10	Cost includes material and labor to install 6" (R-19) kraft paper-faced roll and batt fiberglass insulation, joist or studs 16" O.C.
Install	S.F.	.36	.22		.58	
Demolish and Install	S.F.	.36	.32		.68	
Minimum Charge	Job		150		150	
10" (R-30)						
Demolish	S.F.		.10		.10	Cost includes material and labor to install 10" (R-30) kraft paper-faced roll and batt fiberglass insulation, joist or studs 16" O.C.
Install	S.F.	.66	.26		.92	
Demolish and Install	S.F.	.66	.36		1.02	
Minimum Charge	Job		150		150	
12" (R-38)						
Demolish	S.F.		.10		.10	Cost includes material and labor to install 12" (R-38) kraft paper-faced roll and batt fiberglass insulation, joist or studs 16" O.C.
Install	S.F.	.84	.30		1.14	
Demolish and Install	S.F.	.84	.40		1.24	
Minimum Charge	Job		150		150	
Foil Faced Batt						
3-1/2" (R-11)						
Demolish	S.F.		.10		.10	Cost includes material and labor to install 3-1/2" (R-11) foil faced roll and batt fiberglass insulation, joist or studs 16" O.C.
Install	S.F.	.37	.19		.56	
Demolish and Install	S.F.	.37	.29		.66	
Minimum Charge	Job		150		150	

Ceiling Insulation		Unit	Material	Labor	Equip.	Total	Specification
6" (R-19)							
	Demolish	S.F.		.10		.10	Cost includes material and labor to install 6" (R-19) foil faced roll and batt fiberglass insulation, joist or studs 16" O.C.
	Install	S.F.	.45	.22		.67	
	Demolish and Install	S.F.	.45	.32		.77	
	Minimum Charge	Job		150		150	
Unfaced Batt							
3-1/2" (R-11)							
	Demolish	S.F.		.10		.10	Cost includes material and labor to install 3-1/2" (R-11) unfaced roll and batt fiberglass insulation, joist or studs 16" O.C.
	Install	S.F.	.23	.22		.45	
	Demolish and Install	S.F.	.23	.32		.55	
	Minimum Charge	Job		150		150	
6" (R-19)							
	Demolish	S.F.		.10		.10	Cost includes material and labor to install 6" (R-19) unfaced roll and batt fiberglass insulation, joist or studs 16" O.C.
	Install	S.F.	.37	.26		.63	
	Demolish and Install	S.F.	.37	.36		.73	
	Minimum Charge	Job		150		150	
10" (R-30)							
	Demolish	S.F.		.10		.10	Cost includes material and labor to install 10" (R-30) unfaced roll and batt fiberglass insulation, joist or studs 16" O.C.
	Install	S.F.	.66	.30		.96	
	Demolish and Install	S.F.	.66	.40		1.06	
	Minimum Charge	Job		150		150	
12" (R-38)							
	Demolish	S.F.		.10		.10	Cost includes material and labor to install 12" (R-38) unfaced roll and batt fiberglass insulation, joist or studs 16" O.C.
	Install	S.F.	.84	.30		1.14	
	Demolish and Install	S.F.	.84	.40		1.24	
	Minimum Charge	Job		150		150	
Blown							
5"							
	Demolish	S.F.		.30		.30	Includes material and labor to install fiberglass blown-in insulation.
	Install	S.F.	.18	.18	.06	.42	
	Demolish and Install	S.F.	.18	.48	.06	.72	
	Minimum Charge	Job		340	114	454	
6"							
	Demolish	S.F.		.30		.30	Includes material and labor to install fiberglass blown-in insulation.
	Install	S.F.	.22	.23	.08	.53	
	Demolish and Install	S.F.	.22	.53	.08	.83	
	Minimum Charge	Job		340	114	454	
9"							
	Demolish	S.F.		.30		.30	Includes material and labor to install fiberglass blown-in insulation.
	Install	S.F.	.30	.31	.10	.71	
	Demolish and Install	S.F.	.30	.61	.10	1.01	
	Minimum Charge	Job		340	114	454	
12"							
	Demolish	S.F.		.30		.30	Includes material and labor to install fiberglass blown-in insulation.
	Install	S.F.	.44	.45	.15	1.04	
	Demolish and Install	S.F.	.44	.75	.15	1.34	
	Minimum Charge	Job		340	114	454	
Rigid Board							
1/2"							
	Demolish	S.F.		.17		.17	Includes material and labor to install rigid foam insulation board, foil faced, two sides.
	Install	S.F.	.32	.38		.70	
	Demolish and Install	S.F.	.32	.55		.87	
	Minimum Charge	Job		150		150	

Ceiling Insulation	Unit	Material	Labor	Equip.	Total	Specification
3/4"						
Demolish	S.F.		.17		.17	Includes material and labor to install rigid foam insulation board, foil faced, two sides.
Install	S.F.	.34	.38		.72	
Demolish and Install	S.F.	.34	.55		.89	
Minimum Charge	Job		150		150	
1-1/2"						
Demolish	S.F.		.17		.17	Includes material and labor to install rigid foam insulation board, foil faced, two sides.
Install	S.F.	.42	.41		.83	
Demolish and Install	S.F.	.42	.58		1	
Minimum Charge	Job		150		150	

Exterior Insulation	Unit	Material	Labor	Equip.	Total	Specification
Rigid Board						
1/2"						
Demolish	S.F.		.09		.09	Includes material and labor to install rigid foam insulation board foil faced two sides.
Install	S.F.	.32	.38		.70	
Demolish and Install	S.F.	.32	.47		.79	
Minimum Charge	Job		150		150	
3/4"						
Demolish	S.F.		.09		.09	Includes material and labor to install rigid foam insulation board foil faced two sides.
Install	S.F.	.34	.38		.72	
Demolish and Install	S.F.	.34	.47		.81	
Minimum Charge	Job		150		150	
1-1/2"						
Demolish	S.F.		.09		.09	Includes material and labor to install rigid foam insulation board foil faced two sides.
Install	S.F.	.42	.40		.82	
Demolish and Install	S.F.	.42	.49		.91	
Minimum Charge	Job		150		150	
Vapor Barrier						
Demolish	S.F.		.05		.05	Cost includes material and labor to install 4 mil polyethylene vapor barrier 10' wide sheets with 6" overlaps.
Install	S.F.	.02	.08		.10	
Demolish and Install	S.F.	.02	.13		.15	
Minimum Charge	Job		150		150	

Roof Insulation Board	Unit	Material	Labor	Equip.	Total	Specification
Fiberglass						
3/4" Fiberglass						
Demolish	Sq.		32		32	Includes material and labor to install fiberglass board insulation 3/4" thick, R-2.80 and C-0.36 values.
Install	Sq.	50.50	22.50		73	
Demolish and Install	Sq.	50.50	54.50		105	
Minimum Charge	Job		150		150	
1" Fiberglass						
Demolish	Sq.		32		32	Includes material and labor to install fiberglass board insulation 1" thick, R-4.20 and C-0.24 values.
Install	Sq.	72	25		97	
Demolish and Install	Sq.	72	57		129	
Minimum Charge	Job		150		150	
1-3/8" Fiberglass						
Demolish	Sq.		32		32	Includes material and labor to install fiberglass board insulation 1-3/8" thick, R-5.30 and C-0.19 values.
Install	Sq.	116	27.50		143.50	
Demolish and Install	Sq.	116	59.50		175.50	
Minimum Charge	Job		150		150	

Roof Insulation Board	Unit	Material	Labor	Equip.	Total	Specification
1-5/8" Fiberglass						
Demolish	Sq.		32		32	Includes material and labor to install fiberglass board insulation 1-5/8" thick, R-6.70 and C-0.15 values.
Install	Sq.	118	28.50		146.50	
Demolish and Install	Sq.	118	60.50		178.50	
Minimum Charge	Job		150		150	
2-1/4" Fiberglass						
Demolish	Sq.		32		32	Includes material and labor to install fiberglass board insulation 2-1/4" thick, R-8.30 and C-0.12 values.
Install	Sq.	141	31.50		172.50	
Demolish and Install	Sq.	141	63.50		204.50	
Minimum Charge	Job		150		150	
Perlite						
1" Perlite						
Demolish	Sq.		32		32	Includes material and labor to install perlite board insulation 1" thick, R-2.80 and C-0.36 values.
Install	Sq.	33	25		58	
Demolish and Install	Sq.	33	57		90	
Minimum Charge	Job		150		150	
4" Perlite						
Demolish	Sq.		32		32	Includes material and labor to install perlite board insulation 4" thick, R-10.0 and C-0.10 values.
Install	Sq.	132	48		180	
Demolish and Install	Sq.	132	80		212	
Minimum Charge	Job		150		150	
1-1/2" Perlite						
Demolish	Sq.		32		32	Includes material and labor to install perlite board insulation 1-1/2" thick, R-4.20 and C-0.24 values.
Install	Sq.	43	27.50		70.50	
Demolish and Install	Sq.	43	59.50		102.50	
Minimum Charge	Job		150		150	
2" Perlite						
Demolish	Sq.		32		32	Includes material and labor to install perlite board insulation 2" thick, R-5.30 and C-0.19 values.
Install	Sq.	65	33.50		98.50	
Demolish and Install	Sq.	65	65.50		130.50	
Minimum Charge	Job		150		150	
2-1/2" Perlite						
Demolish	Sq.		32		32	Includes material and labor to install perlite board insulation 2-1/2" thick, R-6.70 and C-0.15 values.
Install	Sq.	84.50	37.50		122	
Demolish and Install	Sq.	84.50	69.50		154	
Minimum Charge	Job		150		150	
3" Perlite						
Demolish	Sq.		32		32	Includes material and labor to install perlite board insulation 3" thick, R-8.30 and C-0.12 values.
Install	Sq.	99	40		139	
Demolish and Install	Sq.	99	72		171	
Minimum Charge	Job		150		150	
Polystyrene						
2" Polystyrene						
Demolish	Sq.		32		32	Includes material and labor to install extruded polystyrene board, R5.0/ 2" in thickness @25PSI compressive strength.
Install	Sq.	29.50	31.50		61	
Demolish and Install	Sq.	29.50	63.50		93	
Minimum Charge	Job		150		150	
1" Polystyrene						
Demolish	Sq.		32		32	Includes material and labor to install extruded polystyrene board, R5.0/ 1" in thickness @25PSI compressive strength.
Install	Sq.	28.50	21.50		50	
Demolish and Install	Sq.	28.50	53.50		82	
Minimum Charge	Job		150		150	
1-1/2" Polystyrene						
Demolish	Sq.		32		32	Includes material and labor to install extruded polystyrene board, R5.0/ 1-1/2" in thickness @25PSI compressive strength.
Install	Sq.	29	25		54	
Demolish and Install	Sq.	29	57		86	
Minimum Charge	Job		150		150	

Roof Insulation Board	Unit	Material	Labor	Equip.	Total	Specification
Urethane						
3/4" Urethane						
Demolish	Sq.		32		32	Includes material and labor to install urethane board insulation 3/4" thick, R-5.30 and C-0.19 values.
Install	Sq.	35	18.80		53.80	
Demolish and Install	Sq.	35	50.80		85.80	
Minimum Charge	Job		150		150	
1" Urethane						
Demolish	Sq.		32		32	Includes material and labor to install urethane board insulation 1" thick, R-6.70 and C-0.15 values.
Install	Sq.	38.50	20		58.50	
Demolish and Install	Sq.	38.50	52		90.50	
Minimum Charge	Job		150		150	
1-1/2" Urethane						
Demolish	Sq.		32		32	Includes material and labor to install urethane board insulation 1-1/2" thick, R-10.1 and C-0.09 values.
Install	Sq.	40.50	23		63.50	
Demolish and Install	Sq.	40.50	55		95.50	
Minimum Charge	Job		150		150	
2" Urethane						
Demolish	Sq.		32		32	Includes material and labor to install urethane board insulation 2" thick, R-13.4 and C-0.06 values.
Install	Sq.	51	27.50		78.50	
Demolish and Install	Sq.	51	59.50		110.50	
Minimum Charge	Job		150		150	

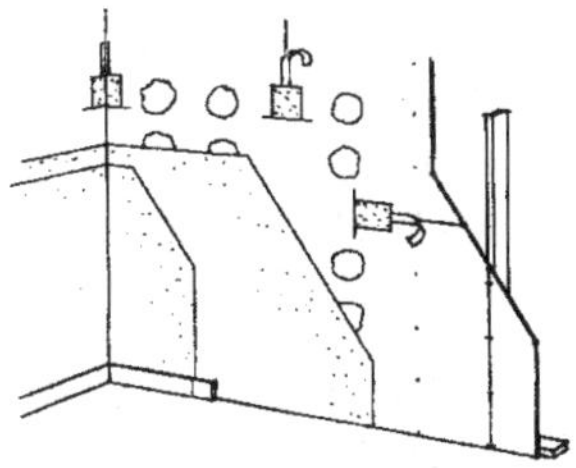

Gypsum Wallboard

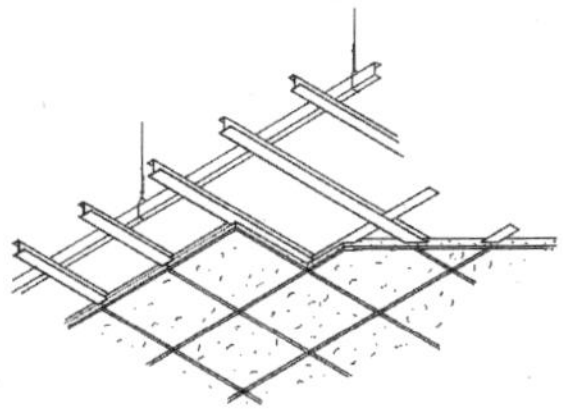

Tile Ceiling

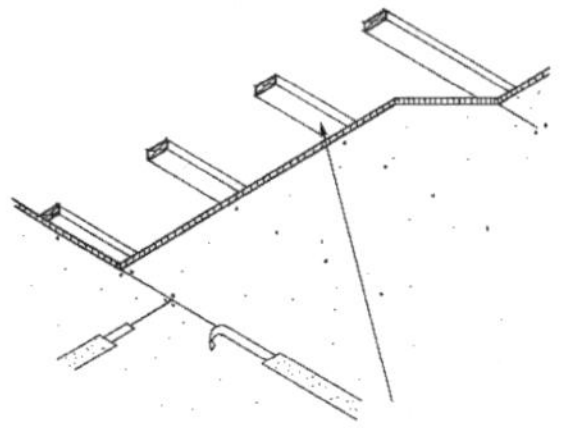

Gypsum Wallboard Ceiling

Wall Plaster	Unit	Material	Labor	Equip.	Total	Specification
Interior						
Cement						
Demolish	S.Y.		4.04		4.04	Includes material and labor to install cement plaster on interior wood framing with metal lath.
Install	S.Y.	13.65	17.35	1.31	32.31	
Demolish and Install	S.Y.	13.65	21.39	1.31	36.35	
Clean	S.Y.		1.98		1.98	
Paint	S.Y.	1.60	3.52		5.12	
Minimum Charge	Job		136		136	
Acoustic						
Demolish	S.Y.		4.04		4.04	Includes material and labor to install acoustical cement plaster on interior wood framing with metal lath.
Install	S.Y.	7.60	21	1.31	29.91	
Demolish and Install	S.Y.	7.60	25.04	1.31	33.95	
Clean	S.Y.		1.98		1.98	
Paint	S.Y.	1.60	3.52		5.12	
Minimum Charge	Job		136		136	
Gypsum						
Demolish	S.Y.		4.04		4.04	Includes material and labor to install gypsum wall or ceiling plaster on interior wood framing with metal lath.
Install	S.Y.	6.55	17.85	1.11	25.51	
Demolish and Install	S.Y.	6.55	21.89	1.11	29.55	
Clean	S.Y.		1.98		1.98	
Paint	S.Y.	1.60	3.52		5.12	
Minimum Charge	Job		136		136	
Stucco						
Demolish	S.Y.		9.70		9.70	Includes material and labor to install stucco on wire mesh over wood framing.
Install	S.Y.	4.38	24.50	1.54	30.42	
Demolish and Install	S.Y.	4.38	34.20	1.54	40.12	
Clean	S.Y.		1.98		1.98	
Paint	S.Y.	1.60	3.52		5.12	
Minimum Charge	Job		136		136	
Thin Coat						
Veneer						
Demolish	S.Y.		4.04		4.04	Includes materials and labor to install thin coat veneer plaster.
Install	S.Y.	.58	3.21	.24	4.03	
Demolish and Install	S.Y.	.58	7.25	.24	8.07	
Clean	S.Y.		1.98		1.98	
Paint	S.Y.	1.60	3.52		5.12	
Minimum Charge	Job		640	48.50	688.50	

Wall Plaster	Unit	Material	Labor	Equip.	Total	Specification
Lath						
Metal						
Demolish	S.Y.		.95		.95	Includes material and labor to install wire lath on steel framed walls.
Install	S.Y.	2.65	3.61		6.26	
Demolish and Install	S.Y.	2.65	4.56		7.21	
Clean	S.Y.		1.98		1.98	
Minimum Charge	Job		63.50		63.50	
Gypsum						
Demolish	S.Y.		.95		.95	Includes material and labor to install gypsum lath on walls.
Install	S.Y.	3.86	3.39		7.25	
Demolish and Install	S.Y.	3.86	4.34		8.20	
Clean	S.Y.		1.98		1.98	
Minimum Charge	Job		136		136	
Repairs						
Fill in Small Cracks						
Install	L.F.	.04	2.18		2.22	Includes labor and material to fill-in hairline cracks up to 1/8″ with filler, sand and prep for paint.
Minimum Charge	Job		136		136	
Repair up to 1″ Crack						
Install	L.F.	.19	2.72		2.91	Includes labor and material to fill-in hairline cracks up to 1″ with filler, sand and prep for paint.
Minimum Charge	Job		136		136	
Patch Small Hole (2″)						
Install	Ea.	.04	2.86		2.90	Includes labor and material to patch up to a 2″ diameter hole in plaster.
Minimum Charge	Job		136		136	
Patch Large Hole						
Install	Ea.	5.80	45.50		51.30	Includes labor and material to patch up to a 20″ diameter hole in plaster.
Minimum Charge	Job		136		136	
Patch Plaster Section						
Install	Ea.	2	2.72		4.72	Includes labor and material to repair section of plaster wall, area less than 20 S.F.
Minimum Charge	Job		136		136	
Repair Inside Corner						
Install	L.F.	2.30	6.80		9.10	Includes material and labor to patch plaster corners.
Minimum Charge	Job		136		136	

Ceiling Plaster	Unit	Material	Labor	Equip.	Total	Specification
Interior						
Cement						
Demolish	S.Y.		4.04		4.04	Includes material and labor to install cement plaster on interior wood framing with metal lath.
Install	S.Y.	13.65	17.35	1.31	32.31	
Demolish and Install	S.Y.	13.65	21.39	1.31	36.35	
Clean	S.Y.		2.08		2.08	
Paint	S.Y.	1.60	3.52		5.12	
Minimum Charge	Job		136		136	
Acoustic						
Demolish	S.Y.		4.04		4.04	Includes material and labor to install acoustical cement plaster on interior wood framing with metal lath.
Install	S.Y.	7.60	21	1.31	29.91	
Demolish and Install	S.Y.	7.60	25.04	1.31	33.95	
Clean	S.Y.		2.08		2.08	
Paint	S.Y.	1.60	3.52		5.12	
Minimum Charge	Job		136		136	

Ceiling Plaster	Unit	Material	Labor	Equip.	Total	Specification
Gypsum						
Demolish	S.Y.		4.04		4.04	Includes material and labor to install gypsum wall or ceiling plaster on interior wood framing with metal lath.
Install	S.Y.	6.55	17.85	1.11	25.51	
Demolish and Install	S.Y.	6.55	21.89	1.11	29.55	
Clean	S.Y.		2.08		2.08	
Paint	S.Y.	1.60	3.52		5.12	
Minimum Charge	Job		136		136	
Stucco						
Demolish	S.Y.		9.70		9.70	Includes material and labor to install stucco on wire mesh over wood framing.
Install	S.Y.	4.38	24.50	1.54	30.42	
Demolish and Install	S.Y.	4.38	34.20	1.54	40.12	
Clean	S.Y.		2.08		2.08	
Paint	S.Y.	1.60	3.52		5.12	
Minimum Charge	Job		136		136	
Thin Coat						
Veneer						
Demolish	S.Y.		4.04		4.04	Includes materials and labor to install thin coat veneer plaster.
Install	S.Y.	.58	3.21	.24	4.03	
Demolish and Install	S.Y.	.58	7.25	.24	8.07	
Clean	S.Y.		2.08		2.08	
Paint	S.Y.	1.60	3.52		5.12	
Minimum Charge	Job		136		136	
Lath						
Metal						
Demolish	S.Y.		.95		.95	Includes material and labor to install wire lath on steel framed ceilings.
Install	S.Y.	2.65	4.51		7.16	
Demolish and Install	S.Y.	2.65	5.46		8.11	
Minimum Charge	Job		136		136	
Gypsum						
Demolish	S.Y.		.95		.95	Includes material and labor to install gypsum lath on ceilings.
Install	S.Y.	3.86	2.82		6.68	
Demolish and Install	S.Y.	3.86	3.77		7.63	
Minimum Charge	Job		136		136	
Repairs						
Fill in Small Cracks						
Install	S.F.	.04	.49		.53	Includes labor and material to fill in hairline cracks up to 1/8″ with filler, sand and prep for paint.
Minimum Charge	Job		272		272	

Gypsum Wallboard	Unit	Material	Labor	Equip.	Total	Specification
Finished						
1/4″						
Demolish	S.F.		.24		.24	Cost includes material and labor to install 1/4″ drywall, including joint taping, premixed compound and screws.
Install	S.F.	.21	.91		1.12	
Demolish and Install	S.F.	.21	1.15		1.36	
Clean	S.F.		.20		.20	
Paint	S.F.	.18	.23		.41	
Minimum Charge	Job		150		150	
3/8″						
Demolish	S.F.		.24		.24	Cost includes material and labor to install 3/8″ drywall including joint tape, premixed compound and screws.
Install	S.F.	.19	.92		1.11	
Demolish and Install	S.F.	.19	1.16		1.35	
Clean	S.F.		.20		.20	
Paint	S.F.	.18	.23		.41	
Minimum Charge	Job		150		150	

Gypsum Wallboard	Unit	Material	Labor	Equip.	Total	Specification
1/2"						
Demolish	S.F.		.24		.24	Cost includes material and labor to install 1/2" drywall including joint tape, premixed compound and screws.
Install	S.F.	.21	.93		1.14	
Demolish and Install	S.F.	.21	1.17		1.38	
Clean	S.F.		.20		.20	
Paint	S.F.	.18	.23		.41	
Minimum Charge	Job		150		150	
1/2" Water Resistant						
Demolish	S.F.		.24		.24	Cost includes material and labor to install 1/2" WR drywall including joint tape, premixed compound and screws.
Install	S.F.	.17	.93		1.10	
Demolish and Install	S.F.	.17	1.17		1.34	
Clean	S.F.		.20		.20	
Paint	S.F.	.18	.23		.41	
Minimum Charge	Job		150		150	
1/2" Fire-rated						
Demolish	S.F.		.24		.24	Cost includes material and labor to install 1/2" FR drywall including screws. Taping and finishing not included.
Install	S.F.	.21	.93		1.14	
Demolish and Install	S.F.	.21	1.17		1.38	
Clean	S.F.		.20		.20	
Paint	S.F.	.18	.23		.41	
Minimum Charge	Job		150		150	
5/8"						
Demolish	S.F.		.24		.24	Cost includes material and labor to install 5/8" drywall including joint tape, premixed compound and screws.
Install	S.F.	.23	1.05		1.28	
Demolish and Install	S.F.	.23	1.29		1.52	
Clean	S.F.		.20		.20	
Paint	S.F.	.18	.23		.41	
Minimum Charge	Job		150		150	
5/8" Fire-rated						
Demolish	S.F.		.24		.24	Cost includes material and labor to install 5/8" FR drywall, including joint taping, premixed compound and screws.
Install	S.F.	.24	1.05		1.29	
Demolish and Install	S.F.	.24	1.29		1.53	
Clean	S.F.		.20		.20	
Paint	S.F.	.18	.23		.41	
Minimum Charge	Job		150		150	
2-hour Fire-rated						
Demolish	S.F.		.61		.61	Cost includes material and labor to install 2 layers of 5/8" type X drywall including joint taping, premixed compound and screws.
Install	S.F.	.52	2.09		2.61	
Demolish and Install	S.F.	.52	2.70		3.22	
Clean	S.F.		.20		.20	
Paint	S.F.	.18	.23		.41	
Minimum Charge	Job		150		150	
Unfinished						
1/4"						
Demolish	S.F.		.24		.24	Cost includes material and labor to install 1/4" drywall including screws. Taping and finishing not included.
Install	S.F.	.19	.87		1.06	
Demolish and Install	S.F.	.19	1.11		1.30	
Clean	S.F.		.20		.20	
Minimum Charge	Job		150		150	
3/8"						
Demolish	S.F.		.24		.24	Cost includes material and labor to install 3/8" drywall including screws. Tape, finish or texture is not included.
Install	S.F.	.17	.89		1.06	
Demolish and Install	S.F.	.17	1.13		1.30	
Clean	S.F.		.20		.20	
Minimum Charge	Job		150		150	
1/2"						
Demolish	S.F.		.24		.24	Cost includes material and labor to install 1/2" drywall including screws. Taping and finishing not included.
Install	S.F.	.18	.90		1.08	
Demolish and Install	S.F.	.18	1.14		1.32	
Clean	S.F.		.20		.20	
Minimum Charge	Job		150		150	

Gypsum Wallboard	Unit	Material	Labor	Equip.	Total	Specification
1/2" Water Resistant						
Demolish	S.F.		.24		.24	Cost includes material and labor to install 1/2" WR drywall including screws. Taping and finishing is not included.
Install	S.F.	.15	.90		1.05	
Demolish and Install	S.F.	.15	1.14		1.29	
Clean	S.F.		.20		.20	
Minimum Charge	Job		150		150	
1/2" Fire-rated						
Demolish	S.F.		.24		.24	Cost includes labor and material to install 1/2" FR drywall including screws. Tape, finish or texture is not included.
Install	S.F.	.18	.22		.40	
Demolish and Install	S.F.	.18	.46		.64	
Clean	S.F.		.20		.20	
Minimum Charge	Job		150		150	
5/8"						
Demolish	S.F.		.24		.24	Cost includes material and labor to install 5/8" drywall including screws. Taping and finishing not included.
Install	S.F.	.20	1.01		1.21	
Demolish and Install	S.F.	.20	1.25		1.45	
Clean	S.F.		.20		.20	
Minimum Charge	Job		150		150	
5/8" Fire-rated						
Demolish	S.F.		.24		.24	Cost includes material and labor to install 5/8" FR drywall including screws. Taping and finishing not included.
Install	S.F.	.21	1.01		1.22	
Demolish and Install	S.F.	.21	1.25		1.46	
Clean	S.F.		.20		.20	
Minimum Charge	Job		150		150	
2-hour Fire-rated						
Demolish	S.F.		.61		.61	Cost includes material and labor to install 2 layers of 5/8" type X drywall including joint taping, premixed compound and screws.
Install	S.F.	.52	2.09		2.61	
Demolish and Install	S.F.	.52	2.70		3.22	
Clean	S.F.		.20		.20	
Minimum Charge	Job		150		150	
Accessories						
Tape & Finish						
Install	S.F.	.03	.30		.33	Includes labor and material for taping, sanding and finish.
Minimum Charge	Job		150		150	
Texture						
Sprayed						
Install	S.F.	.04	.34		.38	Includes labor and material to install by spray, texture finish.
Minimum Charge	Job		150		150	
Trowel						
Install	S.F.		.18		.18	Includes labor and material to apply hand troweled texture.
Minimum Charge	Job		150		150	
Patch Hole in Gypsum Wallboard						
Install	S.F.	.36	40		40.36	Includes labor and material to patch a one square foot area of damaged drywall.
Minimum Charge	Job		150		150	
Furring Strips						
Demolish	S.F.		.06		.06	Cost includes material and labor to install 1" x 2" furring strips, 16" O.C. per S.F. of surface area to be covered.
Install	S.F.	.17	.67		.84	
Demolish and Install	S.F.	.17	.73		.90	
Minimum Charge	Job		75		75	

Ceramic Wall Tile	Unit	Material	Labor	Equip.	Total	Specification
Thin Set						
2" x 2"						
Demolish	S.F.		.81		.81	Cost includes material and labor to install 2" x 2" ceramic tile in organic adhesive including cutting and grout.
Install	S.F.	6.05	4.59		10.64	
Demolish and Install	S.F.	6.05	5.40		11.45	
Clean	S.F.		.42		.42	
Minimum Charge	Job		138		138	
4-1/4" x 4-1/4"						
Demolish	S.F.		.81		.81	Cost includes material and labor to install 4-1/4" x 4-1/4" ceramic tile in organic adhesive including grout.
Install	S.F.	2.30	2.59		4.89	
Demolish and Install	S.F.	2.30	3.40		5.70	
Clean	S.F.		.42		.42	
Minimum Charge	Job		138		138	
6" x 6"						
Demolish	S.F.		.81		.81	Cost includes material and labor to install 6" x 6" ceramic tile in organic adhesive including and grout.
Install	S.F.	2.86	2.46		5.32	
Demolish and Install	S.F.	2.86	3.27		6.13	
Clean	S.F.		.42		.42	
Minimum Charge	Job		138		138	
8" x 8"						
Demolish	S.F.		.81		.81	Cost includes material and labor to install 8" x 8" ceramic tile in organic adhesive including cutting and grout.
Install	S.F.	3.43	2.19		5.62	
Demolish and Install	S.F.	3.43	3		6.43	
Clean	S.F.		.42		.42	
Minimum Charge	Job		138		138	
Thick Set						
2" x 2"						
Demolish	S.F.		.97		.97	Cost includes material and labor to install 2" x 2" ceramic tile in mortar bed including cutting and grout.
Install	S.F.	4.92	5.50		10.42	
Demolish and Install	S.F.	4.92	6.47		11.39	
Clean	S.F.		.42		.42	
Minimum Charge	Job		138		138	
4-1/4" x 4-1/4"						
Demolish	S.F.		.97		.97	Cost includes material and labor to install 4-1/4" x 4-1/4" ceramic tile in mortar bed including cutting and grout.
Install	S.F.	6.65	5.50		12.15	
Demolish and Install	S.F.	6.65	6.47		13.12	
Clean	S.F.		.42		.42	
Minimum Charge	Job		138		138	
6" x 6"						
Demolish	S.F.		.97		.97	Cost includes material and labor to install 6" x 6" ceramic tile in mortar bed including cutting and grout.
Install	S.F.	7.45	5		12.45	
Demolish and Install	S.F.	7.45	5.97		13.42	
Clean	S.F.		.42		.42	
Minimum Charge	Job		138		138	
8" x 8"						
Demolish	S.F.		.97		.97	Cost includes material and labor to install 8" x 8" ceramic tile in mortar bed including cutting and grout.
Install	S.F.	9.75	4.59		14.34	
Demolish and Install	S.F.	9.75	5.56		15.31	
Clean	S.F.		.42		.42	
Minimum Charge	Job		138		138	
Re-grout						
Install	S.F.	.14	2.75		2.89	Includes material and labor to regrout ceramic tile walls.
Minimum Charge	Job		138		138	

Wall Tile		Unit	Material	Labor	Equip.	Total	Specification
Mirror							
Custom Cut Wall							
	Demolish	S.F.		.46		.46	Includes material and labor to install distortion-free float glass mirror, 1/8" to 3/16" thick with cut and polished edges.
	Install	S.F.	1.75	2.99		4.74	
	Demolish and Install	S.F.	1.75	3.45		5.20	
	Clean	S.F.	.02	.13		.15	
	Minimum Charge	Job		146		146	
Beveled Wall							
	Demolish	S.F.		.46		.46	Includes material and labor to install distortion-free float glass mirror, 1/8" to 3/16" thick with beveled edges.
	Install	S.F.	4.24	5.30		9.54	
	Demolish and Install	S.F.	4.24	5.76		10	
	Clean	S.F.	.02	.13		.15	
	Minimum Charge	Job		146		146	
Smoked Wall							
	Demolish	S.F.		.46		.46	Includes material and labor to install distortion-free smoked float glass mirror, 1/8" to 3/16" thick with cut and polished edges.
	Install	S.F.	6.70	5.30		12	
	Demolish and Install	S.F.	6.70	5.76		12.46	
	Clean	S.F.	.02	.13		.15	
	Minimum Charge	Job		146		146	
Marble							
Good Grade							
	Demolish	S.F.		1.51		1.51	Cost includes material and labor to install 3/8" x 12" x 12" marble tile in mortar bed with grout.
	Install	S.F.	7.60	11.95		19.55	
	Demolish and Install	S.F.	7.60	13.46		21.06	
	Clean	S.F.		.42		.42	
	Minimum Charge	Job		138		138	
Better Grade							
	Demolish	S.F.		1.51		1.51	Cost includes material and labor to install 3/8" x 12" x 12" marble, in mortar bed with grout.
	Install	S.F.	9.80	11.95		21.75	
	Demolish and Install	S.F.	9.80	13.46		23.26	
	Clean	S.F.		.42		.42	
	Minimum Charge	Job		138		138	
Premium Grade							
	Demolish	S.F.		1.51		1.51	Cost includes material and labor to install 3/8" x 12" x 12" marble, in mortar bed with grout.
	Install	S.F.	18.90	11.95		30.85	
	Demolish and Install	S.F.	18.90	13.46		32.36	
	Clean	S.F.		.42		.42	
	Minimum Charge	Job		138		138	
Glass Block							
	Demolish	S.F.		1.21		1.21	Cost includes material and labor to install 3-7/8" thick, 6" x 6" smooth face glass block including mortar and cleaning.
	Install	S.F.	15.75	11.95		27.70	
	Demolish and Install	S.F.	15.75	13.16		28.91	
	Clean	S.F.		.42		.42	
	Minimum Charge	Job		267		267	
Plastic							
	Demolish	S.F.		.37		.37	Cost includes material and labor to install 4" plastic tile.
	Install	S.F.	.92	1.57		2.49	
	Demolish and Install	S.F.	.92	1.94		2.86	
	Clean	S.F.		.42		.42	
	Minimum Charge	Job		138		138	
Cork							
	Demolish	S.F.		.19		.19	Cost includes material and labor to install 3/16" thick cork tile and installation.
	Install	S.F.	2.86	1.10		3.96	
	Demolish and Install	S.F.	2.86	1.29		4.15	
	Clean	S.F.		.42		.42	
	Minimum Charge	Job		132		132	

Wall Tile	Unit	Material	Labor	Equip.	Total	Specification
Re-grout						
Install	S.F.	.14	2.75		2.89	Includes material and labor to regrout ceramic tile walls.
Minimum Charge	Job		138		138	

Wall Paneling	Unit	Material	Labor	Equip.	Total	Specification
Plywood						
Good Grade						
Demolish	S.F.		.24		.24	Cost includes material and labor to install 4′ x 8′ x 1/4″ paneling, including nails and adhesive.
Install	S.F.	.81	1.20		2.01	
Demolish and Install	S.F.	.81	1.44		2.25	
Clean	S.F.		.20		.20	
Paint	S.F.	.15	.35		.50	
Minimum Charge	Job		150		150	
Better Grade						
Demolish	S.F.		.24		.24	Cost includes material and labor to install 4′ x 8′ x 1/4″ plywood wall panel, including nails and adhesive.
Install	S.F.	1.23	1.43		2.66	
Demolish and Install	S.F.	1.23	1.67		2.90	
Clean	S.F.		.20		.20	
Paint	S.F.	.15	.35		.50	
Minimum Charge	Job		150		150	
Premium Grade						
Demolish	S.F.		.24		.24	Cost includes material and labor to install 4′ x 8′ x 1/4″ plywood wall panel, including nails and adhesive.
Install	S.F.	1.80	1.72		3.52	
Demolish and Install	S.F.	1.80	1.96		3.76	
Clean	S.F.		.20		.20	
Paint	S.F.	.15	.35		.50	
Minimum Charge	Job		150		150	
Hardboard						
Embossed Paneling						
Demolish	S.F.		.24		.24	Includes material and labor to install hardboard embossed paneling including adhesives.
Install	S.F.	.59	1.20		1.79	
Demolish and Install	S.F.	.59	1.44		2.03	
Clean	S.F.		.20		.20	
Minimum Charge	Job		150		150	
Plastic Molding						
Demolish	L.F.		.24		.24	Includes material and labor to install plastic trim for paneling.
Install	L.F.	.54	.75		1.29	
Demolish and Install	L.F.	.54	.99		1.53	
Clean	L.F.	.02	.13		.15	
Minimum Charge	Job		150		150	
Hardboard						
Embossed Brick						
Demolish	S.F.		.24		.24	Cost includes material and labor to install 4′ x 8′ non-ceramic mineral brick-like veneer wall panels including adhesive.
Install	S.F.	1.97	1.34		3.31	
Demolish and Install	S.F.	1.97	1.58		3.55	
Clean	S.F.		.20		.20	
Paint	S.F.	.18	.40		.58	
Minimum Charge	Job		150		150	
Embossed Ceramic Tile						
Demolish	S.F.		.24		.24	Includes material and labor to install hardboard embossed paneling including adhesives.
Install	S.F.	.59	1.20		1.79	
Demolish and Install	S.F.	.59	1.44		2.03	
Clean	S.F.		.20		.20	
Paint	S.F.	.18	.40		.58	
Minimum Charge	Job		150		150	

Wall Paneling		Unit	Material	Labor	Equip.	Total	Specification
Pegboard							
	Demolish	S.F.		.24		.24	Cost includes material and labor to install 4′ x 8′ x 1/4″ pegboard panel, including adhesive.
	Install	S.F.	.41	1.20		1.61	
	Demolish and Install	S.F.	.41	1.44		1.85	
	Clean	S.F.		.20		.20	
	Paint	S.F.	.18	.40		.58	
	Minimum Charge	Job		150		150	
Furring Strips							
	Demolish	S.F.		.06		.06	Cost includes material and labor to install 1″ x 2″ furring strips, 16″ O.C. per S.F. of surface area to be covered.
	Install	S.F.	.17	.67		.84	
	Demolish and Install	S.F.	.17	.73		.90	
	Minimum Charge	Job		150		150	
Molding							
Unfinished Base							
	Demolish	L.F.		.40		.40	Cost includes material and labor to install 1/2″ x 2-1/2″ pine or birch base molding.
	Install	L.F.	.48	.93		1.41	
	Demolish and Install	L.F.	.48	1.33		1.81	
	Remove and Install	L.F.		.74		.74	
	Clean	L.F.	.02	.13		.15	
	Paint	L.F.	.03	.29		.32	
	Minimum Charge	Job		150		150	
Unfinished Corner							
	Demolish	L.F.		.40		.40	Includes material and labor to install unfinished pine corner molding.
	Install	L.F.	1.13	1.25		2.38	
	Demolish and Install	L.F.	1.13	1.65		2.78	
	Remove and Install	L.F.		1		1	
	Clean	Ea.	.68	19.60		20.28	
	Paint	L.F.	.12	.42		.54	
	Minimum Charge	Job		150		150	
Unfinished Cove							
	Demolish	L.F.		.40		.40	Cost includes material and labor to install 3/4″ x 3-1/2″ unfinished pine cove molding.
	Install	L.F.	1.56	1.11		2.67	
	Demolish and Install	L.F.	1.56	1.51		3.07	
	Remove and Install	L.F.		.89		.89	
	Clean	Ea.	.68	19.60		20.28	
	Paint	L.F.	.12	.42		.54	
	Minimum Charge	Job		150		150	
Unfinished Crown							
	Demolish	L.F.		.48		.48	Includes material and labor to install unfinished pine crown molding.
	Install	L.F.	1.84	1.20		3.04	
	Demolish and Install	L.F.	1.84	1.68		3.52	
	Remove and Install	L.F.		.96		.96	
	Clean	L.F.	.03	.17		.20	
	Paint	L.F.	.04	.29		.33	
	Minimum Charge	Job		150		150	
Prefinished Base							
	Demolish	L.F.		.40		.40	Cost includes material and labor to install 9/16″ x 3-5/16″ prefinished base molding.
	Install	L.F.	1.69	.96		2.65	
	Demolish and Install	L.F.	1.69	1.36		3.05	
	Remove and Install	L.F.		.77		.77	
	Clean	L.F.	.02	.13		.15	
	Paint	L.F.	.03	.29		.32	
	Minimum Charge	Job		150		150	

Wall Paneling	Unit	Material	Labor	Equip.	Total	Specification
Prefinished Corner						
Demolish	L.F.		.40		.40	Includes material and labor to install inside or outside prefinished corner molding.
Install	L.F.	1.98	1.25		3.23	
Demolish and Install	L.F.	1.98	1.65		3.63	
Remove and Install	L.F.		1		1	
Clean	L.F.		.08		.08	
Paint	L.F.	.12	.42		.54	
Minimum Charge	Job		150		150	
Prefinished Crown						
Demolish	L.F.		.48		.48	Cost includes material and labor to install 3/4" x 3-13/16" prefinished crown molding.
Install	L.F.	2.24	1.20		3.44	
Demolish and Install	L.F.	2.24	1.68		3.92	
Remove and Install	L.F.		.96		.96	
Clean	L.F.	.03	.17		.20	
Paint	L.F.	.04	.29		.33	
Minimum Charge	Job		150		150	

Wallcovering	Unit	Material	Labor	Equip.	Total	Specification
Surface Prep						
Patch Walls						
Install	S.F.	.10	2.72		2.82	Includes labor and material to repair section of plaster wall, area less than 20 S.F.
Minimum Charge	Job		132		132	
Sizing						
Install	S.F.	.07	.13		.20	Includes labor and materials to apply wall sizing.
Minimum Charge	Job		132		132	
Good Grade						
Walls						
Demolish	S.F.		.49		.49	Includes material and labor to install pre-pasted, vinyl-coated wallcovering.
Install	S.F.	.80	.55		1.35	
Demolish and Install	S.F.	.80	1.04		1.84	
Clean	S.F.	.01	.05		.06	
Minimum Charge	Job		132		132	
Ceiling						
Demolish	S.F.		.49		.49	Includes material and labor to install pre-pasted, vinyl-coated wallcovering.
Install	S.F.	.80	.55		1.35	
Demolish and Install	S.F.	.80	1.04		1.84	
Clean	S.F.	.02	.13		.15	
Minimum Charge	Job		132		132	
Better Grade						
Walls						
Demolish	S.F.		.49		.49	Includes material and labor to install pre-pasted, vinyl-coated wallcovering.
Install	S.F.	.90	.55		1.45	
Demolish and Install	S.F.	.90	1.04		1.94	
Clean	S.F.	.01	.05		.06	
Minimum Charge	Job		132		132	
Ceiling						
Demolish	S.F.		.49		.49	Includes material and labor to install pre-pasted, vinyl-coated wallcovering.
Install	S.F.	.90	.55		1.45	
Demolish and Install	S.F.	.90	1.04		1.94	
Clean	S.F.	.02	.13		.15	
Minimum Charge	Job		132		132	

Wallcovering	Unit	Material	Labor	Equip.	Total	Specification
Premium Grade						
Walls						
Demolish	S.F.		.61		.61	Includes material and labor to install pre-pasted, vinyl-coated wallcovering.
Install	S.F.	1.14	.55		1.69	
Demolish and Install	S.F.	1.14	1.16		2.30	
Clean	S.F.	.01	.05		.06	
Minimum Charge	Job		132		132	
Ceiling						
Demolish	S.F.		.61		.61	Includes material and labor to install pre-pasted, vinyl-coated wallcovering.
Install	S.F.	1.14	.55		1.69	
Demolish and Install	S.F.	1.14	1.16		2.30	
Clean	S.F.	.02	.13		.15	
Minimum Charge	Job		132		132	
Mural						
Pre-printed						
Demolish	Ea.		75.50		75.50	Includes material and labor to install wallcovering with a pre-printed mural.
Install	Ea.	213	199		412	
Demolish and Install	Ea.	213	274.50		487.50	
Clean	Ea.		5.45		5.45	
Minimum Charge	Job		132		132	
Hand-painted						
Demolish	Ea.		75.50		75.50	Includes material and labor to install wallcovering with a hand-painted mural.
Install	Ea.	430	400		830	
Demolish and Install	Ea.	430	475.50		905.50	
Clean	Ea.		5.45		5.45	
Minimum Charge	Job		132		132	
Vinyl-coated						
Good Grade						
Demolish	Ea.		13.45		13.45	Includes material and labor to install pre-pasted, vinyl-coated wallcovering. Based upon 30 S.F. rolls.
Install	Ea.	17.75	18.50		36.25	
Demolish and Install	Ea.	17.75	31.95		49.70	
Clean	Ea.		5.45		5.45	
Minimum Charge	Job		132		132	
Better Grade						
Demolish	Ea.		13.45		13.45	Includes material and labor to install pre-pasted, vinyl-coated wallcovering. Based upon 30 S.F. rolls.
Install	Ea.	23.50	18.50		42	
Demolish and Install	Ea.	23.50	31.95		55.45	
Clean	Ea.		5.45		5.45	
Minimum Charge	Job		132		132	
Premium Grade						
Demolish	Ea.		13.45		13.45	Includes material and labor to install pre-pasted, vinyl-coated wallcovering. Based upon 30 S.F. rolls.
Install	Ea.	41.50	18.50		60	
Demolish and Install	Ea.	41.50	31.95		73.45	
Clean	Ea.		5.45		5.45	
Minimum Charge	Job		132		132	
Blank Underliner						
Demolish	Ea.		13.45		13.45	Includes material and labor to install blank stock liner paper or bridging paper. Based upon 30 S.F. rolls.
Install	Ea.	10.50	18.90		29.40	
Demolish and Install	Ea.	10.50	32.35		42.85	
Clean	Ea.		5.45		5.45	
Minimum Charge	Job		132		132	
Solid Vinyl						
Good Grade						
Demolish	Roll		16.30		16.30	Includes material and labor to install pre-pasted, vinyl wallcovering. Based upon 30 S.F. rolls.
Install	Ea.	23.50	18.50		42	
Demolish and Install	Ea.	23.50	34.80		58.30	
Clean	Ea.		5.45		5.45	
Minimum Charge	Job		132		132	

Wallcovering	Unit	Material	Labor	Equip.	Total	Specification
Better Grade						
Demolish	Roll		16.30		16.30	Includes material and labor to install pre-pasted, vinyl wallcovering. Based upon 30 S.F. rolls.
Install	Ea.	33	18.50		51.50	
Demolish and Install	Ea.	33	34.80		67.80	
Clean	Ea.		5.45		5.45	
Minimum Charge	Job		132		132	
Premium Grade						
Demolish	Roll		16.30		16.30	Includes material and labor to install pre-pasted, vinyl wallcovering. Based upon 30 S.F. rolls.
Install	Ea.	42.50	18.50		61	
Demolish and Install	Ea.	42.50	34.80		77.30	
Clean	Ea.		5.45		5.45	
Minimum Charge	Job		132		132	
Blank Underliner						
Demolish	Ea.		13.45		13.45	Includes material and labor to install blank stock liner paper or bridging paper. Based upon 30 S.F. rolls.
Install	Ea.	10.50	18.90		29.40	
Demolish and Install	Ea.	10.50	32.35		42.85	
Clean	Ea.		5.45		5.45	
Minimum Charge	Job		132		132	
Grasscloth / String						
Good Grade						
Demolish	Ea.		13.45		13.45	Includes material and labor to install grasscloth or string wallcovering. Based upon 30 S.F. rolls.
Install	Ea.	33.50	24		57.50	
Demolish and Install	Ea.	33.50	37.45		70.95	
Clean	Ea.		7.25		7.25	
Minimum Charge	Job		132		132	
Better Grade						
Demolish	Ea.		13.45		13.45	Includes material and labor to install grasscloth or string wallcovering. Based upon 30 S.F. rolls.
Install	Ea.	43.50	24		67.50	
Demolish and Install	Ea.	43.50	37.45		80.95	
Clean	Ea.		7.25		7.25	
Minimum Charge	Job		132		132	
Premium Grade						
Demolish	Ea.		13.45		13.45	Includes material and labor to install grasscloth or string wallcovering. Based upon 30 S.F. rolls.
Install	Ea.	55.50	24		79.50	
Demolish and Install	Ea.	55.50	37.45		92.95	
Clean	Ea.		7.25		7.25	
Minimum Charge	Job		132		132	
Blank Underliner						
Demolish	Ea.		13.45		13.45	Includes material and labor to install blank stock liner paper or bridging paper. Based upon 30 S.F. rolls.
Install	Ea.	10.50	18.90		29.40	
Demolish and Install	Ea.	10.50	32.35		42.85	
Clean	Ea.		5.45		5.45	
Minimum Charge	Job		132		132	
Designer						
Premium Grade						
Demolish	S.F.		.61		.61	Includes material and labor to install paperback vinyl, untrimmed and hand-painted. Based upon 30 S.F. rolls.
Install	Ea.	103	28		131	
Demolish and Install	Ea.	103	28.61		131.61	
Clean	Ea.		5.45		5.45	
Minimum Charge	Job		132		132	
Blank Underliner						
Demolish	Ea.		13.45		13.45	Includes material and labor to install blank stock liner paper or bridging paper. Based upon 30 S.F. rolls.
Install	Ea.	10.50	18.90		29.40	
Demolish and Install	Ea.	10.50	32.35		42.85	
Clean	Ea.		5.45		5.45	
Minimum Charge	Job		132		132	

Wallcovering

Wallcovering	Unit	Material	Labor	Equip.	Total	Specification
Foil						
Demolish	Ea.		13.45		13.45	Includes material and labor to install foil coated wallcovering. Based upon 30 S.F. rolls.
Install	Ea.	36	24		60	
Demolish and Install	Ea.	36	37.45		73.45	
Clean	Ea.		5.45		5.45	
Minimum Charge	Job		132		132	
Wallcovering Border						
Good Grade						
Demolish	L.F.		.88		.88	Includes material and labor install vinyl-coated border 4″ to 6″ wide.
Install	L.F.	1.77	1.02		2.79	
Demolish and Install	L.F.	1.77	1.90		3.67	
Clean	L.F.		.08		.08	
Minimum Charge	Job		132		132	
Better Grade						
Demolish	L.F.		.88		.88	Includes material and labor install vinyl-coated border 4″ to 6″ wide.
Install	L.F.	2.37	1.02		3.39	
Demolish and Install	L.F.	2.37	1.90		4.27	
Clean	L.F.		.08		.08	
Minimum Charge	Job		132		132	
Premium Grade						
Demolish	L.F.		.88		.88	Includes material and labor install vinyl-coated border 4″ to 6″ wide.
Install	L.F.	2.96	1.02		3.98	
Demolish and Install	L.F.	2.96	1.90		4.86	
Clean	L.F.		.08		.08	
Minimum Charge	Job		132		132	

Acoustical Ceiling

Acoustical Ceiling	Unit	Material	Labor	Equip.	Total	Specification
Complete System						
2′ x 2′ Square Edge						
Demolish	S.F.		.40		.40	Includes material and labor to install T-bar grid acoustical ceiling system, random pinhole 5/8″ thick square edge tile.
Install	S.F.	1.09	.93		2.02	
Demolish and Install	S.F.	1.09	1.33		2.42	
Remove and Install	S.F.		.74		.74	
Clean	S.F.	.03	.17		.20	
Minimum Charge	Job		150		150	
2′ x 2′ Reveal Edge						
Demolish	S.F.		.40		.40	Includes material and labor to install T-bar grid acoustical ceiling system, random pinhole 5/8″ thick reveal edge tile.
Install	S.F.	1.85	1.20		3.05	
Demolish and Install	S.F.	1.85	1.60		3.45	
Remove and Install	S.F.		.96		.96	
Clean	S.F.	.03	.17		.20	
Minimum Charge	Job		150		150	
2′ x 2′ Fire-rated						
Demolish	S.F.		.40		.40	Includes material and labor to install T-bar grid FR acoustical ceiling system, random pinhole 5/8″ thick square edge tile.
Install	S.F.	2.23	.93		3.16	
Demolish and Install	S.F.	2.23	1.33		3.56	
Remove and Install	S.F.		.74		.74	
Clean	S.F.	.03	.17		.20	
Minimum Charge	Job		150		150	
2′ x 4′ Square Edge						
Demolish	S.F.		.40		.40	Cost includes material and labor to install 2′ x 4′ T-bar grid acoustical ceiling system, random pinhole 5/8″ thick square edge tile.
Install	S.F.	.92	.79		1.71	
Demolish and Install	S.F.	.92	1.19		2.11	
Remove and Install	S.F.		.63		.63	
Clean	S.F.	.03	.17		.20	
Minimum Charge	Job		150		150	

Acoustical Ceiling

	Unit	Material	Labor	Equip.	Total	Specification
2' x 4' Reveal Edge						
Demolish	S.F.		.40		.40	Includes material and labor to install T-bar grid acoustical ceiling system, random pinhole 5/8" thick reveal edge tile.
Install	S.F.	1.85	1.09		2.94	
Demolish and Install	S.F.	1.85	1.49		3.34	
Remove and Install	S.F.		.88		.88	
Clean	S.F.	.03	.17		.20	
Minimum Charge	Job		150		150	
2' x 4' Fire-rated						
Demolish	S.F.		.40		.40	Includes material and labor to install T-bar grid FR acoustical ceiling system, random pinhole 5/8" thick square edge tile.
Install	S.F.	2.23	.93		3.16	
Demolish and Install	S.F.	2.23	1.33		3.56	
Remove and Install	S.F.		.74		.74	
Clean	S.F.	.03	.17		.20	
Minimum Charge	Job		150		150	
T-bar Grid						
2' x 2'						
Demolish	S.F.		.47		.47	Includes material and labor to install T-bar grid acoustical ceiling suspension system.
Install	S.F.	.56	.46		1.02	
Demolish and Install	S.F.	.56	.93		1.49	
Remove and Install	S.F.		.37		.37	
Clean	S.F.	.03	.17		.20	
Minimum Charge	Job		136		136	
2' x 4'						
Demolish	S.F.		.47		.47	Includes material and labor to install T-bar grid acoustical ceiling suspension system.
Install	S.F.	.45	.38		.83	
Demolish and Install	S.F.	.45	.85		1.30	
Remove and Install	S.F.		.30		.30	
Clean	S.F.	.03	.17		.20	
Minimum Charge	Job		136		136	
Tile Panels						
2' x 2' Square Edge						
Demolish	S.F.		.13		.13	Includes material and labor to install square edge panels 5/8" thick with pinhole pattern. T-bar grid not included.
Install	S.F.	.47	.48		.95	
Demolish and Install	S.F.	.47	.61		1.08	
Remove and Install	S.F.		.39		.39	
Clean	S.F.	.03	.17		.20	
Minimum Charge	Job		150		150	
2' x 2' Reveal Edge						
Demolish	S.F.		.13		.13	Includes material and labor to install reveal edge panels 5/8" thick with pinhole pattern. T-bar grid not included.
Install	S.F.	1.14	.64		1.78	
Demolish and Install	S.F.	1.14	.77		1.91	
Remove and Install	S.F.		.51		.51	
Clean	S.F.	.03	.17		.20	
Minimum Charge	Job		150		150	
2' x 2' Fire-rated						
Demolish	S.F.		.13		.13	Includes material and labor to install FR square edge panels 5/8" thick with pinhole pattern. T-bar grid not included.
Install	S.F.	.92	.45		1.37	
Demolish and Install	S.F.	.92	.58		1.50	
Remove and Install	S.F.		.36		.36	
Clean	S.F.	.03	.17		.20	
Minimum Charge	Job		150		150	
2' x 4' Square Edge						
Demolish	S.F.		.13		.13	Includes material and labor to install square edge panels 5/8" thick with pinhole pattern. T-bar grid not included.
Install	S.F.	.47	.48		.95	
Demolish and Install	S.F.	.47	.61		1.08	
Remove and Install	S.F.		.39		.39	
Clean	S.F.	.03	.17		.20	
Minimum Charge	Job		150		150	

Acoustical Ceiling	Unit	Material	Labor	Equip.	Total	Specification
2′ x 4′ Reveal Edge						
Demolish	S.F.		.13		.13	Includes material and labor to install reveal edge panels 5/8″ thick with pinhole pattern. T-bar grid not included.
Install	S.F.	1.14	.64		1.78	
Demolish and Install	S.F.	1.14	.77		1.91	
Remove and Install	S.F.		.51		.51	
Clean	S.F.	.03	.17		.20	
Minimum Charge	Job		150		150	
2′ x 4′ Fire-rated						
Demolish	S.F.		.13		.13	Includes material and labor to install FR square edge panels 5/8″ thick with pinhole pattern. T-bar grid not included.
Install	S.F.	.92	.45		1.37	
Demolish and Install	S.F.	.92	.58		1.50	
Remove and Install	S.F.		.36		.36	
Clean	S.F.	.03	.17		.20	
Minimum Charge	Job		150		150	
Bleach Out Stains						
Install	S.F.	.08	.22		.30	Includes labor and material to apply laundry type chlorine bleach to remove light staining of acoustic tile material.
Minimum Charge	Job		150		150	
Adhesive Tile						
Good Grade						
Demolish	S.F.		.54		.54	Cost includes material and labor to install 12″ x 12″ tile with embossed texture pattern. Furring strips not included.
Install	S.F.	.85	1		1.85	
Demolish and Install	S.F.	.85	1.54		2.39	
Clean	S.F.	.02	.13		.15	
Minimum Charge	Job		150		150	
Better Grade						
Demolish	S.F.		.54		.54	Cost includes material and labor to install 12″ x 12″ tile with embossed texture pattern. Furring strips not included.
Install	S.F.	1.13	1		2.13	
Demolish and Install	S.F.	1.13	1.54		2.67	
Clean	S.F.	.02	.13		.15	
Minimum Charge	Job		150		150	
Premium Grade						
Demolish	S.F.		.54		.54	Cost includes material and labor to install 12″ x 12″ tile with embossed texture pattern. Furring strips not included.
Install	S.F.	1.75	1		2.75	
Demolish and Install	S.F.	1.75	1.54		3.29	
Clean	S.F.	.02	.13		.15	
Minimum Charge	Job		150		150	
Bleach Out Stains						
Install	S.F.	.08	.22		.30	Includes labor and material to apply laundry type chlorine bleach to remove light staining of acoustic tile material.
Minimum Charge	Job		150		150	
Stapled Tile						
Good Grade						
Demolish	S.F.		.32		.32	Cost includes material and labor to install 12″ x 12″ tile with embossed texture pattern. Furring strips not included.
Install	S.F.	.85	1		1.85	
Demolish and Install	S.F.	.85	1.32		2.17	
Clean	S.F.	.02	.13		.15	
Minimum Charge	Job		150		150	
Better Grade						
Demolish	S.F.		.32		.32	Cost includes material and labor to install 12″ x 12″ tile with embossed texture pattern. Furring strips not included.
Install	S.F.	1.13	1		2.13	
Demolish and Install	S.F.	1.13	1.32		2.45	
Clean	S.F.	.02	.13		.15	
Minimum Charge	Job		150		150	

Acoustical Ceiling	Unit	Material	Labor	Equip.	Total	Specification
Premium Grade						
Demolish	S.F.		.32		.32	Cost includes material and labor to install 12" x 12" tile with embossed texture pattern. Furring strips not included.
Install	S.F.	1.75	1		2.75	
Demolish and Install	S.F.	1.75	1.32		3.07	
Clean	S.F.	.02	.13		.15	
Minimum Charge	Job		150		150	
Bleach Out Stains						
Install	S.F.	.08	.22		.30	Includes labor and material to apply laundry type chlorine bleach to remove light staining of acoustic tile material.
Minimum Charge	Job		150		150	
Blown						
Scrape Off Existing						
Install	S.F.		.45		.45	Includes labor and material to remove existing blown acoustical ceiling material.
Minimum Charge	Job		150		150	
Seal Ceiling Gypsum Wallboard						
Install	S.F.	.04	.23		.27	Includes labor and material to seal existing drywall with shellac-based material.
Minimum Charge	Job		132		132	
Reblow Existing						
Install	S.F.	.04	.34		.38	Includes labor and material to install by spray, texture finish.
Minimum Charge	Job		360	66	426	
Bleach Out Stains						
Install	S.F.	.08	.22		.30	Includes labor and material to apply laundry type chlorine bleach to remove light staining of acoustic tile material.
Minimum Charge	Job		150		150	
New Blown						
Install	S.F.	.04	.34		.38	Includes labor and material to install by spray, texture finish.
Minimum Charge	Job		132		132	
Apply Glitter To						
Install	S.F.	.09	.12		.21	Includes labor and material to apply glitter material to existing or new drywall.
Minimum Charge	Job		132		132	
Furring Strips						
Demolish	S.F.		.06		.06	Cost includes material and labor to install 1" x 2" furring strips, 16" O.C. per S.F. of surface area to be covered.
Install	S.F.	.17	.67		.84	
Demolish and Install	S.F.	.17	.73		.90	
Minimum Charge	Job		75		75	

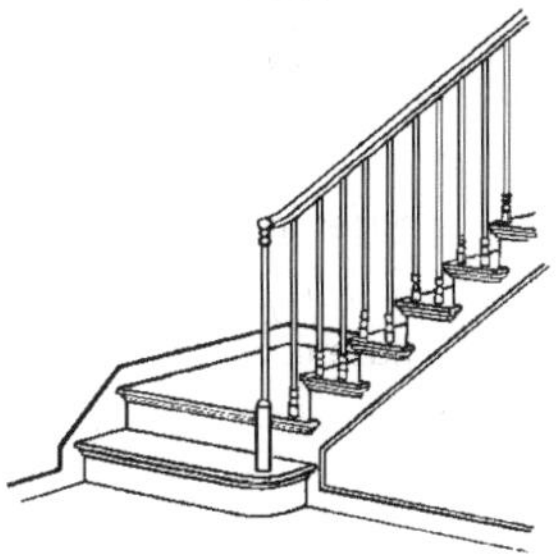

Finished Stair

Bi-Fold Door

Raised Panel Door

Hollow Core Door	Unit	Material	Labor	Equip.	Total	Specification
Pre-hung Wood						
1' 6" x 6' 8"						
Demolish	Ea.		28		28	Cost includes material and labor to install 1' 6" x 6' 8" pre-hung 1-3/8" lauan door with split pine jamb including casing, stops, and hinges.
Install	Ea.	142	14.30		156.30	
Demolish and Install	Ea.	142	42.30		184.30	
Remove and Install	Ea.		14.32		14.32	
Clean	Ea.	2.73	9.10		11.83	
Paint	Ea.	7.55	37.50		45.05	
Minimum Charge	Job		150		150	
2' x 6' 8"						
Demolish	Ea.		28		28	Cost includes material and labor to install 2' x 6' 8" pre-hung 1-3/8" lauan door with split pine jamb including casing, stops, and hinges.
Install	Ea.	145	15.05		160.05	
Demolish and Install	Ea.	145	43.05		188.05	
Remove and Install	Ea.		15.04		15.04	
Clean	Ea.	2.73	9.10		11.83	
Paint	Ea.	7.55	37.50		45.05	
Minimum Charge	Job		150		150	
2' 4" x 6' 8"						
Demolish	Ea.		28		28	Cost includes material and labor to install 2' 4" x 6' 8" pre-hung 1-3/8" lauan door with split pine jamb including casing, stops, and hinges.
Install	Ea.	152	15.05		167.05	
Demolish and Install	Ea.	152	43.05		195.05	
Remove and Install	Ea.		15.04		15.04	
Clean	Ea.	2.73	9.10		11.83	
Paint	Ea.	7.55	37.50		45.05	
Minimum Charge	Job		150		150	
2' 6" x 6' 8"						
Demolish	Ea.		28		28	Cost includes material and labor to install 2' 6" x 6' 8" pre-hung 1-3/8" lauan door with split pine jamb including casing, stops, and hinges.
Install	Ea.	152	15.05		167.05	
Demolish and Install	Ea.	152	43.05		195.05	
Remove and Install	Ea.		15.04		15.04	
Clean	Ea.	2.73	9.10		11.83	
Paint	Ea.	7.55	37.50		45.05	
Minimum Charge	Job		150		150	
2' 8" x 6' 8"						
Demolish	Ea.		28		28	Cost includes material and labor to install 2' 8" x 6' 8" pre-hung 1-3/8" lauan door with split pine jamb, including casing, stop and hinges.
Install	Ea.	156	15.05		171.05	
Demolish and Install	Ea.	156	43.05		199.05	
Remove and Install	Ea.		15.04		15.04	
Clean	Ea.	2.73	9.10		11.83	
Paint	Ea.	7.55	37.50		45.05	
Minimum Charge	Job		150		150	

Hollow Core Door	Unit	Material	Labor	Equip.	Total	Specification
3' x 6' 8"						
Demolish	Ea.		28		28	Cost includes material and labor to install 3' x 6' 8" pre-hung 1-3/8" lauan door with split pine jamb including casing, stops, and hinges.
Install	Ea.	162	15.85		177.85	
Demolish and Install	Ea.	162	43.85		205.85	
Remove and Install	Ea.		15.83		15.83	
Clean	Ea.	2.73	9.10		11.83	
Paint	Ea.	7.55	37.50		45.05	
Minimum Charge	Job		150		150	
Pocket Type						
Demolish	Ea.		42		42	Cost includes material and labor to install 2' 8" x 6' 8" lauan door.
Install	Ea.	32.50	33.50		66	
Demolish and Install	Ea.	32.50	75.50		108	
Remove and Install	Ea.		33.42		33.42	
Clean	Ea.	2.73	9.10		11.83	
Paint	Ea.	7.55	37.50		45.05	
Minimum Charge	Job		150		150	
Stain						
Install	Ea.	3.93	22		25.93	Includes labor and material to stain single door and trim on both sides.
Minimum Charge	Job		132		132	
Pre-hung Masonite						
1' 6" x 6' 8"						
Demolish	Ea.		28		28	Cost includes material and labor to install 1' 6" x 6' 8" pre-hung 1-3/8" masonite door with split pine jamb including casing, stops, and hinges.
Install	Ea.	180	14.30		194.30	
Demolish and Install	Ea.	180	42.30		222.30	
Remove and Install	Ea.		14.32		14.32	
Clean	Ea.	2.73	9.10		11.83	
Paint	Ea.	7.55	37.50		45.05	
Minimum Charge	Job		150		150	
2' x 6' 8"						
Demolish	Ea.		28		28	Cost includes material and labor to install 2' x 6' 8" pre-hung 1-3/8" masonite door with split pine jamb including casing, stops, and hinges.
Install	Ea.	189	15.05		204.05	
Demolish and Install	Ea.	189	43.05		232.05	
Remove and Install	Ea.		15.04		15.04	
Clean	Ea.	2.73	9.10		11.83	
Paint	Ea.	7.55	37.50		45.05	
Minimum Charge	Job		150		150	
2' 4" x 6' 8"						
Demolish	Ea.		28		28	Cost includes material and labor to install 2' 4" x 6' 8" pre-hung 1-3/8" masonite door with split pine jamb including casing, stops, and hinges.
Install	Ea.	191	15.05		206.05	
Demolish and Install	Ea.	191	43.05		234.05	
Remove and Install	Ea.		15.04		15.04	
Clean	Ea.	2.73	9.10		11.83	
Paint	Ea.	7.55	37.50		45.05	
Minimum Charge	Job		150		150	
2' 8" x 6' 8"						
Demolish	Ea.		28		28	Cost includes material and labor to install 2' 8" x 6' 8" pre-hung 1-3/8" masonite door with split pine jamb including casing, stops, and hinges.
Install	Ea.	194	15.05		209.05	
Demolish and Install	Ea.	194	43.05		237.05	
Remove and Install	Ea.		15.04		15.04	
Clean	Ea.	2.73	9.10		11.83	
Paint	Ea.	7.55	37.50		45.05	
Minimum Charge	Job		150		150	
3' x 6' 8"						
Demolish	Ea.		28		28	Cost includes material and labor to install 3' x 6' 8" pre-hung 1-3/8" masonite door with split pine jamb including casing, stops, and hinges.
Install	Ea.	198	15.85		213.85	
Demolish and Install	Ea.	198	43.85		241.85	
Remove and Install	Ea.		15.83		15.83	
Clean	Ea.	2.73	9.10		11.83	
Paint	Ea.	7.55	37.50		45.05	
Minimum Charge	Job		150		150	

Hollow Core Door	Unit	Material	Labor	Equip.	Total	Specification
Pocket Type						
Demolish	Ea.		42		42	Cost includes material and labor to install 2′ 8″ x 6′ 8″ lauan door.
Install	Ea.	32.50	33.50		66	
Demolish and Install	Ea.	32.50	75.50		108	
Remove and Install	Ea.		33.42		33.42	
Clean	Ea.	2.73	9.10		11.83	
Paint	Ea.	7.55	37.50		45.05	
Minimum Charge	Job		150		150	
Stain						
Install	Ea.	3.93	22		25.93	Includes labor and material to stain single door and trim on both sides.
Minimum Charge	Job		132		132	
Hollow Core Door Only						
1′ 6″ x 6′ 8′						
Demolish	Ea.		5.95		5.95	Cost includes material and labor to install 1′ 6″ x 6′ 8″ lauan hollow core door slab.
Install	Ea.	36	15.85		51.85	
Demolish and Install	Ea.	36	21.80		57.80	
Remove and Install	Ea.		15.83		15.83	
Clean	Ea.	1.14	6.80		7.94	
Paint	Ea.	2.33	33		35.33	
Minimum Charge	Job		150		150	
2′ x 6′ 8′						
Demolish	Ea.		5.95		5.95	Cost includes material and labor to install 2′ x 6′ 8″ lauan hollow core door slab.
Install	Ea.	37.50	16.70		54.20	
Demolish and Install	Ea.	37.50	22.65		60.15	
Remove and Install	Ea.		16.71		16.71	
Clean	Ea.	1.14	6.80		7.94	
Paint	Ea.	2.33	33		35.33	
Minimum Charge	Job		150		150	
2′ 4″ x 6′ 8′						
Demolish	Ea.		5.95		5.95	Cost includes material and labor to install 2′ 4″ x 6′ 8″ lauan hollow core door slab.
Install	Ea.	40.50	16.70		57.20	
Demolish and Install	Ea.	40.50	22.65		63.15	
Remove and Install	Ea.		16.71		16.71	
Clean	Ea.	1.14	6.80		7.94	
Paint	Ea.	2.33	33		35.33	
Minimum Charge	Job		150		150	
2′ 8″ x 6′ 8′						
Demolish	Ea.		5.95		5.95	Cost includes material and labor to install 2′ 8″ x 6′ 8″ lauan hollow core door slab.
Install	Ea.	42.50	16.70		59.20	
Demolish and Install	Ea.	42.50	22.65		65.15	
Remove and Install	Ea.		16.71		16.71	
Clean	Ea.	1.14	6.80		7.94	
Paint	Ea.	2.33	33		35.33	
Minimum Charge	Job		150		150	
3′ x 6′ 8′						
Demolish	Ea.		5.95		5.95	Cost includes material and labor to install 3′ x 6′ 8″ lauan hollow core door slab.
Install	Ea.	44	17.70		61.70	
Demolish and Install	Ea.	44	23.65		67.65	
Remove and Install	Ea.		17.69		17.69	
Clean	Ea.	1.14	6.80		7.94	
Paint	Ea.	2.33	33		35.33	
Minimum Charge	Job		150		150	
Pocket Type						
Demolish	Ea.		5.95		5.95	Cost includes material and labor to install 2′ 8″ x 6′ 8″ lauan door.
Install	Ea.	32.50	33.50		66	
Demolish and Install	Ea.	32.50	39.45		71.95	
Remove and Install	Ea.		33.42		33.42	
Clean	Ea.	1.14	6.80		7.94	
Paint	Ea.	2.33	33		35.33	
Minimum Charge	Job		150		150	

Hollow Core Door	Unit	Material	Labor	Equip.	Total	Specification
Stain						
Install	Ea.	2.17	29.50		31.67	Includes labor and materials to stain a flush style door on both sides by brush.
Minimum Charge	Job		132		132	
Casing Trim						
Single Width						
Demolish	Ea.		5.30		5.30	Cost includes material and labor to install 11/16" x 2-1/2" pine ranch style casing for one side of a standard door opening.
Install	Opng.	15.90	51		66.90	
Demolish and Install	Opng.	15.90	56.30		72.20	
Remove and Install	Opng.		40.79		40.79	
Clean	Opng.	.39	2.18		2.57	
Paint	Ea.	3.06	5.30		8.36	
Minimum Charge	Job		150		150	
Double Width						
Demolish	Ea.		6.30		6.30	Cost includes material and labor to install 11/16" x 2-1/2" pine ranch style door casing for one side of a double door opening.
Install	Opng.	18.70	60		78.70	
Demolish and Install	Opng.	18.70	66.30		85	
Remove and Install	Opng.		48.13		48.13	
Clean	Opng.	.45	4.36		4.81	
Paint	Ea.	3.06	6		9.06	
Minimum Charge	Job		150		150	
Hardware						
Doorknob w / Lock						
Demolish	Ea.		11.15		11.15	Includes material and labor to install privacy lockset.
Install	Ea.	19.65	18.80		38.45	
Demolish and Install	Ea.	19.65	29.95		49.60	
Remove and Install	Ea.		15.04		15.04	
Clean	Ea.	.28	6.80		7.08	
Minimum Charge	Job		150		150	
Doorknob						
Demolish	Ea.		11.15		11.15	Includes material and labor to install residential passage lockset, keyless bored type.
Install	Ea.	17.15	18.80		35.95	
Demolish and Install	Ea.	17.15	29.95		47.10	
Remove and Install	Ea.		15.04		15.04	
Clean	Ea.	.28	6.80		7.08	
Minimum Charge	Job		150		150	
Deadbolt						
Demolish	Ea.		10.45		10.45	Includes material and labor to install a deadbolt.
Install	Ea.	36.50	21.50		58	
Demolish and Install	Ea.	36.50	31.95		68.45	
Remove and Install	Ea.		17.19		17.19	
Clean	Ea.	.28	6.80		7.08	
Minimum Charge	Job		150		150	
Lever Handle						
Demolish	Ea.		11.15		11.15	Includes material and labor to install residential passage lockset, keyless bored type with lever handle.
Install	Ea.	27	30		57	
Demolish and Install	Ea.	27	41.15		68.15	
Remove and Install	Ea.		24.06		24.06	
Clean	Ea.	.28	6.80		7.08	
Minimum Charge	Job		150		150	
Closer						
Demolish	Ea.		6.25		6.25	Includes material and labor to install pneumatic light duty closer for interior type doors.
Install	Ea.	86.50	46.50		133	
Demolish and Install	Ea.	86.50	52.75		139.25	
Remove and Install	Ea.		37.02		37.02	
Clean	Ea.	.13	9.10		9.23	
Minimum Charge	Job		150		150	

Hollow Core Door	Unit	Material	Labor	Equip.	Total	Specification
Push Plate						
Demolish	Ea.		9.30		9.30	Includes material and labor to install bronze push-pull plate.
Install	Ea.	12.55	25		37.55	
Demolish and Install	Ea.	12.55	34.30		46.85	
Remove and Install	Ea.		20.05		20.05	
Clean	Ea.	.28	6.80		7.08	
Minimum Charge	Job		150		150	
Kickplate						
Demolish	Ea.		9.30		9.30	Includes material and labor to install aluminum kickplate, 10″ x 28″.
Install	Ea.	16.45	20		36.45	
Demolish and Install	Ea.	16.45	29.30		45.75	
Remove and Install	Ea.		16.04		16.04	
Clean	Ea.	.28	6.80		7.08	
Minimum Charge	Job		150		150	
Exit Sign						
Demolish	Ea.		16.65		16.65	Includes material and labor to install a wall mounted interior electric exit sign.
Install	Ea.	43	41		84	
Demolish and Install	Ea.	43	57.65		100.65	
Remove and Install	Ea.		32.92		32.92	
Clean	Ea.	.02	4.90		4.92	
Minimum Charge	Job		150		150	
Jamb						
Demolish	Ea.		21		21	Includes material and labor to install flat pine jamb with square cut heads and rabbeted sides for 6′ 8″ high and 3-9/16″ door including trim sets for both sides.
Install	Ea.	77	8.15		85.15	
Demolish and Install	Ea.	77	29.15		106.15	
Clean	Ea.	.09	5.45		5.54	
Paint	Ea.	5.55	17.60		23.15	
Minimum Charge	Job		150		150	
Shave & Refit						
Install	Ea.		25		25	Includes labor to shave and rework door to fit opening at the job site.
Minimum Charge	Job		150		150	

Solid Core Door	Unit	Material	Labor	Equip.	Total	Specification
Pre-hung Wood						
2′ 4″ x 6′ 8″						
Demolish	Ea.		28		28	Cost includes material and labor to install 2′ 4″ x 6′ 8″ pre-hung 1-3/4″ door, 4-1/2″ split jamb, including casing and stop, hinges, aluminum sill, weatherstripping.
Install	Ea.	174	15.05		189.05	
Demolish and Install	Ea.	174	43.05		217.05	
Remove and Install	Ea.		15.04		15.04	
Clean	Ea.	2.73	9.10		11.83	
Paint	Ea.	8.85	37.50		46.35	
Minimum Charge	Job		150		150	
2′ 6″ x 6′ 8″						
Demolish	Ea.		28		28	Cost includes material and labor to install 2′ 6″ x 6′ 8″ pre-hung 1-3/4″ door, 4-1/2″ split jamb, including casing and stop, hinges, aluminum sill, weatherstripping.
Install	Ea.	166	15.05		181.05	
Demolish and Install	Ea.	166	43.05		209.05	
Remove and Install	Ea.		15.04		15.04	
Clean	Ea.	2.73	9.10		11.83	
Paint	Ea.	8.85	37.50		46.35	
Minimum Charge	Job		150		150	

Solid Core Door	Unit	Material	Labor	Equip.	Total	Specification
2' 8" x 6' 8"						
Demolish	Ea.		28		28	Cost includes material and labor to install 2' 8" x 6' 8" pre-hung 1-3/4" door, 4-1/2" split jamb, including casing and stop, hinges, aluminum sill, weatherstripping.
Install	Ea.	166	15.05		181.05	
Demolish and Install	Ea.	166	43.05		209.05	
Remove and Install	Ea.		15.04		15.04	
Clean	Ea.	2.73	9.10		11.83	
Paint	Ea.	8.85	37.50		46.35	
Minimum Charge	Job		150		150	
3' x 6' 8"						
Demolish	Ea.		28		28	Cost includes material and labor to install 3' x 6' 8" pre-hung 1-3/4" door, including casing and stop, hinges, jamb, aluminum sill, weatherstripping.
Install	Ea.	173	15.85		188.85	
Demolish and Install	Ea.	173	43.85		216.85	
Remove and Install	Ea.		15.83		15.83	
Clean	Ea.	2.73	9.10		11.83	
Paint	Ea.	8.85	37.50		46.35	
Minimum Charge	Job		150		150	
3' 6" x 6' 8"						
Demolish	Ea.		28		28	Cost includes material and labor to install 3' 6" x 6' 8" pre-hung 1-3/4" door, 4-1/2" split jamb, including casing and stop, hinges, aluminum sill, weatherstripping.
Install	Ea.	215	16.70		231.70	
Demolish and Install	Ea.	215	44.70		259.70	
Remove and Install	Ea.		16.71		16.71	
Clean	Ea.	2.73	9.10		11.83	
Paint	Ea.	8.85	37.50		46.35	
Minimum Charge	Job		150		150	
Raised Panel						
Demolish	Ea.		28		28	Cost includes material and labor to install 3' x 6' 8" pre-hung 1-3/4" door, jamb, including casing and stop, hinges, aluminum sill, weatherstripping.
Install	Ea.	340	16.70		356.70	
Demolish and Install	Ea.	340	44.70		384.70	
Remove and Install	Ea.		16.71		16.71	
Clean	Ea.	2.73	9.10		11.83	
Paint	Ea.	8.85	88		96.85	
Minimum Charge	Job		150		150	
Stain						
Install	Ea.	3.93	22		25.93	Includes labor and material to stain single door and trim on both sides.
Minimum Charge	Job		132		132	

Bifold Door	Unit	Material	Labor	Equip.	Total	Specification
Wood, Single						
3'						
Demolish	Ea.		21		21	Cost includes material and labor to install 3' wide solid wood bi-fold door not including jamb, trim, track, and hardware.
Install	Ea.	53	46.50		99.50	
Demolish and Install	Ea.	53	67.50		120.50	
Remove and Install	Ea.		46.28		46.28	
Clean	Ea.	2.73	9.10		11.83	
Paint	Ea.	7.55	37.50		45.05	
Minimum Charge	Job		150		150	
5'						
Demolish	Ea.		21		21	Includes material and labor to install pair of 2' 6" wide solid wood bi-fold doors not including jamb, trim, track, and hardware.
Install	Ea.	97.50	54.50		152	
Demolish and Install	Ea.	97.50	75.50		173	
Remove and Install	Ea.		54.69		54.69	
Clean	Ea.	1.20	15.55		16.75	
Paint	Ea.	16.25	66		82.25	
Minimum Charge	Job		150		150	

Bifold Door	Unit	Material	Labor	Equip.	Total	Specification
6'						
Demolish	Ea.		21		21	Includes material and labor to install pair of 3' wide solid wood bi-fold doors not including jamb, trim, track, and hardware.
Install	Ea.	106	60		166	
Demolish and Install	Ea.	106	81		187	
Remove and Install	Ea.		60.16		60.16	
Clean	Ea.	1.36	18.15		19.51	
Paint	Ea.	16.25	66		82.25	
Minimum Charge	Job		150		150	
8'						
Demolish	Ea.		21		21	Includes material and labor to install pair of 4' wide solid wood bi-fold doors not including jamb, trim, track, and hardware.
Install	Ea.	185	33.50		218.50	
Demolish and Install	Ea.	185	54.50		239.50	
Remove and Install	Ea.		33.42		33.42	
Clean	Ea.	1.88	22		23.88	
Paint	Ea.	8.45	66		74.45	
Minimum Charge	Job		150		150	
Mirrored						
5'						
Demolish	Ea.		21		21	Includes material and labor to install pair of 2' 6" wide mirrored bi-fold doors including jamb, trim, track, and hardware.
Install	Ea.	375	27.50		402.50	
Demolish and Install	Ea.	375	48.50		423.50	
Remove and Install	Ea.		27.35		27.35	
Clean	Ea.	1.20	7.80		9	
Minimum Charge	Job		150		150	
6'						
Demolish	Ea.		21		21	Includes material and labor to install pair of 3' wide mirrored bi-fold doors including jamb, trim, track, and hardware.
Install	Ea.	415	33.50		448.50	
Demolish and Install	Ea.	415	54.50		469.50	
Remove and Install	Ea.		33.42		33.42	
Clean	Ea.	1.36	9.10		10.46	
Minimum Charge	Job		150		150	
8'						
Demolish	Ea.		21		21	Includes material and labor to install pair of 4' wide mirrored bi-fold doors including jamb, trim, track, and hardware.
Install	Ea.	665	50		715	
Demolish and Install	Ea.	665	71		736	
Remove and Install	Ea.		50.13		50.13	
Clean	Ea.	1.88	10.90		12.78	
Minimum Charge	Job		150		150	
Louvered						
3'						
Demolish	Ea.		21		21	Cost includes material and labor to install 3' wide louvered bi-fold door not including jamb, trim, track, and hardware.
Install	Ea.	124	46.50		170.50	
Demolish and Install	Ea.	124	67.50		191.50	
Remove and Install	Ea.		46.28		46.28	
Clean	Ea.	2.73	9.10		11.83	
Paint	Ea.	7.55	37.50		45.05	
Minimum Charge	Job		150		150	
5'						
Demolish	Ea.		21		21	Includes material and labor to install pair of 2' 6" wide louvered wood bi-fold doors not including jamb, trim, track, and hardware.
Install	Ea.	200	54.50		254.50	
Demolish and Install	Ea.	200	75.50		275.50	
Remove and Install	Ea.		54.69		54.69	
Clean	Ea.	1.20	15.55		16.75	
Paint	Ea.	16.25	66		82.25	
Minimum Charge	Job		150		150	

Bifold Door	Unit	Material	Labor	Equip.	Total	Specification
6'						
Demolish	Ea.		21		21	Includes material and labor to install pair of 3' wide louvered wood bi-fold doors not including jamb, trim, track, and hardware.
Install	Ea.	222	60		282	
Demolish and Install	Ea.	222	81		303	
Remove and Install	Ea.		60.16		60.16	
Clean	Ea.	1.36	18.15		19.51	
Paint	Ea.	16.25	66		82.25	
Minimum Charge	Job		150		150	
8'						
Demolish	Ea.		21		21	Includes material and labor to install pair of 4' wide solid wood bi-fold doors not including jamb, trim, track, and hardware.
Install	Ea.	390	30		420	
Demolish and Install	Ea.	390	51		441	
Remove and Install	Ea.		30.08		30.08	
Clean	Ea.	6.50	15.55		22.05	
Paint	Ea.	8.45	66		74.45	
Minimum Charge	Job		150		150	
Accordion Type						
Demolish	S.F.		.71		.71	Includes material and labor to install vinyl folding accordion closet doors, including track and frame.
Install	S.F.	1.76	1.50		3.26	
Demolish and Install	S.F.	1.76	2.21		3.97	
Remove and Install	S.F.		1.50		1.50	
Clean	S.F.		.26		.26	
Minimum Charge	Job		150		150	
Casing Trim						
Single Width						
Demolish	Ea.		5.30		5.30	Cost includes material and labor to install 11/16" x 2-1/2" pine ranch style casing for one side of a standard door opening.
Install	Opng.	15.90	51		66.90	
Demolish and Install	Opng.	15.90	56.30		72.20	
Remove and Install	Opng.		40.79		40.79	
Clean	Opng.	.39	2.18		2.57	
Paint	Ea.	3.06	5.30		8.36	
Minimum Charge	Job		150		150	
Double Width						
Demolish	Ea.		6.30		6.30	Cost includes material and labor to install 11/16" x 2-1/2" pine ranch style door casing for one side of a double door opening.
Install	Opng.	18.70	60		78.70	
Demolish and Install	Opng.	18.70	66.30		85	
Remove and Install	Opng.		48.13		48.13	
Clean	Opng.	.45	4.36		4.81	
Paint	Ea.	3.06	6		9.06	
Minimum Charge	Job		150		150	
Jamb						
Demolish	Ea.		21		21	Includes material and labor to install flat pine jamb with square cut heads and rabbeted sides for 6' 8" high and 3-9/16" door including trim sets for both sides.
Install	Ea.	77	8.15		85.15	
Demolish and Install	Ea.	77	29.15		106.15	
Clean	Ea.	.09	5.45		5.54	
Paint	Ea.	5.55	17.60		23.15	
Minimum Charge	Job		75		75	
Stain						
Install	Ea.	3.93	22		25.93	Includes labor and material to stain single door and trim on both sides.
Minimum Charge	Job		132		132	
Shave & Refit						
Install	Ea.		25		25	Includes labor to shave and rework door to fit opening at the job site.
Minimum Charge	Job		75		75	

Louver Door	Unit	Material	Labor	Equip.	Total	Specification
Full						
2' x 6' 8"						
Demolish	Ea.		5.95		5.95	Cost includes material and labor to install 2' x 6' 8" pine full louver door, jamb, hinges.
Install	Ea.	168	30		198	
Demolish and Install	Ea.	168	35.95		203.95	
Remove and Install	Ea.		30.08		30.08	
Clean	Ea.	5.05	27.50		32.55	
Paint	Ea.	7.55	37.50		45.05	
Minimum Charge	Job		150		150	
2' 8" x 6' 8"						
Demolish	Ea.		5.95		5.95	Cost includes material and labor to install 2' 8" x 6' 8" pine full louver door, jamb, hinges.
Install	Ea.	187	30		217	
Demolish and Install	Ea.	187	35.95		222.95	
Remove and Install	Ea.		30.08		30.08	
Clean	Ea.	5.05	27.50		32.55	
Paint	Ea.	7.55	37.50		45.05	
Minimum Charge	Job		150		150	
3' x 6' 8"						
Demolish	Ea.		5.95		5.95	Cost includes material and labor to install 3' x 6' 8" pine full louver door, jamb, hinges.
Install	Ea.	197	31.50		228.50	
Demolish and Install	Ea.	197	37.45		234.45	
Remove and Install	Ea.		31.66		31.66	
Clean	Ea.	5.05	27.50		32.55	
Paint	Ea.	7.55	37.50		45.05	
Minimum Charge	Job		150		150	
Half						
2' x 6' 8"						
Demolish	Ea.		5.95		5.95	Cost includes material and labor to install 2' x 6' 8" pine half louver door, jamb, hinges.
Install	Ea.	278	30		308	
Demolish and Install	Ea.	278	35.95		313.95	
Remove and Install	Ea.		30.08		30.08	
Clean	Ea.	5.05	27.50		32.55	
Paint	Ea.	7.55	37.50		45.05	
Minimum Charge	Job		150		150	
2' 8" x 6' 8"						
Demolish	Ea.		5.95		5.95	Cost includes material and labor to install 2' 8" x 6' 8" pine half louver door, jamb, hinges.
Install	Ea.	300	30		330	
Demolish and Install	Ea.	300	35.95		335.95	
Remove and Install	Ea.		30.08		30.08	
Clean	Ea.	5.05	27.50		32.55	
Paint	Ea.	7.55	37.50		45.05	
Minimum Charge	Job		150		150	
3' x 6' 8"						
Demolish	Ea.		5.95		5.95	Cost includes material and labor to install 3' x 6' 8" pine half louver door, jamb, hinges.
Install	Ea.	315	31.50		346.50	
Demolish and Install	Ea.	315	37.45		352.45	
Remove and Install	Ea.		31.66		31.66	
Clean	Ea.	5.05	27.50		32.55	
Paint	Ea.	7.55	37.50		45.05	
Minimum Charge	Job		150		150	
Jamb						
Demolish	Ea.		21		21	Includes material and labor to install flat pine jamb with square cut heads and rabbeted sides for 6' 8" high and 3-9/16" door including trim sets for both sides.
Install	Ea.	77	8.15		85.15	
Demolish and Install	Ea.	77	29.15		106.15	
Clean	Ea.	.09	5.45		5.54	
Paint	Ea.	5.55	17.60		23.15	
Minimum Charge	Job		75		75	

Louver Door	Unit	Material	Labor	Equip.	Total	Specification
Stain						
Install	Ea.	3.93	22		25.93	Includes labor and material to stain single door and trim on both sides.
Minimum Charge	Job		132		132	
Shave & Refit						
Install	Ea.		25		25	Includes labor to shave and rework door to fit opening at the job site.
Minimum Charge	Job		132		132	

Bypass Sliding Door	Unit	Material	Labor	Equip.	Total	Specification
Wood						
5′						
Demolish	Ea.		28		28	Includes material and labor to install pair of 2′ 6″ wide lauan hollow core wood bi-pass doors with jamb, track hardware.
Install	Opng.	185	54.50		239.50	
Demolish and Install	Opng.	185	82.50		267.50	
Remove and Install	Opng.		54.69		54.69	
Clean	Ea.	1.20	15.55		16.75	
Paint	Ea.	16.25	66		82.25	
Minimum Charge	Job		150		150	
6′						
Demolish	Ea.		28		28	Includes material and labor to install pair of 3′ wide lauan hollow core wood bi-pass doors with jamb, track and hardware.
Install	Opng.	198	60		258	
Demolish and Install	Opng.	198	88		286	
Remove and Install	Opng.		60.16		60.16	
Clean	Ea.	1.36	18.15		19.51	
Paint	Ea.	16.25	66		82.25	
Minimum Charge	Job		150		150	
8′						
Demolish	Ea.		28		28	Includes material and labor to install pair of 4′ wide lauan hollow core wood bi-pass doors with jamb, track and hardware.
Install	Opng.	275	75		350	
Demolish and Install	Opng.	275	103		378	
Remove and Install	Opng.		75.20		75.20	
Clean	Ea.	1.88	22		23.88	
Paint	Ea.	8.45	66		74.45	
Minimum Charge	Job		150		150	
Mirrored						
5′						
Demolish	Ea.		28		28	Includes material and labor to install pair of 2′ 6″ wide mirrored bi-pass doors with jamb, trim, track hardware.
Install	Opng.	208	60		268	
Demolish and Install	Opng.	208	88		296	
Remove and Install	Opng.		60.16		60.16	
Clean	Ea.	1.20	7.80		9	
Minimum Charge	Job		150		150	
6′						
Demolish	Ea.		28		28	Includes material and labor to install pair of 3′ wide mirrored bi-pass doors with jamb, trim, track hardware.
Install	Opng.	238	60		298	
Demolish and Install	Opng.	238	88		326	
Remove and Install	Opng.		60.16		60.16	
Clean	Ea.	1.36	9.10		10.46	
Minimum Charge	Job		150		150	
8′						
Demolish	Ea.		28		28	Includes material and labor to install pair of 4′ wide mirrored bi-pass doors with jamb, trim, track hardware.
Install	Opng.	390	67		457	
Demolish and Install	Opng.	390	95		485	
Remove and Install	Opng.		66.84		66.84	
Clean	Ea.	1.88	10.90		12.78	
Minimum Charge	Job		150		150	

Bypass Sliding Door	Unit	Material	Labor	Equip.	Total	Specification
Casing Trim						
Single Width						
Demolish	Ea.		5.30		5.30	Cost includes material and labor to install 11/16" x 2-1/2" pine ranch style casing for one side of a standard door opening.
Install	Opng.	15.90	51		66.90	
Demolish and Install	Opng.	15.90	56.30		72.20	
Remove and Install	Opng.		40.79		40.79	
Clean	Opng.	.39	2.18		2.57	
Paint	Ea.	3.06	5.30		8.36	
Minimum Charge	Job		150		150	
Double Width						
Demolish	Ea.		6.30		6.30	Cost includes material and labor to install 11/16" x 2-1/2" pine ranch style door casing for one side of a double door opening.
Install	Opng.	18.70	60		78.70	
Demolish and Install	Opng.	18.70	66.30		85	
Remove and Install	Opng.		48.13		48.13	
Clean	Opng.	.45	4.36		4.81	
Paint	Ea.	3.06	6		9.06	
Minimum Charge	Job		150		150	
Jamb						
Demolish	Ea.		21		21	Includes material and labor to install flat pine jamb with square cut heads and rabbeted sides for 6' 8" high and 3-9/16" door including trim sets for both sides.
Install	Ea.	77	8.15		85.15	
Demolish and Install	Ea.	77	29.15		106.15	
Clean	Ea.	.09	5.45		5.54	
Paint	Ea.	5.55	17.60		23.15	
Minimum Charge	Job		75		75	
Stain						
Install	Ea.	3.93	22		25.93	Includes labor and material to stain single door and trim on both sides.
Minimum Charge	Job		132		132	
Shave & Refit						
Install	Ea.		25		25	Includes labor to shave and rework door to fit opening at the job site.
Minimum Charge	Job		75		75	

Base Molding	Unit	Material	Labor	Equip.	Total	Specification
One Piece						
1-5/8"						
Demolish	L.F.		.40		.40	Includes material and labor to install pine or birch base molding.
Install	L.F.	.23	.89		1.12	
Demolish and Install	L.F.	.23	1.29		1.52	
Remove and Install	L.F.		.71		.71	
Clean	L.F.	.02	.13		.15	
Paint	L.F.	.03	.29		.32	
Minimum Charge	Job		150		150	
2-1/2"						
Demolish	L.F.		.40		.40	Cost includes material and labor to install 1/2" x 2-1/2" pine or birch base molding.
Install	L.F.	.48	.93		1.41	
Demolish and Install	L.F.	.48	1.33		1.81	
Remove and Install	L.F.		.74		.74	
Clean	L.F.	.02	.13		.15	
Paint	L.F.	.03	.29		.32	
Minimum Charge	Job		150		150	

Base Molding	Unit	Material	Labor	Equip.	Total	Specification
3-1/2"						
Demolish	L.F.		.40		.40	Cost includes material and labor to install 9/16" x 3-1/2" pine or birch base molding.
Install	L.F.	.93	.96		1.89	
Demolish and Install	L.F.	.93	1.36		2.29	
Remove and Install	L.F.		.77		.77	
Clean	L.F.	.02	.13		.15	
Paint	L.F.	.03	.31		.34	
Minimum Charge	Job		150		150	
6"						
Demolish	L.F.		.40		.40	Cost includes material and labor to install 1" x 6" pine or birch base molding.
Install	L.F.	.69	1.05		1.74	
Demolish and Install	L.F.	.69	1.45		2.14	
Remove and Install	L.F.		.84		.84	
Clean	L.F.	.02	.18		.20	
Paint	L.F.	.04	.40		.44	
Minimum Charge	Job		150		150	
8"						
Demolish	L.F.		.48		.48	Cost includes material and labor to install 1" x 8" pine or birch base molding.
Install	L.F.	1.08	1.16		2.24	
Demolish and Install	L.F.	1.08	1.64		2.72	
Remove and Install	L.F.		.93		.93	
Clean	L.F.	.03	.22		.25	
Paint	L.F.	.06	.44		.50	
Minimum Charge	Job		150		150	
Premium / Custom Grade						
Demolish	L.F.		.48		.48	Includes material and labor to install custom 2 or 3 piece oak or other hardwood, stain grade trim.
Install	L.F.	3.63	2.89		6.52	
Demolish and Install	L.F.	3.63	3.37		7	
Remove and Install	L.F.		2.31		2.31	
Clean	L.F.	.03	.22		.25	
Paint	L.F.	.06	.44		.50	
Minimum Charge	Job		150		150	
One Piece Oak						
1-5/8"						
Demolish	L.F.		.40		.40	Cost includes material and labor to install 1/2" x 1-5/8" oak base molding.
Install	L.F.	.69	.89		1.58	
Demolish and Install	L.F.	.69	1.29		1.98	
Remove and Install	L.F.		.71		.71	
Clean	L.F.	.02	.13		.15	
Paint	L.F.	.08	.39		.47	
Minimum Charge	Job		150		150	
2-1/2"						
Demolish	L.F.		.40		.40	Cost includes material and labor to install 1/2" x 2-1/2" oak base molding.
Install	L.F.	1.06	.91		1.97	
Demolish and Install	L.F.	1.06	1.31		2.37	
Remove and Install	L.F.		.73		.73	
Clean	L.F.	.02	.13		.15	
Paint	L.F.	.08	.40		.48	
Minimum Charge	Job		150		150	
3-1/2"						
Demolish	L.F.		.40		.40	Cost includes material and labor to install 1/2" x 3-1/2" oak base molding.
Install	L.F.	2.09	.93		3.02	
Demolish and Install	L.F.	2.09	1.33		3.42	
Remove and Install	L.F.		.74		.74	
Clean	L.F.	.02	.13		.15	
Paint	L.F.	.08	.41		.49	
Minimum Charge	Job		150		150	

Base Molding	Unit	Material	Labor	Equip.	Total	Specification
6"						
Demolish	L.F.		.40		.40	Cost includes material and labor to install 1" x 6" oak base molding.
Install	L.F.	2.26	1.05		3.31	
Demolish and Install	L.F.	2.26	1.45		3.71	
Remove and Install	L.F.		.84		.84	
Clean	L.F.	.03	.22		.25	
Paint	L.F.	.08	.42		.50	
Minimum Charge	Job		150		150	
8"						
Demolish	L.F.		.48		.48	Cost includes material and labor to install 1" x 8" oak base molding.
Install	L.F.	2.76	1.16		3.92	
Demolish and Install	L.F.	2.76	1.64		4.40	
Remove and Install	L.F.		.93		.93	
Clean	L.F.	.03	.22		.25	
Paint	L.F.	.08	.44		.52	
Minimum Charge	Job		150		150	
Vinyl Base Molding						
2-1/2"						
Demolish	L.F.		.40		.40	Cost includes material and labor to install 2-1/2" vinyl or rubber cove base including adhesive.
Install	L.F.	.46	.87		1.33	
Demolish and Install	L.F.	.46	1.27		1.73	
Remove and Install	L.F.		.87		.87	
Clean	L.F.		.20		.20	
Minimum Charge	Job		138		138	
4"						
Demolish	L.F.		.40		.40	Cost includes material and labor to install 4" vinyl or rubber cove base including adhesive.
Install	L.F.	.53	.87		1.40	
Demolish and Install	L.F.	.53	1.27		1.80	
Remove and Install	L.F.		.87		.87	
Clean	L.F.		.20		.20	
Minimum Charge	Job		138		138	
6"						
Demolish	L.F.		.40		.40	Cost includes material and labor to install 6" vinyl or rubber cove base including adhesive.
Install	L.F.	.85	.87		1.72	
Demolish and Install	L.F.	.85	1.27		2.12	
Remove and Install	L.F.		.87		.87	
Clean	L.F.		.20		.20	
Minimum Charge	Job		138		138	
Prefinished						
Demolish	L.F.		.40		.40	Cost includes material and labor to install 9/16" x 3-5/16" prefinished base molding.
Install	L.F.	1.69	.96		2.65	
Demolish and Install	L.F.	1.69	1.36		3.05	
Remove and Install	L.F.		.77		.77	
Clean	L.F.	.02	.13		.15	
Paint	L.F.	.03	.31		.34	
Minimum Charge	Job		150		150	
Shoe						
Demolish	L.F.		.40		.40	Includes material and labor to install pine or birch base molding.
Install	L.F.	.23	.89		1.12	
Demolish and Install	L.F.	.23	1.29		1.52	
Remove and Install	L.F.		.71		.71	
Clean	L.F.	.02	.13		.15	
Paint	L.F.	.12	.42		.54	
Minimum Charge	Job		150		150	

Base Molding	Unit	Material	Labor	Equip.	Total	Specification
Carpeted						
Demolish	L.F.		.16		.16	Includes material and labor to install carpet baseboard with adhesive.
Install	L.F.	1.17	1.67		2.84	
Demolish and Install	L.F.	1.17	1.83		3	
Remove and Install	L.F.		1.33		1.33	
Clean	L.F.		.08		.08	
Minimum Charge	Job		138		138	
Ceramic Tile						
Demolish	L.F.		.88		.88	Includes material and labor to install thin set ceramic tile cove base including grout.
Install	L.F.	3.36	3.84		7.20	
Demolish and Install	L.F.	3.36	4.72		8.08	
Clean	L.F.		.42		.42	
Minimum Charge	Job		138		138	
Quarry Tile						
Demolish	L.F.		.88		.88	Includes material and labor to install quarry tile base including grout.
Install	L.F.	4.20	4.47		8.67	
Demolish and Install	L.F.	4.20	5.35		9.55	
Clean	L.F.		.42		.42	
Minimum Charge	Job		138		138	
Caulk and Renail						
Install	L.F.	.01	.47		.48	Includes labor and material to renail base and caulk where necessary.
Minimum Charge	Job		150		150	

Molding	Unit	Material	Labor	Equip.	Total	Specification
Oak Base						
Demolish	L.F.		.40		.40	Cost includes material and labor to install 1/2" x 2-1/2" oak base molding.
Install	L.F.	1.06	.91		1.97	
Demolish and Install	L.F.	1.06	1.31		2.37	
Remove and Install	L.F.		.73		.73	
Clean	L.F.	.02	.13		.15	
Paint	L.F.	.08	.40		.48	
Minimum Charge	Job		150		150	
Shoe						
Demolish	L.F.		.40		.40	Includes material and labor to install pine or birch base molding.
Install	L.F.	.23	.89		1.12	
Demolish and Install	L.F.	.23	1.29		1.52	
Remove and Install	L.F.		.71		.71	
Clean	L.F.	.02	.13		.15	
Paint	L.F.	.12	.42		.54	
Minimum Charge	Job		150		150	
Quarter Round						
Demolish	L.F.		.40		.40	Includes material and labor to install pine base quarter round molding.
Install	L.F.	.40	1.18		1.58	
Demolish and Install	L.F.	.40	1.58		1.98	
Remove and Install	L.F.		.94		.94	
Clean	L.F.	.02	.13		.15	
Paint	L.F.	.06	.66		.72	
Minimum Charge	Job		150		150	
Chair Rail						
Demolish	L.F.		.40		.40	Cost includes material and labor to install 5/8" x 2-1/2" oak chair rail molding.
Install	L.F.	.94	1.11		2.05	
Demolish and Install	L.F.	.94	1.51		2.45	
Remove and Install	L.F.		.89		.89	
Clean	L.F.	.02	.13		.15	
Paint	L.F.	.06	.66		.72	
Minimum Charge	Job		150		150	

Molding	Unit	Material	Labor	Equip.	Total	Specification
Crown						
Demolish	L.F.		.40		.40	Includes material and labor to install unfinished pine crown molding.
Install	L.F.	1.84	1.20		3.04	
Demolish and Install	L.F.	1.84	1.60		3.44	
Remove and Install	L.F.		.96		.96	
Clean	L.F.	.03	.17		.20	
Paint	L.F.	.06	.66		.72	
Minimum Charge	Job		150		150	
Cove						
Demolish	L.F.		.40		.40	Cost includes material and labor to install 1/2" x 2-3/4" pine cove molding.
Install	L.F.	1.38	1.18		2.56	
Demolish and Install	L.F.	1.38	1.58		2.96	
Remove and Install	L.F.		.94		.94	
Clean	L.F.	.02	.13		.15	
Paint	L.F.	.06	.66		.72	
Minimum Charge	Job		150		150	
Corner						
Demolish	L.F.		.40		.40	Includes material and labor to install inside or outside pine corner molding.
Install	L.F.	.56	.47		1.03	
Demolish and Install	L.F.	.56	.87		1.43	
Remove and Install	L.F.		.38		.38	
Clean	L.F.	.02	.13		.15	
Paint	L.F.	.06	.66		.72	
Minimum Charge	Job		150		150	
Picture						
Demolish	L.F.		.40		.40	Cost includes material and labor to install 9/16" x 2-1/2" pine casing.
Install	L.F.	.94	1.25		2.19	
Demolish and Install	L.F.	.94	1.65		2.59	
Remove and Install	L.F.		1		1	
Clean	L.F.	.02	.13		.15	
Paint	L.F.	.06	.66		.72	
Minimum Charge	Job		150		150	
Single Cased Opening						
Demolish	Ea.		21		21	Includes material and labor to install flat pine jamb with square cut heads and rabbeted sides for 6' 8" high opening including trim sets for both sides.
Install	Ea.	13.95	18.80		32.75	
Demolish and Install	Ea.	13.95	39.80		53.75	
Clean	Ea.	.19	4.54		4.73	
Paint	Ea.	.99	6.60		7.59	
Minimum Charge	Job		150		150	
Double Cased Opening						
Demolish	Ea.		21		21	Includes material and labor to install flat pine jamb with square cut heads and rabbeted sides for 6' 8" high opening including trim sets for both sides.
Install	Ea.	28	21.50		49.50	
Demolish and Install	Ea.	28	42.50		70.50	
Clean	Ea.	.22	5.30		5.52	
Paint	Ea.	1.14	7.75		8.89	
Minimum Charge	Job		150		150	

Window Trim Set	Unit	Material	Labor	Equip.	Total	Specification
Single						
Demolish	Ea.		5.40		5.40	Cost includes material and labor to install 11/16" x 2-1/2" pine ranch style window casing.
Install	Opng.	26	30		56	
Demolish and Install	Opng.	26	35.40		61.40	
Remove and Install	Opng.		24.06		24.06	
Clean	Ea.	.28	3.11		3.39	
Paint	Ea.	.92	6.30		7.22	
Minimum Charge	Job		150		150	

Finish Carpentry

Window Trim Set	Unit	Material	Labor	Equip.	Total	Specification
Double						
Demolish	Ea.		6.05		6.05	Cost includes material and labor to install 11/16" x 2-1/2" pine ranch style trim for casing one side of a double window opening.
Install	Ea.	35	37.50		72.50	
Demolish and Install	Ea.	35	43.55		78.55	
Remove and Install	Ea.		30.08		30.08	
Clean	Ea.	.48	3.63		4.11	
Paint	Ea.	1.56	10.55		12.11	
Minimum Charge	Job		150		150	
Triple						
Demolish	Ea.		6.90		6.90	Cost includes material and labor to install 11/16" x 2-1/2" pine ranch style trim for casing one side of a triple window opening.
Install	Opng.	60.50	50		110.50	
Demolish and Install	Opng.	60.50	56.90		117.40	
Remove and Install	Opng.		40.11		40.11	
Clean	Ea.	.68	4.36		5.04	
Paint	Ea.	2.18	14.65		16.83	
Minimum Charge	Job		150		150	
Window Casing Per L.F.						
Demolish	L.F.		.40		.40	Cost includes material and labor to install 11/16" x 2-1/2" pine ranch style trim for casing.
Install	L.F.	.94	1.25		2.19	
Demolish and Install	L.F.	.94	1.65		2.59	
Remove and Install	L.F.		1		1	
Clean	L.F.	.02	.13		.15	
Paint	L.F.	.06	.66		.72	
Minimum Charge	Job		150		150	

Window Sill	Unit	Material	Labor	Equip.	Total	Specification
Wood						
Demolish	L.F.		.40		.40	Includes material and labor to install flat, wood window stool.
Install	L.F.	1.68	1.50		3.18	
Demolish and Install	L.F.	1.68	1.90		3.58	
Remove and Install	L.F.		1.20		1.20	
Clean	L.F.	.02	.13		.15	
Paint	L.F.	.06	.66		.72	
Minimum Charge	Job		150		150	
Marble						
Demolish	L.F.		.61		.61	Includes material and labor to install marble window sills, 6" x 3/4" thick.
Install	L.F.	7.80	6.30		14.10	
Demolish and Install	L.F.	7.80	6.91		14.71	
Remove and Install	L.F.		5.03		5.03	
Clean	L.F.	.02	.13		.15	
Minimum Charge	Job		138		138	

Shelving	Unit	Material	Labor	Equip.	Total	Specification
Pine						
18"						
Demolish	L.F.		.69		.69	Cost includes material and labor to install 1" thick custom shelving board 18" wide.
Install	L.F.	6.90	3.17		10.07	
Demolish and Install	L.F.	6.90	3.86		10.76	
Remove and Install	L.F.		2.53		2.53	
Clean	L.F.	.04	.36		.40	
Paint	L.F.	.72	.42		1.14	
Minimum Charge	Job		150		150	

Shelving	Unit	Material	Labor	Equip.	Total	Specification
24″						
Demolish	L.F.		.69		.69	Cost includes material and labor to install 1″ thick custom shelving board.
Install	L.F.	9.20	3.54		12.74	
Demolish and Install	L.F.	9.20	4.23		13.43	
Remove and Install	L.F.		2.83		2.83	
Clean	L.F.	.06	.55		.61	
Paint	L.F.	.97	.56		1.53	
Minimum Charge	Job		150		150	
Particle Board						
Demolish	L.F.		.64		.64	Cost includes material and labor to install 3/4″ thick particle shelving board 18″ wide.
Install	L.F.	1.10	1.58		2.68	
Demolish and Install	L.F.	1.10	2.22		3.32	
Remove and Install	L.F.		1.27		1.27	
Clean	L.F.	.04	.36		.40	
Paint	S.F.	.10	.61		.71	
Minimum Charge	Job		150		150	
Plywood						
Demolish	L.F.		.64		.64	Cost includes material and labor to install 3/4″ thick plywood shelving board.
Install	L.F.	2.01	4.63		6.64	
Demolish and Install	L.F.	2.01	5.27		7.28	
Remove and Install	L.F.		3.70		3.70	
Clean	L.F.	.04	.36		.40	
Paint	S.F.	.10	.61		.71	
Minimum Charge	Job		150		150	
Hardwood (oak)						
Demolish	L.F.		.64		.64	Includes material and labor to install oak shelving, 1″ x 12″, incl. cleats and bracing.
Install	L.F.	8.25	2.07		10.32	
Demolish and Install	L.F.	8.25	2.71		10.96	
Remove and Install	L.F.		1.66		1.66	
Clean	L.F.	.04	.36		.40	
Paint	S.F.	.10	.61		.71	
Minimum Charge	Job		150		150	
Custom Bookcase						
Demolish	L.F.		8.05		8.05	Includes material and labor to install custom modular bookcase unit with clear pine faced frames, shelves 12″ O.C., 7′ high, 8″ deep.
Install	L.F.	26	30		56	
Demolish and Install	L.F.	26	38.05		64.05	
Remove and Install	L.F.		24.06		24.06	
Clean	L.F.	1.09	7.80		8.89	
Paint	L.F.	.94	20.50		21.44	
Minimum Charge	Job		150		150	
Glass						
Demolish	L.F.		.64		.64	Includes material and labor to install glass shelving, 1″ x 12″, incl. cleats and bracing.
Install	L.F.	22.50	2.07		24.57	
Demolish and Install	L.F.	22.50	2.71		25.21	
Remove and Install	L.F.		1.66		1.66	
Clean	L.F.	.04	.34		.38	
Minimum Charge	Job		146		146	

Closet Shelving	Unit	Material	Labor	Equip.	Total	Specification
Closet Shelf and Rod						
Demolish	L.F.		.64		.64	Cost includes material and labor to install 1″ thick custom shelving board 18″ wide and 1″ diameter clothes pole with brackets 3′ O.C.
Install	L.F.	2.98	5		7.98	
Demolish and Install	L.F.	2.98	5.64		8.62	
Remove and Install	L.F.		4.01		4.01	
Clean	L.F.	.02	.27		.29	
Paint	L.F.	.11	.66		.77	
Minimum Charge	Job		150		150	

Closet Shelving	Unit	Material	Labor	Equip.	Total	Specification
Clothing Rod						
Demolish	L.F.		.42		.42	Includes material and labor to install fir closet pole, 1-5/8" diameter.
Install	L.F.	1.31	1.50		2.81	
Demolish and Install	L.F.	1.31	1.92		3.23	
Remove and Install	L.F.		1.20		1.20	
Clean	L.F.	.02	.38		.40	
Paint	L.F.	.12	.72		.84	
Minimum Charge	Job		150		150	

Stair Assemblies	Unit	Material	Labor	Equip.	Total	Specification
Straight Hardwood						
Demolish	Ea.		290		290	Includes material and labor to install factory cut and assembled straight closed box stairs with oak treads and prefinished stair rail with balusters.
Install	Flight	855	201		1056	
Demolish and Install	Flight	855	491		1346	
Clean	Flight	2.73	18.90		21.63	
Paint	Flight	10	35		45	
Minimum Charge	Job		150		150	
Clear Oak Tread						
Demolish	Ea.		7.55		7.55	Includes material and labor to install clear oak treads per riser.
Install	Ea.	26	16.70		42.70	
Demolish and Install	Ea.	26	24.25		50.25	
Clean	Ea.	.09	1.89		1.98	
Paint	Ea.	.33	1.96		2.29	
Minimum Charge	Job		150		150	
Landing						
Demolish	S.F.		2.52		2.52	Includes material and labor to install clear oak landing.
Install	S.F.	10.35	16.70		27.05	
Demolish and Install	S.F.	10.35	19.22		29.57	
Clean	S.F.		.27		.27	
Paint	S.F.	.25	.65		.90	
Minimum Charge	Job		150		150	
Refinish / Stain						
Install	S.F.	1.14	1.68		2.82	Includes labor and material to sand and finish (3 passes) with two coats of urethane on a new floor.
Minimum Charge	Job		132		132	
Spiral Hardwood						
Demolish	Ea.		290		290	Cost includes material and labor to install 4' - 6' diameter spiral stairs with oak treads, factory cut and assembled with double handrails.
Install	Flight	4400	400		4800	
Demolish and Install	Flight	4400	690		5090	
Clean	Flight	2.73	18.90		21.63	
Paint	Flight	10	35		45	
Minimum Charge	Job		150		150	
Curved Hardwood						
Demolish	Ea.		290		290	Includes material and labor to install curved stairs open on one side with clear oak treads and handrail.
Install	Flight	7025	860		7885	
Demolish and Install	Flight	7025	1150		8175	
Clean	Flight	2.73	18.90		21.63	
Paint	Flight	10	35		45	
Minimum Charge	Job		150		150	

Stairs	Unit	Material	Labor	Equip.	Total	Specification
Disappearing						
Demolish	Ea.		15.15		15.15	Includes material and labor to install a folding staircase.
Install	Ea.	110	86		196	
Demolish and Install	Ea.	110	101.15		211.15	
Remove and Install	Ea.		68.75		68.75	
Clean	Ea.	.41	13.65		14.06	
Minimum Charge	Job		150		150	

Stair Components	Unit	Material	Labor	Equip.	Total	Specification
Handrail w / Balusters						
Demolish	L.F.		2.02		2.02	Includes material and labor to install prefinished assembled stair rail with brackets, turned balusters and newel.
Install	L.F.	29.50	6.25		35.75	
Demolish and Install	L.F.	29.50	8.27		37.77	
Clean	L.F.	.55	1.50		2.05	
Paint	L.F.	1.33	3.11		4.44	
Minimum Charge	Job		150		150	
Bannister						
Demolish	L.F.		1.21		1.21	Includes material and labor to install built-up oak railings.
Install	L.F.	33	5		38	
Demolish and Install	L.F.	33	6.21		39.21	
Clean	L.F.	.55	1.36		1.91	
Paint	L.F.	1.33	2.40		3.73	
Minimum Charge	Job		150		150	

Wall and Base Cabinets

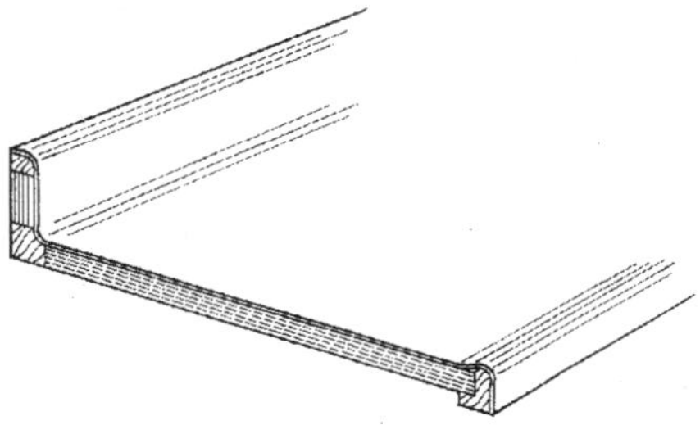

Countertop

Wall Cabinets	Unit	Material	Labor	Equip.	Total	Specification
Good Grade Laminated						
12" Wide, 1 Door						
Demolish	Ea.		6.05		6.05	Includes material and labor to install good modular unit with melamine laminated to particle board, in textured colors or wood grain print finish, hinges and pulls.
Install	Ea.	101	27.50		128.50	
Demolish and Install	Ea.	101	33.55		134.55	
Remove and Install	Ea.		21.88		21.88	
Clean	Ea.	.41	4.19		4.60	
Paint	Ea.	.40	9.45		9.85	
Minimum Charge	Job		150		150	
15" Wide, 1 Door						
Demolish	Ea.		7.55		7.55	Includes material and labor to install good modular unit with melamine laminated to particle board, in textured colors or wood grain print finish, hinges and pulls.
Install	Ea.	114	28		142	
Demolish and Install	Ea.	114	35.55		149.55	
Remove and Install	Ea.		22.49		22.49	
Clean	Ea.	.52	5.20		5.72	
Paint	Ea.	.50	12		12.50	
Minimum Charge	Job		150		150	
18" Wide, 1 Door						
Demolish	Ea.		9.15		9.15	Includes material and labor to install good modular unit with melamine laminated to particle board, in textured colors or wood grain print finish, hinges and pulls.
Install	Ea.	124	29		153	
Demolish and Install	Ea.	124	38.15		162.15	
Remove and Install	Ea.		23.03		23.03	
Clean	Ea.	.62	6.25		6.87	
Paint	Ea.	.59	13.90		14.49	
Minimum Charge	Job		150		150	
21" Wide, 1 Door						
Demolish	Ea.		10.55		10.55	Includes material and labor to install good modular unit with melamine laminated to particle board, in textured colors or wood grain print finish, hinges and pulls.
Install	Ea.	146	29		175	
Demolish and Install	Ea.	146	39.55		185.55	
Remove and Install	Ea.		23.36		23.36	
Clean	Ea.	.72	7.25		7.97	
Paint	Ea.	.70	16.50		17.20	
Minimum Charge	Job		150		150	
24" w / Blind Corner						
Demolish	Ea.		12.10		12.10	Includes material and labor to install good modular unit with melamine laminated to particle board, in textured colors or wood grain print finish, hinges and pulls.
Install	Ea.	141	29.50		170.50	
Demolish and Install	Ea.	141	41.60		182.60	
Remove and Install	Ea.		23.71		23.71	
Clean	Ea.	.83	8.40		9.23	
Paint	Ea.	.79	18.85		19.64	
Minimum Charge	Job		150		150	

Wall Cabinets	Unit	Material	Labor	Equip.	Total	Specification
27" Wide, 2 Door						
Demolish	Ea.		13.45		13.45	Includes material and labor to install good modular unit with melamine laminated to particle board, in textured colors or wood grain print finish, hinges and pulls.
Install	Ea.	174	30.50		204.50	
Demolish and Install	Ea.	174	43.95		217.95	
Remove and Install	Ea.		24.31		24.31	
Clean	Ea.	.92	9.50		10.42	
Paint	Ea.	.90	22		22.90	
Minimum Charge	Job		150		150	
30" Wide, 2 Door						
Demolish	Ea.		15.15		15.15	Includes material and labor to install good modular unit with melamine laminated to particle board, in textured colors or wood grain print finish, hinges and pulls.
Install	Ea.	168	31		199	
Demolish and Install	Ea.	168	46.15		214.15	
Remove and Install	Ea.		24.94		24.94	
Clean	Ea.	1.02	10.40		11.42	
Paint	Ea.	1	24		25	
Minimum Charge	Job		150		150	
36" Wide, 2 Door						
Demolish	Ea.		17.95		17.95	Includes material and labor to install good modular unit with melamine laminated to particle board, in textured colors or wood grain print finish, hinges and pulls.
Install	Ea.	193	32		225	
Demolish and Install	Ea.	193	49.95		242.95	
Remove and Install	Ea.		25.60		25.60	
Clean	Ea.	1.23	12.80		14.03	
Paint	Ea.	1.21	29.50		30.71	
Minimum Charge	Job		150		150	
48" Wide, 2 Door						
Demolish	Ea.		24		24	Includes material and labor to install good modular unit with melamine laminated to particle board, in textured colors or wood grain print finish, hinges and pulls.
Install	Ea.	235	32.50		267.50	
Demolish and Install	Ea.	235	56.50		291.50	
Remove and Install	Ea.		26.16		26.16	
Clean	Ea.	1.64	16.75		18.39	
Paint	Ea.	1.61	37.50		39.11	
Minimum Charge	Job		150		150	
Above-Appliance						
Demolish	Ea.		15.15		15.15	Includes material and labor to install good modular unit with melamine laminated to particle board, in textured colors or wood grain print finish, hinges and pulls.
Install	Ea.	111	24.50		135.50	
Demolish and Install	Ea.	111	39.65		150.65	
Remove and Install	Ea.		19.41		19.41	
Clean	Ea.	1.02	10.40		11.42	
Paint	Ea.	1	24		25	
Minimum Charge	Job		150		150	
Better Grade Wood						
12" Wide, 1 Door						
Demolish	Ea.		6.05		6.05	Includes material and labor to install custom modular unit with solid hardwood faced frames, hardwood door frames and drawer fronts, hardwood veneer on raised door panels, hinges and pulls.
Install	Ea.	134	27.50		161.50	
Demolish and Install	Ea.	134	33.55		167.55	
Remove and Install	Ea.		21.88		21.88	
Clean	Ea.	.41	4.19		4.60	
Paint	Ea.	.40	10.15		10.55	
Minimum Charge	Job		150		150	
15" Wide, 1 Door						
Demolish	Ea.		7.55		7.55	Includes material and labor to install custom modular unit with solid hardwood faced frames, hardwood door frames and drawer fronts, hardwood veneer on raised door panels, hinges and pulls.
Install	Ea.	152	28		180	
Demolish and Install	Ea.	152	35.55		187.55	
Remove and Install	Ea.		22.49		22.49	
Clean	Ea.	.52	5.20		5.72	
Paint	Ea.	.50	12.55		13.05	
Minimum Charge	Job		150		150	

Wall Cabinets	Unit	Material	Labor	Equip.	Total	Specification
18" Wide, 1 Door						
Demolish	Ea.		9.15		9.15	Includes material and labor to install custom modular unit with solid hardwood faced frames, hardwood door frames and drawer fronts, hardwood veneer on raised door panels, hinges and pulls.
Install	Ea.	166	29		195	
Demolish and Install	Ea.	166	38.15		204.15	
Remove and Install	Ea.		23.03		23.03	
Clean	Ea.	.62	6.25		6.87	
Paint	Ea.	.59	15.55		16.14	
Minimum Charge	Job		150		150	
21" Wide, 1 Door						
Demolish	Ea.		10.55		10.55	Includes material and labor to install custom modular unit with solid hardwood faced frames, hardwood door frames and drawer fronts, hardwood veneer on raised door panels, hinges and pulls.
Install	Ea.	189	29		218	
Demolish and Install	Ea.	189	39.55		228.55	
Remove and Install	Ea.		23.36		23.36	
Clean	Ea.	.72	7.25		7.97	
Paint	Ea.	.70	17.60		18.30	
Minimum Charge	Job		150		150	
24" w / Blind Corner						
Demolish	Ea.		12.10		12.10	Includes material and labor to install custom modular unit with solid hardwood faced frames, hardwood door frames and drawer fronts, hardwood veneer on raised door panels, hinges and pulls.
Install	Ea.	187	29.50		216.50	
Demolish and Install	Ea.	187	41.60		228.60	
Remove and Install	Ea.		23.71		23.71	
Clean	Ea.	.83	8.40		9.23	
Paint	Ea.	.79	20.50		21.29	
Minimum Charge	Job		150		150	
27" Wide, 2 Door						
Demolish	Ea.		13.45		13.45	Includes material and labor to install custom modular unit with solid hardwood faced frames, hardwood door frames and drawer fronts, hardwood veneer on raised door panels, hinges and pulls.
Install	Ea.	211	30.50		241.50	
Demolish and Install	Ea.	211	43.95		254.95	
Remove and Install	Ea.		24.31		24.31	
Clean	Ea.	.92	9.50		10.42	
Paint	Ea.	.90	22		22.90	
Minimum Charge	Job		150		150	
30" Wide, 2 Door						
Demolish	Ea.		15.15		15.15	Includes material and labor to install custom modular unit with solid hardwood faced frames, hardwood door frames and drawer fronts, hardwood veneer on raised door panels, hinges and pulls.
Install	Ea.	224	31		255	
Demolish and Install	Ea.	224	46.15		270.15	
Remove and Install	Ea.		24.94		24.94	
Clean	Ea.	1.02	10.40		11.42	
Paint	Ea.	1	26.50		27.50	
Minimum Charge	Job		150		150	
36" Wide, 2 Door						
Demolish	Ea.		17.95		17.95	Includes material and labor to install custom modular unit with solid hardwood faced frames, hardwood door frames and drawer fronts, hardwood veneer on raised door panels, hinges and pulls.
Install	Ea.	256	32		288	
Demolish and Install	Ea.	256	49.95		305.95	
Remove and Install	Ea.		25.60		25.60	
Clean	Ea.	1.23	12.80		14.03	
Paint	Ea.	1.21	29.50		30.71	
Minimum Charge	Job		150		150	
48" Wide, 2 Door						
Demolish	Ea.		24		24	Includes material and labor to install custom modular unit with solid hardwood faced frames, hardwood door frames and drawer fronts, hardwood veneer on raised door panels, hinges and pulls.
Install	Ea.	315	32.50		347.50	
Demolish and Install	Ea.	315	56.50		371.50	
Remove and Install	Ea.		26.16		26.16	
Clean	Ea.	1.64	16.75		18.39	
Paint	Ea.	1.61	37.50		39.11	
Minimum Charge	Job		150		150	

Wall Cabinets	Unit	Material	Labor	Equip.	Total	Specification
Above-appliance						
Demolish	Ea.		15.15		15.15	Includes material and labor to install custom modular unit with solid hardwood faced frames, hardwood door frames and drawer fronts, hardwood veneer on raised door panels, hinges and pulls.
Install	Ea.	153	24.50		177.50	
Demolish and Install	Ea.	153	39.65		192.65	
Remove and Install	Ea.		19.41		19.41	
Clean	Ea.	1.02	10.40		11.42	
Paint	Ea.	1	26.50		27.50	
Minimum Charge	Job		150		150	
Premium Grade Wood						
12" Wide, 1 Door						
Demolish	Ea.		6.05		6.05	Includes material and labor to install premium modular unit with solid hardwood faced frames, hardwood door frames and drawer fronts, hardwood veneer on raised door panels, hinges and pulls.
Install	Ea.	174	27.50		201.50	
Demolish and Install	Ea.	174	33.55		207.55	
Remove and Install	Ea.		21.88		21.88	
Clean	Ea.	.41	4.19		4.60	
Paint	Ea.	.40	11		11.40	
Minimum Charge	Job		150		150	
15" Wide, 1 Door						
Demolish	Ea.		7.55		7.55	Includes material and labor to install premium modular unit with solid hardwood faced frames, hardwood door frames and drawer fronts, hardwood veneer on raised door panels, hinges and pulls.
Install	Ea.	202	28		230	
Demolish and Install	Ea.	202	35.55		237.55	
Remove and Install	Ea.		22.49		22.49	
Clean	Ea.	.52	5.20		5.72	
Paint	Ea.	.50	13.90		14.40	
Minimum Charge	Job		150		150	
18" Wide, 1 Door						
Demolish	Ea.		9.15		9.15	Includes material and labor to install premium modular unit with solid hardwood faced frames, hardwood door frames and drawer fronts, hardwood veneer on raised door panels, hinges and pulls.
Install	Ea.	213	29		242	
Demolish and Install	Ea.	213	38.15		251.15	
Remove and Install	Ea.		23.03		23.03	
Clean	Ea.	.62	6.25		6.87	
Paint	Ea.	.59	16.50		17.09	
Minimum Charge	Job		150		150	
21" Wide, 1 Door						
Demolish	Ea.		10.55		10.55	Includes material and labor to install premium modular unit with solid hardwood faced frames, hardwood door frames and drawer fronts, hardwood veneer on raised door panels, hinges and pulls.
Install	Ea.	227	29		256	
Demolish and Install	Ea.	227	39.55		266.55	
Remove and Install	Ea.		23.36		23.36	
Clean	Ea.	.72	7.25		7.97	
Paint	Ea.	.70	18.85		19.55	
Minimum Charge	Job		150		150	
24" w / Blind Corner						
Demolish	Ea.		12.10		12.10	Includes material and labor to install premium modular unit with solid hardwood faced frames, hardwood door frames and drawer fronts, hardwood veneer on raised door panels, hinges and pulls.
Install	Ea.	249	29.50		278.50	
Demolish and Install	Ea.	249	41.60		290.60	
Remove and Install	Ea.		23.71		23.71	
Clean	Ea.	.83	8.40		9.23	
Paint	Ea.	.79	22		22.79	
Minimum Charge	Job		150		150	
27" Wide, 2 Door						
Demolish	Ea.		13.45		13.45	Includes material and labor to install premium modular unit with solid hardwood faced frames, hardwood door frames and drawer fronts, hardwood veneer on raised door panels, hinges and pulls.
Install	Ea.	232	30.50		262.50	
Demolish and Install	Ea.	232	43.95		275.95	
Remove and Install	Ea.		24.31		24.31	
Clean	Ea.	.92	9.50		10.42	
Paint	Ea.	.90	24		24.90	
Minimum Charge	Job		150		150	

Wall Cabinets	Unit	Material	Labor	Equip.	Total	Specification
30" Wide, 2 Door						
Demolish	Ea.		15.15		15.15	Includes material and labor to install premium modular unit with solid hardwood faced frames, hardwood door frames and drawer fronts, hardwood veneer on raised door panels, hinges and pulls.
Install	Ea.	283	31		314	
Demolish and Install	Ea.	283	46.15		329.15	
Remove and Install	Ea.		24.94		24.94	
Clean	Ea.	1.02	10.40		11.42	
Paint	Ea.	1	26.50		27.50	
Minimum Charge	Job		150		150	
36" Wide, 2 Door						
Demolish	Ea.		17.95		17.95	Includes material and labor to install premium modular unit with solid hardwood faced frames, hardwood door frames and drawer fronts, hardwood veneer on raised door panels, hinges and pulls.
Install	Ea.	325	32		357	
Demolish and Install	Ea.	325	49.95		374.95	
Remove and Install	Ea.		25.60		25.60	
Clean	Ea.	1.23	12.80		14.03	
Paint	Ea.	1.21	33		34.21	
Minimum Charge	Job		150		150	
48" Wide, 2 Door						
Demolish	Ea.		24		24	Includes material and labor to install premium modular unit with solid hardwood faced frames, hardwood door frames and drawer fronts, hardwood veneer on raised door panels, hinges and pulls.
Install	Ea.	400	32.50		432.50	
Demolish and Install	Ea.	400	56.50		456.50	
Remove and Install	Ea.		26.16		26.16	
Clean	Ea.	1.64	16.75		18.39	
Paint	Ea.	1.61	44		45.61	
Minimum Charge	Job		150		150	
Above-appliance						
Demolish	Ea.		15.15		15.15	Includes material and labor to install premium modular unit with solid hardwood faced frames, hardwood door frames and drawer fronts, hardwood veneer on raised door panels, hinges, pulls.
Install	Ea.	189	24.50		213.50	
Demolish and Install	Ea.	189	39.65		228.65	
Remove and Install	Ea.		19.41		19.41	
Clean	Ea.	1.02	10.40		11.42	
Paint	Ea.	1	26.50		27.50	
Minimum Charge	Job		150		150	
Premium Hardwood						
12" Wide, 1 Door						
Demolish	Ea.		6.05		6.05	Includes material and labor to install premium hardwood modular unit with solid hardwood faced frames, drawer fronts, and door panels, steel drawer guides, hinges, pulls, laminated interior.
Install	Ea.	202	27.50		229.50	
Demolish and Install	Ea.	202	33.55		235.55	
Remove and Install	Ea.		21.88		21.88	
Clean	Ea.	.41	4.19		4.60	
Paint	Ea.	.40	12		12.40	
Minimum Charge	Job		150		150	
15" Wide, 1 Door						
Demolish	Ea.		7.55		7.55	Includes material and labor to install premium hardwood modular unit with solid hardwood faced frames, drawer fronts, and door panels, steel drawer guides, hinges, pulls, laminated interior.
Install	Ea.	266	28		294	
Demolish and Install	Ea.	266	35.55		301.55	
Remove and Install	Ea.		22.49		22.49	
Clean	Ea.	.52	5.20		5.72	
Paint	Ea.	.50	14.65		15.15	
Minimum Charge	Job		150		150	
18" Wide, 1 Door						
Demolish	Ea.		9.15		9.15	Includes material and labor to install premium hardwood modular unit with solid hardwood faced frames, drawer fronts, and door panels, steel drawer guides, hinges, pulls, laminated interior.
Install	Ea.	242	29		271	
Demolish and Install	Ea.	242	38.15		280.15	
Remove and Install	Ea.		23.03		23.03	
Clean	Ea.	.62	6.25		6.87	
Paint	Ea.	.59	17.60		18.19	
Minimum Charge	Job		150		150	

Wall Cabinets	Unit	Material	Labor	Equip.	Total	Specification
21" Wide, 1 Door						
Demolish	Ea.		10.55		10.55	Includes material and labor to install premium hardwood modular unit with solid hardwood faced frames, drawer fronts, and door panels, steel drawer guides, hinges, pulls, laminated interior.
Install	Ea.	249	29		278	
Demolish and Install	Ea.	249	39.55		288.55	
Remove and Install	Ea.		23.36		23.36	
Clean	Ea.	.72	7.25		7.97	
Paint	Ea.	.70	20.50		21.20	
Minimum Charge	Job		150		150	
24" w / Blind Corner						
Demolish	Ea.		12.10		12.10	Includes material and labor to install premium hardwood modular unit with solid hardwood faced frames, drawer fronts, and door panels, steel drawer guides, hinges, pulls, laminated interior.
Install	Ea.	277	29.50		306.50	
Demolish and Install	Ea.	277	41.60		318.60	
Remove and Install	Ea.		23.71		23.71	
Clean	Ea.	.83	8.40		9.23	
Paint	Ea.	.79	24		24.79	
Minimum Charge	Job		150		150	
27" Wide, 2 Door						
Demolish	Ea.		13.45		13.45	Includes material and labor to install premium hardwood modular unit with solid hardwood faced frames, drawer fronts, and door panels, steel drawer guides, hinges, pulls, laminated interior.
Install	Ea.	470	30.50		500.50	
Demolish and Install	Ea.	470	43.95		513.95	
Remove and Install	Ea.		24.31		24.31	
Clean	Ea.	.92	9.50		10.42	
Paint	Ea.	.90	26.50		27.40	
Minimum Charge	Job		150		150	
30" Wide, 2 Door						
Demolish	Ea.		15.15		15.15	Includes material and labor to install premium hardwood modular unit with solid hardwood faced frames, drawer fronts, and door panels, steel drawer guides, hinges, pulls, laminated interior.
Install	Ea.	310	31		341	
Demolish and Install	Ea.	310	46.15		356.15	
Remove and Install	Ea.		24.94		24.94	
Clean	Ea.	1.02	10.40		11.42	
Paint	Ea.	1	29.50		30.50	
Minimum Charge	Job		150		150	
36" Wide, 2 Door						
Demolish	Ea.		17.95		17.95	Includes material and labor to install premium hardwood modular unit with solid hardwood faced frames, drawer fronts, and door panels, steel drawer guides, hinges, pulls, laminated interior.
Install	Ea.	350	32		382	
Demolish and Install	Ea.	350	49.95		399.95	
Remove and Install	Ea.		25.60		25.60	
Clean	Ea.	1.23	12.80		14.03	
Paint	Ea.	1.21	37.50		38.71	
Minimum Charge	Job		150		150	
48" Wide, 2 Door						
Demolish	Ea.		24		24	Includes material and labor to install premium hardwood modular unit with solid hardwood faced frames, drawer fronts, and door panels, steel drawer guides, hinges, pulls, laminated interior.
Install	Ea.	430	32.50		462.50	
Demolish and Install	Ea.	430	56.50		486.50	
Remove and Install	Ea.		26.16		26.16	
Clean	Ea.	1.64	16.75		18.39	
Paint	Ea.	1.61	44		45.61	
Minimum Charge	Job		150		150	
Above-appliance						
Demolish	Ea.		15.15		15.15	Includes material and labor to install premium hardwood modular unit with solid hardwood faced frames, drawer fronts, and door panels, steel drawer guides, hinges, pulls, laminated interior.
Install	Ea.	218	24.50		242.50	
Demolish and Install	Ea.	218	39.65		257.65	
Remove and Install	Ea.		19.41		19.41	
Clean	Ea.	1.02	10.40		11.42	
Paint	Ea.	1	29.50		30.50	
Minimum Charge	Job		150		150	

Wall Cabinets	Unit	Material	Labor	Equip.	Total	Specification
Medicine Cabinet						
Good Grade						
Demolish	Ea.		10.10		10.10	Cost includes material and labor to install 14" x 18" recessed medicine cabinet with stainless steel frame, mirror. Light not included.
Install	Ea.	75	21.50		96.50	
Demolish and Install	Ea.	75	31.60		106.60	
Remove and Install	Ea.		17.19		17.19	
Clean	Ea.	.02	9.80		9.82	
Paint	Ea.	.97	16.50		17.47	
Minimum Charge	Job		150		150	
Good Grade Lighted						
Demolish	Ea.		10.10		10.10	Cost includes material and labor to install 14" x 18" recessed medicine cabinet with stainless steel frame, mirror and light.
Install	Ea.	234	25		259	
Demolish and Install	Ea.	234	35.10		269.10	
Remove and Install	Ea.		20.05		20.05	
Clean	Ea.	.02	9.80		9.82	
Paint	Ea.	.97	16.50		17.47	
Minimum Charge	Job		150		150	
Better Grade						
Demolish	Ea.		10.10		10.10	Includes material and labor to install recessed medicine cabinet with polished chrome frame and mirror. Light not included.
Install	Ea.	103	21.50		124.50	
Demolish and Install	Ea.	103	31.60		134.60	
Remove and Install	Ea.		17.19		17.19	
Clean	Ea.	.02	9.80		9.82	
Paint	Ea.	.97	16.50		17.47	
Minimum Charge	Job		150		150	
Better Grade Lighted						
Demolish	Ea.		10.10		10.10	Includes material and labor to install recessed medicine cabinet with polished chrome frame, mirror and light.
Install	Ea.	263	25		288	
Demolish and Install	Ea.	263	35.10		298.10	
Remove and Install	Ea.		20.05		20.05	
Clean	Ea.	.02	9.80		9.82	
Paint	Ea.	.97	16.50		17.47	
Minimum Charge	Job		150		150	
Premium Grade						
Demolish	Ea.		10.10		10.10	Cost includes material and labor to install 14" x 24" recessed medicine cabinet with beveled mirror, frameless swing door. Light not included.
Install	Ea.	132	21.50		153.50	
Demolish and Install	Ea.	132	31.60		163.60	
Remove and Install	Ea.		17.19		17.19	
Clean	Ea.	.02	9.80		9.82	
Paint	Ea.	.97	16.50		17.47	
Minimum Charge	Job		150		150	
Premium Grade Light						
Demolish	Ea.		10.10		10.10	Cost includes material and labor to install 36" x 30" surface mounted medicine cabinet with beveled tri-view mirror, frameless swing door and light.
Install	Ea.	292	25		317	
Demolish and Install	Ea.	292	35.10		327.10	
Remove and Install	Ea.		20.05		20.05	
Clean	Ea.	.02	9.80		9.82	
Paint	Ea.	.97	16.50		17.47	
Minimum Charge	Job		150		150	
Bath Mirror						
Demolish	Ea.		7.55		7.55	Cost includes material and labor to install 18" x 24" surface mount mirror with stainless steel frame.
Install	Ea.	62.50	15.05		77.55	
Demolish and Install	Ea.	62.50	22.60		85.10	
Remove and Install	Ea.		12.03		12.03	
Clean	Ea.	.02	.65		.67	
Minimum Charge	Job		150		150	

Cabinets and Countertops

Wall Cabinets		Unit	Material	Labor	Equip.	Total	Specification
Garage							
	Demolish	L.F.		6.05		6.05	Includes material and labor to install good modular unit with melamine laminated to particle board, in textured colors or wood grain print finish, hinges and pulls.
	Install	L.F.	49	15.05		64.05	
	Demolish and Install	L.F.	49	21.10		70.10	
	Remove and Install	L.F.		12.03		12.03	
	Clean	L.F.	.41	4.19		4.60	
	Paint	L.F.	.40	9.45		9.85	
	Minimum Charge	Job		150		150	
Strip and Refinish							
	Install	L.F.	1.07	4.80		5.87	Includes labor and material to strip, prep and refinish exterior of cabinets.
	Minimum Charge	Job		132		132	

Tall Cabinets		Unit	Material	Labor	Equip.	Total	Specification
Good Grade Laminated							
	Demolish	L.F.		10.75		10.75	Includes material and labor to install custom modular unit with solid hardwood faced frames, hardwood door frames and drawer fronts, hardwood veneer on raised door panels, hinges and pulls.
	Install	L.F.	199	169		368	
	Demolish and Install	L.F.	199	179.75		378.75	
	Remove and Install	L.F.		135.57		135.57	
	Clean	L.F.	1.09	7.80		8.89	
	Paint	L.F.	.94	18.85		19.79	
	Minimum Charge	Job		150		150	
Better Grade Wood							
	Demolish	L.F.		10.75		10.75	Includes material and labor to install custom modular unit with solid hardwood faced frames, hardwood door frames and drawer fronts, hardwood veneer on raised door panels, hinges and pulls.
	Install	L.F.	265	169		434	
	Demolish and Install	L.F.	265	179.75		444.75	
	Remove and Install	L.F.		135.57		135.57	
	Clean	L.F.	1.09	7.80		8.89	
	Paint	L.F.	.94	20.50		21.44	
	Minimum Charge	Job		150		150	
Premium Grade Wood							
	Demolish	L.F.		10.75		10.75	Includes material and labor to install premium modular unit with solid hardwood faced frames, hardwood door frames and drawer fronts, hardwood veneer on raised door panels, hinges and pulls.
	Install	L.F.	350	169		519	
	Demolish and Install	L.F.	350	179.75		529.75	
	Remove and Install	L.F.		135.57		135.57	
	Clean	L.F.	1.09	7.80		8.89	
	Paint	L.F.	.94	22		22.94	
	Minimum Charge	Job		150		150	
Premium Hardwood							
	Demolish	L.F.		10.75		10.75	Includes material and labor to install premium hardwood modular unit with solid hardwood faced frames, drawer fronts, and door panels, steel drawer guides, hinges and pulls, laminated interior.
	Install	L.F.	460	169		629	
	Demolish and Install	L.F.	460	179.75		639.75	
	Remove and Install	L.F.		135.57		135.57	
	Clean	L.F.	1.09	7.80		8.89	
	Paint	L.F.	.94	24		24.94	
	Minimum Charge	Job		150		150	

Tall Cabinets	Unit	Material	Labor	Equip.	Total	Specification
Single Oven, 27" Wide						
Good Grade Laminated						
Demolish	Ea.		24		24	Includes material and labor to install good modular unit with melamine laminated to particle board, in textured colors or wood grain print finish, hinges and pulls.
Install	Ea.	445	75		520	
Demolish and Install	Ea.	445	99		544	
Remove and Install	Ea.		60.16		60.16	
Clean	Ea.	3.42	13.65		17.07	
Paint	Ea.	1.56	42		43.56	
Minimum Charge	Job		150		150	
Better Grade Veneer						
Demolish	Ea.		24		24	Includes material and labor to install custom modular unit with solid hardwood faced frames, hardwood door frames and drawer fronts, hardwood veneer on raised door panels, hinges and pulls.
Install	Ea.	790	75		865	
Demolish and Install	Ea.	790	99		889	
Remove and Install	Ea.		60.16		60.16	
Clean	Ea.	3.42	13.65		17.07	
Paint	Ea.	1.56	45.50		47.06	
Minimum Charge	Job		150		150	
Premium Grade Veneer						
Demolish	Ea.		24		24	Includes material and labor to install premium modular unit with solid hardwood faced frames, hardwood door frames and drawer fronts, hardwood veneer on raised door panels, hinges and pulls.
Install	Ea.	705	75		780	
Demolish and Install	Ea.	705	99		804	
Remove and Install	Ea.		60.16		60.16	
Clean	Ea.	3.42	13.65		17.07	
Paint	Ea.	1.56	50		51.56	
Minimum Charge	Job		150		150	
Premium Hardwood						
Demolish	Ea.		24		24	Includes material and labor to install premium hardwood modular unit with solid hardwood faced frames, drawer fronts, and door panels, steel drawer guides, hinges and pulls, laminated interior.
Install	Ea.	900	75		975	
Demolish and Install	Ea.	900	99		999	
Remove and Install	Ea.		60.16		60.16	
Clean	Ea.	3.42	13.65		17.07	
Paint	Ea.	1.56	54		55.56	
Minimum Charge	Job		150		150	
Single Oven, 30" Wide						
Good Grade Laminated						
Demolish	Ea.		24		24	Includes material and labor to install good modular unit with melamine laminated to particle board, in textured colors or wood grain print finish, hinges and pulls.
Install	Ea.	405	75		480	
Demolish and Install	Ea.	405	99		504	
Remove and Install	Ea.		60.16		60.16	
Clean	Ea.	3.42	13.65		17.07	
Paint	Ea.	1.56	42		43.56	
Minimum Charge	Job		150		150	
Better Grade Veneer						
Demolish	Ea.		24		24	Includes material and labor to install custom modular unit with solid hardwood faced frames, hardwood door frames and drawer fronts, hardwood veneer on raised door panels, hinges and pulls.
Install	Ea.	595	75		670	
Demolish and Install	Ea.	595	99		694	
Remove and Install	Ea.		60.16		60.16	
Clean	Ea.	3.42	13.65		17.07	
Paint	Ea.	1.56	42		43.56	
Minimum Charge	Job		150		150	

Tall Cabinets	Unit	Material	Labor	Equip.	Total	Specification
Premium Grade Veneer						
Demolish	Ea.		24		24	Includes material and labor to install premium modular unit with solid hardwood faced frames, hardwood door frames and drawer fronts, hardwood veneer on raised door panels, hinges and pulls.
Install	Ea.	695	75		770	
Demolish and Install	Ea.	695	99		794	
Remove and Install	Ea.		60.16		60.16	
Clean	Ea.	3.42	13.65		17.07	
Paint	Ea.	1.56	42		43.56	
Minimum Charge	Job		150		150	
Premium Hardwood						
Demolish	Ea.		24		24	Includes material and labor to install premium hardwood modular unit with solid hardwood faced frames, drawer fronts, and door panels, steel drawer guides, hinges and pulls, laminated interior.
Install	Ea.	825	75		900	
Demolish and Install	Ea.	825	99		924	
Remove and Install	Ea.		60.16		60.16	
Clean	Ea.	3.42	13.65		17.07	
Paint	Ea.	1.56	42		43.56	
Minimum Charge	Job		150		150	
Utility 18" W x 12" D						
Good Grade Laminated						
Demolish	Ea.		24		24	Includes material and labor to install good modular unit with melamine laminated to particle board, in textured colors or wood grain print finish, hinges and pulls.
Install	Ea.	365	60		425	
Demolish and Install	Ea.	365	84		449	
Remove and Install	Ea.		48.13		48.13	
Clean	Ea.	.80	6.60		7.40	
Paint	Ea.	.59	16.50		17.09	
Minimum Charge	Job		150		150	
Better Grade Veneer						
Demolish	Ea.		24		24	Includes material and labor to install custom modular unit with solid hardwood faced frames, hardwood door frames and drawer fronts, hardwood veneer on raised door panels, hinges and pulls.
Install	Ea.	395	60		455	
Demolish and Install	Ea.	395	84		479	
Remove and Install	Ea.		48.13		48.13	
Clean	Ea.	.80	6.60		7.40	
Paint	Ea.	.59	17.60		18.19	
Minimum Charge	Job		150		150	
Premium Grade Veneer						
Demolish	Ea.		24		24	Includes material and labor to install premium modular unit with solid hardwood faced frames, hardwood door frames and drawer fronts, hardwood veneer on raised door panels, hinges and pulls.
Install	Ea.	505	60		565	
Demolish and Install	Ea.	505	84		589	
Remove and Install	Ea.		48.13		48.13	
Clean	Ea.	.80	6.60		7.40	
Paint	Ea.	.59	20.50		21.09	
Minimum Charge	Job		150		150	
Premium Hardwood						
Demolish	Ea.		24		24	Includes material and labor to install premium hardwood modular unit with solid hardwood faced frames, drawer fronts, and door panels, steel drawer guides, hinges and pulls, laminated interior.
Install	Ea.	645	60		705	
Demolish and Install	Ea.	645	84		729	
Remove and Install	Ea.		48.13		48.13	
Clean	Ea.	.80	6.60		7.40	
Paint	Ea.	.59	22		22.59	
Minimum Charge	Job		150		150	

Tall Cabinets	Unit	Material	Labor	Equip.	Total	Specification
Utility 24" W x 24" D						
Good Grade Laminated						
Demolish	Ea.		24		24	Includes material and labor to install good modular unit with melamine laminated to particle board, in textured colors or wood grain print finish, hinges and pulls.
Install	Ea.	440	75		515	
Demolish and Install	Ea.	440	99		539	
Remove and Install	Ea.		60.16		60.16	
Clean	Ea.	1.06	8.70		9.76	
Paint	Ea.	.79	22		22.79	
Minimum Charge	Job		150		150	
Better Grade Veneer						
Demolish	Ea.		24		24	Includes material and labor to install custom modular unit with solid hardwood faced frames, hardwood door frames and drawer fronts, hardwood veneer on raised door panels, hinges and pulls.
Install	Ea.	535	75		610	
Demolish and Install	Ea.	535	99		634	
Remove and Install	Ea.		60.16		60.16	
Clean	Ea.	1.06	8.70		9.76	
Paint	Ea.	.79	24		24.79	
Minimum Charge	Job		150		150	
Premium Grade Veneer						
Demolish	Ea.		24		24	Includes material and labor to install premium modular unit with solid hardwood faced frames, hardwood door frames and drawer fronts, hardwood veneer on raised door panels, hinges and pulls.
Install	Ea.	675	75		750	
Demolish and Install	Ea.	675	99		774	
Remove and Install	Ea.		60.16		60.16	
Clean	Ea.	1.06	8.70		9.76	
Paint	Ea.	.79	26.50		27.29	
Minimum Charge	Job		150		150	
Premium Hardwood						
Demolish	Ea.		24		24	Includes material and labor to install premium hardwood modular unit with solid hardwood faced frames, drawer fronts, and door panels, steel drawer guides, hinges and pulls, laminated interior.
Install	Ea.	825	75		900	
Demolish and Install	Ea.	825	99		924	
Remove and Install	Ea.		60.16		60.16	
Clean	Ea.	1.06	8.70		9.76	
Paint	Ea.	.79	29.50		30.29	
Minimum Charge	Job		150		150	
Plain Shelves						
Install	Ea.	21.50	18.80		40.30	Includes labor and material to install tall cabinet shelving, per set.
Rotating Shelves						
Install	Ea.	72.50	15.05		87.55	Includes labor and material to install lazy susan shelving for wall cabinet.
Strip and Refinish						
Install	L.F.	1.68	12		13.68	Includes labor and material to strip, prep and refinish exterior of cabinets.
Minimum Charge	Job		132		132	
Stain						
Install	L.F.	.15	2.40		2.55	Includes labor and material to replace normal prep and stain on doors and exterior cabinets.
Minimum Charge	Job		132		132	

Base Cabinets	Unit	Material	Labor	Equip.	Total	Specification
Good Grade Laminated						
12" w, 1 Door, 1 Drawer						
Demolish	Ea.		6.05		6.05	Includes material and labor to install good modular unit with melamine laminate, textured colors or wood grain print finish including hinges and pulls.
Install	Ea.	105	24.50		129.50	
Demolish and Install	Ea.	105	30.55		135.55	
Remove and Install	Ea.		19.41		19.41	
Clean	Ea.	.55	4.36		4.91	
Paint	Ea.	.40	11		11.40	
Minimum Charge	Job		150		150	
15" w, 1 Door, 1 Drawer						
Demolish	Ea.		7.55		7.55	Includes material and labor to install good modular unit with melamine laminate, textured colors or wood grain print finish including hinges and pulls.
Install	Ea.	143	25		168	
Demolish and Install	Ea.	143	32.55		175.55	
Remove and Install	Ea.		20.05		20.05	
Clean	Ea.	.68	5.45		6.13	
Paint	Ea.	.50	13.90		14.40	
Minimum Charge	Job		150		150	
18" w, 1 Door, 1 Drawer						
Demolish	Ea.		8.95		8.95	Includes material and labor to install good modular unit with melamine laminate, textured colors or wood grain print finish including hinges and pulls.
Install	Ea.	155	26		181	
Demolish and Install	Ea.	155	34.95		189.95	
Remove and Install	Ea.		20.66		20.66	
Clean	Ea.	.80	6.60		7.40	
Paint	Ea.	.59	16.50		17.09	
Minimum Charge	Job		150		150	
21" w, 1 Door, 1 Drawer						
Demolish	Ea.		10.55		10.55	Includes material and labor to install good modular unit with melamine laminate, textured colors or wood grain print finish including hinges and pulls.
Install	Ea.	162	26.50		188.50	
Demolish and Install	Ea.	162	37.05		199.05	
Remove and Install	Ea.		21.20		21.20	
Clean	Ea.	.98	7.80		8.78	
Paint	Ea.	.70	18.85		19.55	
Minimum Charge	Job		150		150	
24" w, 1 Door, 1 Drawer						
Demolish	Ea.		12.10		12.10	Includes material and labor to install good modular unit with melamine laminate, textured colors or wood grain print finish including hinges and pulls.
Install	Ea.	186	27		213	
Demolish and Install	Ea.	186	39.10		225.10	
Remove and Install	Ea.		21.58		21.58	
Clean	Ea.	1.06	8.70		9.76	
Paint	Ea.	.79	22		22.79	
Minimum Charge	Job		150		150	
30" Wide Sink Base						
Demolish	Ea.		15.15		15.15	Includes material and labor to install good modular unit with melamine laminate, textured colors or wood grain print finish including hinges and pulls.
Install	Ea.	183	28		211	
Demolish and Install	Ea.	183	43.15		226.15	
Remove and Install	Ea.		22.49		22.49	
Clean	Ea.	1.36	10.90		12.26	
Paint	Ea.	1	26.50		27.50	
Minimum Charge	Job		150		150	
36" Wide Sink Base						
Demolish	Ea.		18.65		18.65	Includes material and labor to install good modular unit with melamine laminate, textured colors or wood grain print finish including hinges and pulls.
Install	Ea.	205	29.50		234.50	
Demolish and Install	Ea.	205	48.15		253.15	
Remove and Install	Ea.		23.71		23.71	
Clean	Ea.	1.52	12.80		14.32	
Paint	Ea.	1.21	33		34.21	
Minimum Charge	Job		150		150	

Base Cabinets	Unit	Material	Labor	Equip.	Total	Specification
36" Blind Corner						
Demolish	Ea.		18.65		18.65	Includes material and labor to install good modular unit with melamine laminate, textured colors or wood grain print finish including hinges and pulls.
Install	Ea.	300	33.50		333.50	
Demolish and Install	Ea.	300	52.15		352.15	
Remove and Install	Ea.		26.74		26.74	
Clean	Ea.	1.52	12.80		14.32	
Paint	Ea.	1.21	33		34.21	
Minimum Charge	Job		150		150	
42" w, 2 Door, 2 Drawer						
Demolish	Ea.		22		22	Includes material and labor to install good modular unit with melamine laminate, textured colors or wood grain print finish including hinges and pulls.
Install	Ea.	254	30.50		284.50	
Demolish and Install	Ea.	254	52.50		306.50	
Remove and Install	Ea.		24.31		24.31	
Clean	Ea.	1.95	15.55		17.50	
Paint	Ea.	1.39	37.50		38.89	
Minimum Charge	Job		150		150	
Clean Interior						
Clean	L.F.	.55	3.30		3.85	Includes labor and materials to clean the interior of a 27" cabinet.
Minimum Charge	Job		109		109	
Better Grade Wood						
12" w, 1 Door, 1 Drawer						
Demolish	Ea.		6.05		6.05	Includes material and labor to install custom modular unit including solid hardwood faced frames, hardwood door frames and drawer fronts, hardwood veneer on raised door panels, hinges and pulls.
Install	Ea.	140	24.50		164.50	
Demolish and Install	Ea.	140	30.55		170.55	
Remove and Install	Ea.		19.41		19.41	
Clean	Ea.	.55	4.36		4.91	
Paint	Ea.	.40	12		12.40	
Minimum Charge	Job		150		150	
15" w, 1 Door, 1 Drawer						
Demolish	Ea.		7.55		7.55	Includes material and labor to install custom modular unit including solid hardwood faced frames, hardwood door frames and drawer fronts, hardwood veneer on raised door panels, hinges and pulls.
Install	Ea.	190	25		215	
Demolish and Install	Ea.	190	32.55		222.55	
Remove and Install	Ea.		20.05		20.05	
Clean	Ea.	.68	5.45		6.13	
Paint	Ea.	.50	15.55		16.05	
Minimum Charge	Job		150		150	
18" w, 1 Door, 1 Drawer						
Demolish	Ea.		8.95		8.95	Includes material and labor to install custom modular unit including solid hardwood faced frames, hardwood door frames and drawer fronts, hardwood veneer on raised door panels, hinges and pulls.
Install	Ea.	207	26		233	
Demolish and Install	Ea.	207	34.95		241.95	
Remove and Install	Ea.		20.66		20.66	
Clean	Ea.	.80	6.60		7.40	
Paint	Ea.	.59	17.60		18.19	
Minimum Charge	Job		150		150	
21" w, 1 Door, 1 Drawer						
Demolish	Ea.		10.55		10.55	Includes material and labor to install custom modular unit including solid hardwood faced frames, hardwood door frames and drawer fronts, hardwood veneer on raised door panels, hinges and pulls.
Install	Ea.	216	26.50		242.50	
Demolish and Install	Ea.	216	37.05		253.05	
Remove and Install	Ea.		21.20		21.20	
Clean	Ea.	.98	7.80		8.78	
Paint	Ea.	.70	20.50		21.20	
Minimum Charge	Job		150		150	

Cabinets and Countertops

Base Cabinets	Unit	Material	Labor	Equip.	Total	Specification
24" w, 1 Door, 1 Drawer						
Demolish	Ea.		12.10		12.10	Includes material and labor to install custom modular unit including solid hardwood faced frames, hardwood door frames and drawer fronts, hardwood veneer on raised door panels, hinges and pulls.
Install	Ea.	248	27		275	
Demolish and Install	Ea.	248	39.10		287.10	
Remove and Install	Ea.		21.58		21.58	
Clean	Ea.	1.06	8.70		9.76	
Paint	Ea.	.79	24		24.79	
Minimum Charge	Job		150		150	
30" Wide Sink Base						
Demolish	Ea.		15.15		15.15	Includes material and labor to install custom modular unit including solid hardwood faced frames, hardwood door frames and drawer fronts, hardwood veneer on raised door panels, hinges and pulls.
Install	Ea.	243	28		271	
Demolish and Install	Ea.	243	43.15		286.15	
Remove and Install	Ea.		22.49		22.49	
Clean	Ea.	1.36	10.90		12.26	
Paint	Ea.	1	29.50		30.50	
Minimum Charge	Job		150		150	
36" Wide Sink Base						
Demolish	Ea.		18.65		18.65	Includes material and labor to install custom modular unit including solid hardwood faced frames, hardwood door frames and drawer fronts, hardwood veneer on raised door panels, hinges and pulls.
Install	Ea.	273	29.50		302.50	
Demolish and Install	Ea.	273	48.15		321.15	
Remove and Install	Ea.		23.71		23.71	
Clean	Ea.	1.52	12.80		14.32	
Paint	Ea.	1.21	37.50		38.71	
Minimum Charge	Job		150		150	
36" Blind Corner						
Demolish	Ea.		18.65		18.65	Includes material and labor to install custom modular unit including solid hardwood faced frames, hardwood door frames and drawer fronts, hardwood veneer on raised door panels, hinges and pulls.
Install	Ea.	400	33.50		433.50	
Demolish and Install	Ea.	400	52.15		452.15	
Remove and Install	Ea.		26.74		26.74	
Clean	Ea.	1.52	12.80		14.32	
Paint	Ea.	1.21	37.50		38.71	
Minimum Charge	Job		150		150	
42" w, 2 Door, 2 Drawer						
Demolish	Ea.		22		22	Includes material and labor to install modular unit including solid hardwood faced frames, hardwood door frames and drawer fronts, hardwood veneer on raised door panels, hinges and pulls.
Install	Ea.	340	30.50		370.50	
Demolish and Install	Ea.	340	52.50		392.50	
Remove and Install	Ea.		24.31		24.31	
Clean	Ea.	1.95	15.55		17.50	
Paint	Ea.	1.39	44		45.39	
Minimum Charge	Job		150		150	
Clean Interior						
Clean	L.F.	.55	3.30		3.85	Includes labor and materials to clean the interior of a 27" cabinet.
Minimum Charge	Job		109		109	
Premium Grade Wood						
12" w, 1 Door, 1 Drawer						
Demolish	Ea.		6.05		6.05	Includes material and labor to install premium hardwood modular unit with solid hardwood faced frames, drawer fronts, and door panels, steel drawer guides, hinges, pulls and laminated interior.
Install	Ea.	240	12.15		252.15	
Demolish and Install	Ea.	240	18.20		258.20	
Remove and Install	Ea.		9.70		9.70	
Clean	Ea.	.55	4.36		4.91	
Paint	Ea.	.40	13.20		13.60	
Minimum Charge	Job		150		150	

Base Cabinets	Unit	Material	Labor	Equip.	Total	Specification
15" w, 1 Door, 1 Drawer						
Demolish	Ea.		7.55		7.55	Includes material and labor to install premium modular unit including solid hardwood faced frames, hardwood door frames and drawer fronts, hardwood veneer on raised door panels, hinges and pulls.
Install	Ea.	249	25		274	
Demolish and Install	Ea.	249	32.55		281.55	
Remove and Install	Ea.		20.05		20.05	
Clean	Ea.	.68	5.45		6.13	
Paint	Ea.	.50	16.50		17	
Minimum Charge	Job		150		150	
18" w, 1 Door, 1 Drawer						
Demolish	Ea.		8.95		8.95	Includes material and labor to install premium modular unit including solid hardwood faced frames, hardwood door frames and drawer fronts, hardwood veneer on raised door panels, hinges and pulls.
Install	Ea.	268	26		294	
Demolish and Install	Ea.	268	34.95		302.95	
Remove and Install	Ea.		20.66		20.66	
Clean	Ea.	.80	6.60		7.40	
Paint	Ea.	.59	20.50		21.09	
Minimum Charge	Job		150		150	
21" w, 1 Door, 1 Drawer						
Demolish	Ea.		10.55		10.55	Includes material and labor to install premium modular unit including solid hardwood faced frames, hardwood door frames and drawer fronts, hardwood veneer on raised door panels, hinges and pulls.
Install	Ea.	279	26.50		305.50	
Demolish and Install	Ea.	279	37.05		316.05	
Remove and Install	Ea.		21.20		21.20	
Clean	Ea.	.98	7.80		8.78	
Paint	Ea.	.70	22		22.70	
Minimum Charge	Job		150		150	
24" w, 1 Door, 1 Drawer						
Demolish	Ea.		12.10		12.10	Includes material and labor to install premium modular unit including solid hardwood faced frames, hardwood door frames and drawer fronts, hardwood veneer on raised door panels, hinges and pulls.
Install	Ea.	320	27		347	
Demolish and Install	Ea.	320	39.10		359.10	
Remove and Install	Ea.		21.58		21.58	
Clean	Ea.	1.06	8.70		9.76	
Paint	Ea.	.79	26.50		27.29	
Minimum Charge	Job		150		150	
30" Wide Sink Base						
Demolish	Ea.		15.15		15.15	Includes material and labor to install premium modular unit including solid hardwood faced frames, hardwood door frames and drawer fronts, hardwood veneer on raised door panels, hinges and pulls.
Install	Ea.	283	28		311	
Demolish and Install	Ea.	283	43.15		326.15	
Remove and Install	Ea.		22.49		22.49	
Clean	Ea.	1.36	10.90		12.26	
Paint	Ea.	1	33		34	
Minimum Charge	Job		150		150	
36" Wide Sink Base						
Demolish	Ea.		18.65		18.65	Includes material and labor to install premium modular unit including solid hardwood faced frames, hardwood door frames and drawer fronts, hardwood veneer on raised door panels, hinges and pulls.
Install	Ea.	320	29.50		349.50	
Demolish and Install	Ea.	320	48.15		368.15	
Remove and Install	Ea.		23.71		23.71	
Clean	Ea.	1.52	12.80		14.32	
Paint	Ea.	1.21	37.50		38.71	
Minimum Charge	Job		150		150	
36" Blind Corner						
Demolish	Ea.		18.65		18.65	Includes material and labor to install premium modular unit including solid hardwood faced frames, hardwood door frames and drawer fronts, hardwood veneer on raised door panels, hinges and pulls.
Install	Ea.	665	33.50		698.50	
Demolish and Install	Ea.	665	52.15		717.15	
Remove and Install	Ea.		26.74		26.74	
Clean	Ea.	1.52	12.80		14.32	
Paint	Ea.	1.21	37.50		38.71	
Minimum Charge	Job		150		150	

Base Cabinets	Unit	Material	Labor	Equip.	Total	Specification
42" w, 2 Door, 2 Drawer						
Demolish	Ea.		22		22	Includes material and labor to install premium modular unit including solid hardwood faced frames, hardwood door frames and drawer fronts, hardwood veneer on raised door panels, hinges and pulls.
Install	Ea.	450	30.50		480.50	
Demolish and Install	Ea.	450	52.50		502.50	
Remove and Install	Ea.		24.31		24.31	
Clean	Ea.	1.95	15.55		17.50	
Paint	Ea.	1.39	44		45.39	
Minimum Charge	Job		150		150	
Clean Interior						
Clean	L.F.	.55	3.30		3.85	Includes labor and materials to clean the interior of a 27" cabinet.
Minimum Charge	Job		109		109	
Premium Hardwood						
12" w, 1 Door, 1 Drawer						
Demolish	Ea.		6.05		6.05	Includes material and labor to install premium hardwood modular unit with solid hardwood faced frames, drawer fronts, and door panels, steel drawer guides, hinges, pulls and laminated interior.
Install	Ea.	294	24.50		318.50	
Demolish and Install	Ea.	294	30.55		324.55	
Remove and Install	Ea.		19.41		19.41	
Clean	Ea.	.55	4.36		4.91	
Paint	Ea.	.40	14.65		15.05	
Minimum Charge	Job		150		150	
15" w, 1 Door, 1 Drawer						
Demolish	Ea.		7.55		7.55	Includes material and labor to install premium hardwood modular unit with solid hardwood faced frames, drawer fronts, and door panels, steel drawer guides, hinges, pulls and laminated interior.
Install	Ea.	298	25		323	
Demolish and Install	Ea.	298	32.55		330.55	
Remove and Install	Ea.		20.05		20.05	
Clean	Ea.	.68	5.45		6.13	
Paint	Ea.	.50	18.85		19.35	
Minimum Charge	Job		150		150	
18" w, 1 Door, 1 Drawer						
Demolish	Ea.		8.95		8.95	Includes material and labor to install premium hardwood modular unit with solid hardwood faced frames, drawer fronts, and door panels, steel drawer guides, hinges, pulls and laminated interior.
Install	Ea.	325	26		351	
Demolish and Install	Ea.	325	34.95		359.95	
Remove and Install	Ea.		20.66		20.66	
Clean	Ea.	.80	6.60		7.40	
Paint	Ea.	.59	22		22.59	
Minimum Charge	Job		150		150	
21" w, 1 Door, 1 Drawer						
Demolish	Ea.		10.55		10.55	Includes material and labor to install premium hardwood modular unit with solid hardwood faced frames, drawer fronts, and door panels, steel drawer guides, hinges, pulls and laminated interior.
Install	Ea.	335	26.50		361.50	
Demolish and Install	Ea.	335	37.05		372.05	
Remove and Install	Ea.		21.20		21.20	
Clean	Ea.	.98	7.80		8.78	
Paint	Ea.	.70	24		24.70	
Minimum Charge	Job		150		150	
24" w, 1 Door, 1 Drawer						
Demolish	Ea.		12.10		12.10	Includes material and labor to install premium hardwood modular unit with solid hardwood faced frames, drawer fronts, and door panels, steel drawer guides, hinges and pulls, laminated interior.
Install	Ea.	375	27		402	
Demolish and Install	Ea.	375	39.10		414.10	
Remove and Install	Ea.		21.58		21.58	
Clean	Ea.	1.06	8.70		9.76	
Paint	Ea.	.79	29.50		30.29	
Minimum Charge	Job		150		150	

Base Cabinets	Unit	Material	Labor	Equip.	Total	Specification
30" Wide Sink Base						
Demolish	Ea.		15.15		15.15	Includes material and labor to install premium hardwood modular unit with solid hardwood faced frames, drawer fronts, and door panels, steel drawer guides, hinges, pulls and laminated interior.
Install	Ea.	335	28		363	
Demolish and Install	Ea.	335	43.15		378.15	
Remove and Install	Ea.		22.49		22.49	
Clean	Ea.	1.36	10.90		12.26	
Paint	Ea.	1	33		34	
Minimum Charge	Job		150		150	
36" Wide Sink Base						
Demolish	Ea.		18.65		18.65	Includes material and labor to install premium hardwood modular unit with solid hardwood faced frames, drawer fronts, and door panels, steel drawer guides, hinges, pulls and laminated interior.
Install	Ea.	375	29.50		404.50	
Demolish and Install	Ea.	375	48.15		423.15	
Remove and Install	Ea.		23.71		23.71	
Clean	Ea.	1.52	12.80		14.32	
Paint	Ea.	1.21	44		45.21	
Minimum Charge	Job		150		150	
36" Blind Corner						
Demolish	Ea.		18.65		18.65	Includes material and labor to install premium hardwood modular unit with solid hardwood faced frames, drawer fronts, and door panels, steel drawer guides, hinges, pulls and laminated interior.
Install	Ea.	720	33.50		753.50	
Demolish and Install	Ea.	720	52.15		772.15	
Remove and Install	Ea.		26.74		26.74	
Clean	Ea.	1.52	12.80		14.32	
Paint	Ea.	1.21	44		45.21	
Minimum Charge	Job		150		150	
42" w, 2 Door, 2 Drawer						
Demolish	Ea.		22		22	Includes material and labor to install premium hardwood modular unit with solid hardwood faced frames, drawer fronts, and door panels, steel drawer guides, hinges, pulls and laminated interior.
Install	Ea.	595	30.50		625.50	
Demolish and Install	Ea.	595	52.50		647.50	
Remove and Install	Ea.		24.31		24.31	
Clean	Ea.	1.95	15.55		17.50	
Paint	Ea.	1.39	53		54.39	
Minimum Charge	Job		150		150	
Clean Interior						
Clean	L.F.	.55	3.30		3.85	Includes labor and materials to clean the interior of a 27" cabinet.
Minimum Charge	Job		109		109	
Island						
24" Wide						
Demolish	Ea.		12.10		12.10	Includes material and labor to install custom modular unit with solid hardwood faced frames, hardwood door frames and drawer fronts, hardwood veneer on raised door panels, hinges and pulls.
Install	Ea.	350	25		375	
Demolish and Install	Ea.	350	37.10		387.10	
Remove and Install	Ea.		20.05		20.05	
Clean	Ea.	1.52	12.80		14.32	
Paint	Ea.	1.20	37.50		38.70	
Minimum Charge	Job		150		150	
30" Wide						
Demolish	Ea.		15.15		15.15	Includes material and labor to install custom modular unit with solid hardwood faced frames, hardwood door frames and drawer fronts, hardwood veneer on raised door panels, hinges and pulls.
Install	Ea.	415	31.50		446.50	
Demolish and Install	Ea.	415	46.65		461.65	
Remove and Install	Ea.		25.07		25.07	
Clean	Ea.	1.95	15.55		17.50	
Paint	Ea.	1.50	44		45.50	
Minimum Charge	Job		150		150	

Cabinets and Countertops

Base Cabinets	Unit	Material	Labor	Equip.	Total	Specification
36″ Wide						
Demolish	Ea.		18.65		18.65	Includes material and labor to install custom modular unit with solid hardwood faced frames, hardwood door frames and drawer fronts, hardwood veneer on raised door panels, hinges and pulls.
Install	Ea.	475	37.50		512.50	
Demolish and Install	Ea.	475	56.15		531.15	
Remove and Install	Ea.		30.08		30.08	
Clean	Ea.	2.28	19.80		22.08	
Paint	Ea.	1.82	53		54.82	
Minimum Charge	Job		150		150	
48″ Wide						
Demolish	Ea.		24		24	Includes material and labor to install custom modular unit with solid hardwood faced frames, hardwood door frames and drawer fronts, hardwood veneer on raised door panels, hinges and pulls.
Install	Ea.	705	50		755	
Demolish and Install	Ea.	705	74		779	
Remove and Install	Ea.		40.11		40.11	
Clean	Ea.	2.73	24		26.73	
Paint	Ea.	2.08	66		68.08	
Minimum Charge	Job		150		150	
60″ Wide						
Demolish	Ea.		27		27	Includes material and labor to install custom modular unit with solid hardwood faced frames, hardwood door frames and drawer fronts, hardwood veneer on raised door panels, hinges and pulls.
Install	Ea.	830	62.50		892.50	
Demolish and Install	Ea.	830	89.50		919.50	
Remove and Install	Ea.		50.13		50.13	
Clean	Ea.	3.42	31		34.42	
Paint	Ea.	2.08	88		90.08	
Minimum Charge	Job		150		150	
Clean Interior						
Clean	L.F.	.55	3.30		3.85	Includes labor and materials to clean the interior of a 27″ cabinet.
Minimum Charge	Job		109		109	
Liquor Bar						
Demolish	L.F.		6.05		6.05	Includes material and labor to install custom modular unit with solid hardwood faced frames, hardwood door frames and drawer fronts, hardwood veneer on raised door panels, hinges and pulls.
Install	L.F.	128	30		158	
Demolish and Install	L.F.	128	36.05		164.05	
Remove and Install	L.F.		24.06		24.06	
Clean	L.F.	.55	4.54		5.09	
Paint	L.F.	.40	12		12.40	
Minimum Charge	Job		150		150	
Clean Interior						
Clean	L.F.	.55	3.30		3.85	Includes labor and materials to clean the interior of a 27″ cabinet.
Minimum Charge	Job		109		109	
Seal Interior						
Install	L.F.	.28	4.33		4.61	Includes labor and materials to seal a cabinet interior.
Minimum Charge	Job		132		132	
Back Bar						
Demolish	L.F.		6.05		6.05	Includes material and labor to install custom modular unit with solid hardwood faced frames, hardwood door frames and drawer fronts, hardwood veneer on raised door panels, hinges and pulls.
Install	L.F.	128	30		158	
Demolish and Install	L.F.	128	36.05		164.05	
Remove and Install	L.F.		24.06		24.06	
Clean	L.F.	.55	4.54		5.09	
Paint	L.F.	.40	12		12.40	
Minimum Charge	Job		150		150	
Clean Interior						
Clean	L.F.	.55	3.30		3.85	Includes labor and materials to clean the interior of a 27″ cabinet.
Minimum Charge	Job		109		109	

Base Cabinets	Unit	Material	Labor	Equip.	Total	Specification
Seal Interior						
Install	L.F.	.28	4.33		4.61	Includes labor and materials to seal a cabinet interior.
Minimum Charge	Job		132		132	
Storage (garage)						
Demolish	L.F.		6.05		6.05	Includes material and labor to install good modular unit with melamine laminate, textured colors or wood grain print finish including hinges and pulls.
Install	L.F.	95.50	30		125.50	
Demolish and Install	L.F.	95.50	36.05		131.55	
Remove and Install	L.F.		24.06		24.06	
Clean	L.F.	.55	4.54		5.09	
Paint	L.F.	.40	12		12.40	
Minimum Charge	Job		150		150	
Clean Interior						
Clean	L.F.	.55	3.30		3.85	Includes labor and materials to clean the interior of a 27″ cabinet.
Minimum Charge	Job		109		109	
Seal Interior						
Install	L.F.	.28	4.33		4.61	Includes labor and materials to seal a cabinet interior.
Minimum Charge	Job		132		132	
Built-in Desk						
Demolish	L.F.		6.05		6.05	Includes material and labor to install custom modular unit with solid hardwood faced frames, hardwood door frames and drawer fronts, hardwood veneer on raised door panels, hinges and pulls.
Install	L.F.	128	30		158	
Demolish and Install	L.F.	128	36.05		164.05	
Remove and Install	L.F.		24.06		24.06	
Clean	L.F.	.55	4.54		5.09	
Paint	L.F.	.40	12		12.40	
Minimum Charge	Job		150		150	
Built-in Bookcase						
Demolish	L.F.		8.05		8.05	Includes material and labor to install custom modular bookcase unit with clear pine faced frames, shelves 12″ O.C., 7′ high, 8″ deep.
Install	L.F.	26	30		56	
Demolish and Install	L.F.	26	38.05		64.05	
Remove and Install	L.F.		24.06		24.06	
Clean	L.F.	1.09	7.80		8.89	
Paint	L.F.	.94	20.50		21.44	
Minimum Charge	Job		150		150	
Strip and Refinish						
Install	L.F.	1.07	4.80		5.87	Includes labor and material to strip, prep and refinish exterior of cabinets.
Minimum Charge	Job		132		132	
Stain						
Install	L.F.	.34	1.23		1.57	Includes labor and material to stain door and exterior of cabinets.
Minimum Charge	Job		150		150	

Countertop	Unit	Material	Labor	Equip.	Total	Specification
Laminated						
With Splash						
Demolish	L.F.		4.04		4.04	Includes material and labor to install one piece laminated top with 4″ backsplash.
Install	L.F.	27.50	10.75		38.25	
Demolish and Install	L.F.	27.50	14.79		42.29	
Remove and Install	L.F.		8.59		8.59	
Clean	L.F.	.07	.55		.62	
Minimum Charge	Job		150		150	

Countertop	Unit	Material	Labor	Equip.	Total	Specification
Roll Top						
Demolish	L.F.		4.04		4.04	Includes material and labor to install one piece laminated top with rolled drip edge (post formed) and backsplash.
Install	L.F.	9.55	10.05		19.60	
Demolish and Install	L.F.	9.55	14.09		23.64	
Remove and Install	L.F.		8.02		8.02	
Clean	L.F.	.10	.83		.93	
Minimum Charge	Job		150		150	
Ceramic Tile						
Demolish	S.F.		2.02		2.02	Cost includes material and labor to install 4-1/4" x 4-1/4" to 6" x 6" glazed tile set in mortar bed and grout on particle board substrate.
Install	S.F.	13.85	6		19.85	
Demolish and Install	S.F.	13.85	8.02		21.87	
Clean	S.F.	.04	.42		.46	
Minimum Charge	Job		138		138	
Cultured Marble						
Demolish	L.F.		4.04		4.04	Includes material and labor to install marble countertop, 24" wide, no backsplash.
Install	L.F.	25	30.50		55.50	
Demolish and Install	L.F.	25	34.54		59.54	
Remove and Install	L.F.		24.32		24.32	
Clean	L.F.	.10	.83		.93	
Minimum Charge	Job		138		138	
With Splash						
Demolish	L.F.		4.04		4.04	Includes material and labor to install marble countertop, 24" wide, with backsplash.
Install	L.F.	28	30.50		58.50	
Demolish and Install	L.F.	28	34.54		62.54	
Remove and Install	L.F.		24.32		24.32	
Clean	L.F.	.10	.83		.93	
Minimum Charge	Job		138		138	
Quarry Tile						
Demolish	S.F.		2.02		2.02	Includes material and labor to install quarry tile countertop with no backsplash.
Install	S.F.	7.85	6		13.85	
Demolish and Install	S.F.	7.85	8.02		15.87	
Clean	S.F.	.04	.42		.46	
Minimum Charge	Job		138		138	
With Splash						
Demolish	S.F.		2.02		2.02	Includes material and labor to install quarry tile countertop with backsplash.
Install	S.F.	9.95	6.70		16.65	
Demolish and Install	S.F.	9.95	8.72		18.67	
Clean	S.F.	.04	.42		.46	
Minimum Charge	Job		138		138	
Butcher Block						
Demolish	S.F.		2.02		2.02	Includes material and labor to install solid laminated maple countertop with no backsplash.
Install	S.F.	28.50	5.35		33.85	
Demolish and Install	S.F.	28.50	7.37		35.87	
Remove and Install	S.F.		4.30		4.30	
Clean	S.F.	.02	.17		.19	
Minimum Charge	Job		150		150	
Solid Surface						
Demolish	L.F.		4.04		4.04	Includes material and labor to install solid surface counter top 22" deep.
Install	L.F.	48	21.50		69.50	
Demolish and Install	L.F.	48	25.54		73.54	
Remove and Install	L.F.		17.19		17.19	
Clean	L.F.	.10	.83		.93	
Minimum Charge	Job		150		150	

Vanity Cabinets	Unit	Material	Labor	Equip.	Total	Specification
Good Grade Laminated						
24" Wide, 2 Door						
Demolish	Ea.		12.10		12.10	Includes material and labor to install modular unit with melamine laminated to particle board, textured colors or wood grain print finish, hinges and pulls.
Install	Ea.	139	30		169	
Demolish and Install	Ea.	139	42.10		181.10	
Remove and Install	Ea.		24.06		24.06	
Clean	Ea.	1.06	8.70		9.76	
Paint	Ea.	.79	22		22.79	
Minimum Charge	Job		150		150	
30" Wide, 2 Door						
Demolish	Ea.		15.15		15.15	Includes material and labor to install modular unit with melamine laminated to particle board, textured colors or wood grain print finish, hinges and pulls.
Install	Ea.	158	37.50		195.50	
Demolish and Install	Ea.	158	52.65		210.65	
Remove and Install	Ea.		30.08		30.08	
Clean	Ea.	1.36	10.90		12.26	
Paint	Ea.	1	26.50		27.50	
Minimum Charge	Job		150		150	
36" Wide, 2 Door						
Demolish	Ea.		18.65		18.65	Includes material and labor to install modular unit with melamine laminated to particle board, textured colors or wood grain print finish, hinges and pulls.
Install	Ea.	212	45		257	
Demolish and Install	Ea.	212	63.65		275.65	
Remove and Install	Ea.		36.11		36.11	
Clean	Ea.	1.52	12.80		14.32	
Paint	Ea.	1.21	33		34.21	
Minimum Charge	Job		150		150	
42" w, 3 Doors / Drawer						
Demolish	Ea.		22		22	Includes material and labor to install modular unit with melamine laminated to particle board, textured colors or wood grain print finish, hinges and pulls.
Install	Ea.	305	48.50		353.50	
Demolish and Install	Ea.	305	70.50		375.50	
Remove and Install	Ea.		38.88		38.88	
Clean	Ea.	1.95	15.55		17.50	
Paint	Ea.	1.39	37.50		38.89	
Minimum Charge	Job		150		150	
48" w, 3 Doors / Drawer						
Demolish	Ea.		24		24	Includes material and labor to install modular unit with melamine laminated to particle board, textured colors or wood grain print finish, hinges and pulls.
Install	Ea.	253	52.50		305.50	
Demolish and Install	Ea.	253	76.50		329.50	
Remove and Install	Ea.		42.11		42.11	
Clean	Ea.	2.73	24		26.73	
Paint	Ea.	2.08	66		68.08	
Minimum Charge	Job		150		150	
Clean Interior						
Install	L.F.	.55	4.54		5.09	Includes labor and materials to clean base cabinetry.
Minimum Charge	Job		109		109	
Better Grade Wood						
24" Wide, 2 Door						
Demolish	Ea.		12.10		12.10	Includes material and labor to install custom modular unit with solid hardwood faced frames, hardwood door frames and drawer fronts, hardwood veneer on raised door panels, hinges and pulls.
Install	Ea.	184	30		214	
Demolish and Install	Ea.	184	42.10		226.10	
Remove and Install	Ea.		24.06		24.06	
Clean	Ea.	1.06	8.70		9.76	
Paint	Ea.	.79	24		24.79	
Minimum Charge	Job		150		150	

Vanity Cabinets	Unit	Material	Labor	Equip.	Total	Specification
30" Wide, 2 Door						
Demolish	Ea.		15.15		15.15	Includes material and labor to install custom modular unit with solid hardwood faced frames, hardwood door frames and drawer fronts, hardwood veneer on raised door panels, hinges and pulls.
Install	Ea.	211	37.50		248.50	
Demolish and Install	Ea.	211	52.65		263.65	
Remove and Install	Ea.		30.08		30.08	
Clean	Ea.	1.36	10.90		12.26	
Paint	Ea.	1	29.50		30.50	
Minimum Charge	Job		150		150	
36" Wide, 2 Door						
Demolish	Ea.		18.65		18.65	Includes material and labor to install custom modular unit with solid hardwood faced frames, hardwood door frames and drawer fronts, hardwood veneer on raised door panels, hinges and pulls.
Install	Ea.	283	45		328	
Demolish and Install	Ea.	283	63.65		346.65	
Remove and Install	Ea.		36.11		36.11	
Clean	Ea.	1.52	12.80		14.32	
Paint	Ea.	1.21	37.50		38.71	
Minimum Charge	Job		150		150	
42" w, 3 Doors / Drawer						
Demolish	Ea.		22		22	Includes material and labor to install custom modular unit with solid hardwood faced frames, hardwood door frames and drawer fronts, hardwood veneer on raised door panels, hinges and pulls.
Install	Ea.	305	48.50		353.50	
Demolish and Install	Ea.	305	70.50		375.50	
Remove and Install	Ea.		38.88		38.88	
Clean	Ea.	1.95	15.55		17.50	
Paint	Ea.	1.39	44		45.39	
Minimum Charge	Job		150		150	
48" w, 3 Doors / Drawer						
Demolish	Ea.		24		24	Includes material and labor to install custom modular unit with solid hardwood faced frames, hardwood door frames and drawer fronts, hardwood veneer on raised door panels, hinges and pulls.
Install	Ea.	335	52.50		387.50	
Demolish and Install	Ea.	335	76.50		411.50	
Remove and Install	Ea.		42.11		42.11	
Clean	Ea.	2.73	24		26.73	
Paint	Ea.	2.08	66		68.08	
Minimum Charge	Job		150		150	
Clean Interior						
Install	L.F.	.55	4.54		5.09	Includes labor and materials to clean base cabinetry.
Minimum Charge	Job		109		109	
Premium Grade Wood						
24" Wide, 2 Door						
Demolish	Ea.		12.10		12.10	Includes material and labor to install premium modular unit with solid hardwood faced frames, hardwood door frames and drawer fronts, hardwood veneer on raised door panels, hinges and pulls.
Install	Ea.	219	30		249	
Demolish and Install	Ea.	219	42.10		261.10	
Remove and Install	Ea.		24.06		24.06	
Clean	Ea.	1.06	8.70		9.76	
Paint	Ea.	.79	26.50		27.29	
Minimum Charge	Job		150		150	
30" Wide, 2 Door						
Demolish	Ea.		15.15		15.15	Includes material and labor to install premium modular unit with solid hardwood faced frames, hardwood door frames and drawer fronts, hardwood veneer on raised door panels, hinges and pulls.
Install	Ea.	252	37.50		289.50	
Demolish and Install	Ea.	252	52.65		304.65	
Remove and Install	Ea.		30.08		30.08	
Clean	Ea.	1.36	10.90		12.26	
Paint	Ea.	1	33		34	
Minimum Charge	Job		150		150	

Vanity Cabinets	Unit	Material	Labor	Equip.	Total	Specification
36" Wide, 2 Door						
Demolish	Ea.		18.65		18.65	Includes material and labor to install premium modular unit with solid hardwood faced frames, hardwood door frames and drawer fronts, hardwood veneer on raised door panels, hinges and pulls.
Install	Ea.	390	45		435	
Demolish and Install	Ea.	390	63.65		453.65	
Remove and Install	Ea.		36.11		36.11	
Clean	Ea.	1.52	12.80		14.32	
Paint	Ea.	1.21	37.50		38.71	
Minimum Charge	Job		150		150	
42" w, 3 Doors / Drawer						
Demolish	Ea.		22		22	Includes material and labor to install premium modular unit with solid hardwood faced frames, hardwood door frames and drawer fronts, hardwood veneer on raised door panels, hinges and pulls.
Install	Ea.	375	48.50		423.50	
Demolish and Install	Ea.	375	70.50		445.50	
Remove and Install	Ea.		38.88		38.88	
Clean	Ea.	1.95	15.55		17.50	
Paint	Ea.	1.39	44		45.39	
Minimum Charge	Job		150		150	
48" w, 3 Doors / Drawer						
Demolish	Ea.		24		24	Includes material and labor to install premium modular unit with solid hardwood faced frames, hardwood door frames and drawer fronts, hardwood veneer on raised door panels, hinges and pulls.
Install	Ea.	455	52.50		507.50	
Demolish and Install	Ea.	455	76.50		531.50	
Remove and Install	Ea.		42.11		42.11	
Clean	Ea.	2.73	24		26.73	
Paint	Ea.	2.08	66		68.08	
Minimum Charge	Job		150		150	
Clean Interior						
Install	L.F.	.55	4.54		5.09	Includes labor and materials to clean base cabinetry.
Minimum Charge	Job		109		109	
Premium Hardwood						
24" Wide, 2 Door						
Demolish	Ea.		12.10		12.10	Includes material and labor to install premium modular unit with solid hardwood faced frames, hardwood door frames and drawer fronts, hardwood veneer on raised door panels, hinges and pulls.
Install	Ea.	275	30		305	
Demolish and Install	Ea.	275	42.10		317.10	
Remove and Install	Ea.		24.06		24.06	
Clean	Ea.	1.06	8.70		9.76	
Paint	Ea.	.79	29.50		30.29	
Minimum Charge	Job		150		150	
30" Wide, 2 Door						
Demolish	Ea.		15.15		15.15	Includes material and labor to install premium modular unit with solid hardwood faced frames, hardwood door frames and drawer fronts, hardwood veneer on raised door panels, hinges and pulls.
Install	Ea.	310	37.50		347.50	
Demolish and Install	Ea.	310	52.65		362.65	
Remove and Install	Ea.		30.08		30.08	
Clean	Ea.	1.36	10.90		12.26	
Paint	Ea.	1	33		34	
Minimum Charge	Job		150		150	
36" Wide, 2 Door						
Demolish	Ea.		18.65		18.65	Includes material and labor to install premium modular unit with solid hardwood faced frames, hardwood door frames and drawer fronts, hardwood veneer on raised door panels, hinges and pulls.
Install	Ea.	450	45		495	
Demolish and Install	Ea.	450	63.65		513.65	
Remove and Install	Ea.		36.11		36.11	
Clean	Ea.	1.52	12.80		14.32	
Paint	Ea.	1.21	44		45.21	
Minimum Charge	Job		150		150	

Vanity Cabinets	Unit	Material	Labor	Equip.	Total	Specification
42" w, 3 Doors / Drawer						
Demolish	Ea.		22		22	Includes material and labor to install premium modular unit with solid hardwood faced frames, hardwood door frames and drawer fronts, hardwood veneer on raised door panels, hinges and pulls.
Install	Ea.	420	48.50		468.50	
Demolish and Install	Ea.	420	70.50		490.50	
Remove and Install	Ea.		38.88		38.88	
Clean	Ea.	1.95	15.55		17.50	
Paint	Ea.	1.39	53		54.39	
Minimum Charge	Job		150		150	
48" w, 3 Doors / Drawer						
Demolish	Ea.		24		24	Includes material and labor to install premium modular unit with solid hardwood faced frames, hardwood door frames and drawer fronts, hardwood veneer on raised door panels, hinges and pulls.
Install	Ea.	515	52.50		567.50	
Demolish and Install	Ea.	515	76.50		591.50	
Remove and Install	Ea.		42.11		42.11	
Clean	Ea.	2.73	24		26.73	
Paint	Ea.	2.08	66		68.08	
Minimum Charge	Job		150		150	
Clean Interior						
Install	L.F.	.55	4.54		5.09	Includes labor and materials to clean base cabinetry.
Minimum Charge	Job		109		109	
Strip and Refinish						
Install	L.F.	1.07	4.80		5.87	Includes labor and material to strip, prep and refinish exterior of cabinets.
Minimum Charge	Job		132		132	

Vanity Cabinet Tops	Unit	Material	Labor	Equip.	Total	Specification
Cultured Marble Top						
Demolish	L.F.		4.04		4.04	Includes material and labor to install cultured marble vanity top with integral sink, 22" deep.
Install	L.F.	33.50	17.90		51.40	
Demolish and Install	L.F.	33.50	21.94		55.44	
Remove and Install	L.F.		14.31		14.31	
Clean	L.F.	.10	.83		.93	
Minimum Charge	Job		138		138	
Laminated Top						
Demolish	L.F.		2.42		2.42	Includes material and labor to install one piece laminated top with 4" backsplash.
Install	L.F.	27.50	10.75		38.25	
Demolish and Install	L.F.	27.50	13.17		40.67	
Remove and Install	L.F.		8.59		8.59	
Clean	L.F.	.07	.55		.62	
Minimum Charge	Job		150		150	
Solid Surface						
Demolish	L.F.		4.04		4.04	Includes material and labor to install solid surface vanity top with integral sink installed, 22" deep.
Install	L.F.	67	22.50		89.50	
Demolish and Install	L.F.	67	26.54		93.54	
Remove and Install	L.F.		17.83		17.83	
Clean	L.F.	.10	.83		.93	
Minimum Charge	Job		150		150	

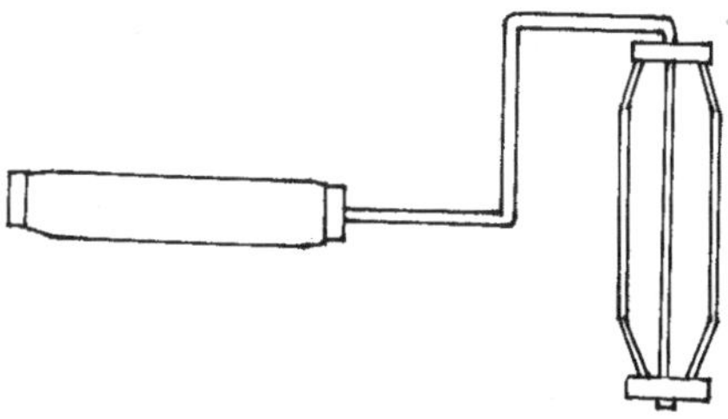

Roller Handle

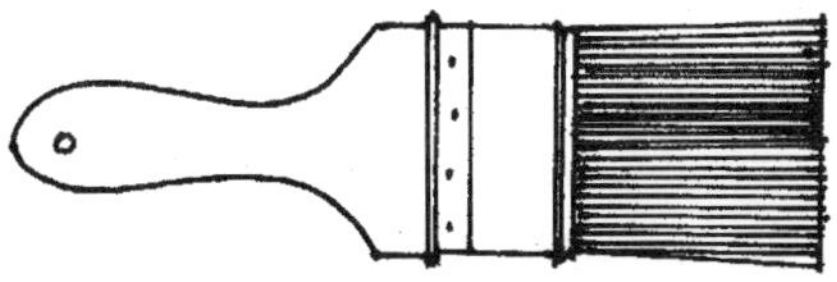

Paint Brush

Paint Preparation, Walls	Unit	Material	Labor	Equip.	Total	Specification
Cover / Protect Floors						
Install	S.F.	.01	.10		.11	Includes material and labor to install plastic masking sheet on large uninterrupted areas.
Minimum Charge	Job		132		132	
Cover / Protect Walls						
Install	S.F.	.01	.10		.11	Includes material and labor to install plastic masking sheet on walls.
Minimum Charge	Job		132		132	
Clean Walls						
Light						
Clean	S.F.		.08		.08	Includes labor to wash gypsum drywall or plaster wall surfaces.
Minimum Charge	Job		132		132	
Heavy						
Clean	S.F.		.11		.11	Includes labor and material for heavy cleaning (multiple applications) with detergent and solvent.
Minimum Charge	Job		132		132	
Prep Walls						
Paint	S.F.	.03	.13		.16	Includes labor and material to prepare for painting including scraping, patching and puttying.
Minimum Charge	Job		132		132	
Sand Walls						
Paint	S.F.		.09		.09	Includes labor to sand gypsum drywall wall surfaces.
Minimum Charge	Job		132		132	
Seal Walls						
Large Area						
Paint	S.F.	.04	.13		.17	Includes labor and material to paint with primer / sealer by roller.
Minimum Charge	Job		132		132	
Spot						
Paint	S.F.	.04	.23		.27	Includes labor and material to seal existing drywall with shellac-based material.
Minimum Charge	Job		132		132	

Paint Preparation, Ceilings	Unit	Material	Labor	Equip.	Total	Specification
Cover / Protect Floors						
Install	S.F.	.01	.10		.11	Includes material and labor to install plastic masking sheet on large uninterrupted areas.
Minimum Charge	Job		132		132	
Cover / Protect Walls						
Install	S.F.	.01	.10		.11	Includes material and labor to install plastic masking sheet on walls.
Minimum Charge	Job		132		132	
Prep Ceiling						
Paint	S.F.	.04	.49		.53	Includes labor and material to fill in hairline cracks up to 1/8" with filler, sand and prep for paint.
Minimum Charge	Job		132		132	
Sand Ceiling						
Paint	S.F.		.12		.12	Includes labor to sand gypsum drywall ceiling surfaces.
Minimum Charge	Job		132		132	
Seal Ceiling						
Large Area						
Paint	S.F.	.06	.20		.26	Includes labor and material to paint ceiling, one coat flat latex.
Minimum Charge	Job		132		132	
Spot						
Paint	S.F.	.04	.23		.27	Includes labor and material to seal existing drywall with shellac-based material.
Minimum Charge	Job		132		132	

Paint / Texture, Walls	Unit	Material	Labor	Equip.	Total	Specification
Texture Walls						
Spray						
Install	S.F.	.04	.34		.38	Includes labor and material to install by spray, texture finish.
Minimum Charge	Job		132		132	
Trowel						
Install	S.F.		.18		.18	Includes labor and material to apply hand troweled texture.
Minimum Charge	Job		132		132	
Paint Walls 1 Coat						
Paint	S.F.	.06	.20		.26	Includes labor and material to paint, one coat flat latex.
Minimum Charge	Job		132		132	
Paint Walls 2 Coats						
Paint	S.F.	.10	.33		.43	Includes labor and material to paint, two coats flat latex.
Minimum Charge	Job		132		132	
Paint Walls 3 Coats						
Paint	S.F.	.15	.41		.56	Includes labor and material to paint, three coats flat latex.
Minimum Charge	Job		132		132	
Clean and Seal						
Paint	S.F.	.21	.70		.91	Includes labor and material to do light cleaning, sealing with 2 coats of latex paint.
Minimum Charge	Job		132		132	

Paint / Texture, Ceilings		Unit	Material	Labor	Equip.	Total	Specification
Texture Ceiling							
Spray							
	Install	S.F.	.04	.34		.38	Includes labor and material to install by spray, texture finish.
	Minimum Charge	Job		132		132	
Trowel							
	Install	S.F.		.18		.18	Includes labor and material to apply hand troweled texture.
	Minimum Charge	Job		132		132	
Paint Ceiling, 1 Coat							
	Paint	S.F.	.06	.20		.26	Includes labor and material to paint ceiling, one coat flat latex.
	Minimum Charge	Job		132		132	
Paint Ceiling, 2 Coats							
	Paint	S.F.	.10	.33		.43	Includes labor and material to paint two coats flat latex paint.
	Minimum Charge	Job		132		132	
Paint Ceiling, 3 Coats							
	Paint	S.F.	.15	.39		.54	Includes labor and material to paint ceiling, three coats flat latex.
	Minimum Charge	Job		132		132	
Clean and Seal							
	Paint	S.F.	.15	.56		.71	Includes labor and material for light cleaning, sealing and paint 3 coats of flat latex paint.
	Minimum Charge	Job		132		132	

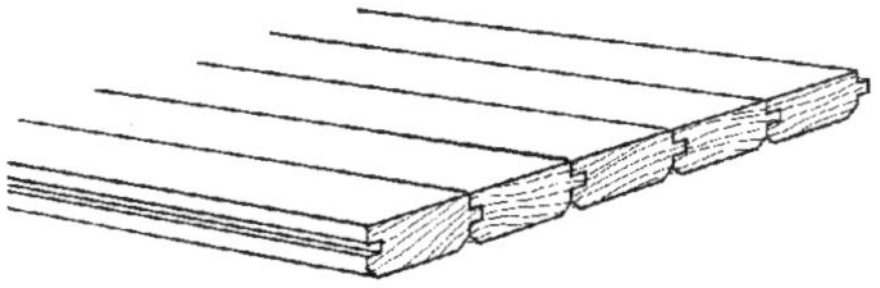

Plank Flooring

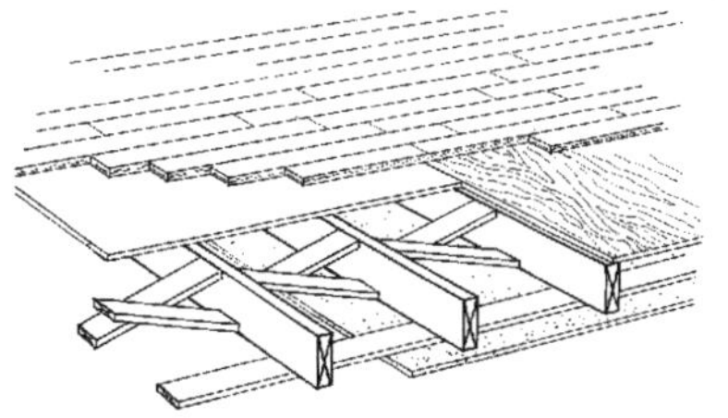

Wood Strip Flooring

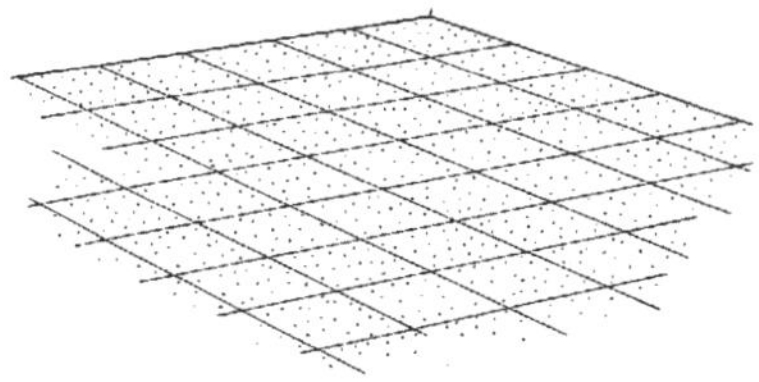

Tile Flooring

Wood Plank Flooring	Unit	Material	Labor	Equip.	Total	Specification
Maple						
2-1/4″						
Demolish	S.F.		.97		.97	Includes material and labor to install unfinished T&G maple flooring, 25/32″ thick, 2-1/4″ wide, random 3′ to 16′ lengths including felt underlayment, nailed in place over prepared subfloor.
Install	S.F.	4.19	1.50		5.69	
Demolish and Install	S.F.	4.19	2.47		6.66	
Remove and Install	S.F.		1.20		1.20	
Clean	S.F.		.27		.27	
Paint	S.F.	1.14	1.68		2.82	
Minimum Charge	Job		150		150	
3-1/4″						
Demolish	S.F.		.97		.97	Includes material and labor to install unfinished T&G maple flooring, 33/32″ thick, 3-1/4″ wide, random 3′ to 16′ lengths including felt underlayment, nailed in place over prepared subfloor.
Install	S.F.	3.11	1.50		4.61	
Demolish and Install	S.F.	3.11	2.47		5.58	
Remove and Install	S.F.		1.20		1.20	
Clean	S.F.		.27		.27	
Paint	S.F.	1.14	1.68		2.82	
Minimum Charge	Job		150		150	
Sand and Finish						
Install	S.F.	.76	.55		1.31	Includes labor and material to sand (3 passes) and finish and two coats of urethane existing floor.
Minimum Charge	Job		150		150	
Oak						
2-1/4″						
Demolish	S.F.		.97		.97	Includes material and labor to install unfinished T&G oak flooring, 25/32″ thick, 2-1/4″ wide, random 3′ to 16′ lengths including felt underlayment, nailed in place over prepared subfloor.
Install	S.F.	3.82	1.50		5.32	
Demolish and Install	S.F.	3.82	2.47		6.29	
Remove and Install	S.F.		1.20		1.20	
Clean	S.F.		.27		.27	
Paint	S.F.	1.14	1.68		2.82	
Minimum Charge	Job		150		150	
3-1/4″						
Demolish	S.F.		.97		.97	Includes material and labor to install unfinished T&G oak flooring, 25/32″ thick, 3-1/4″ wide, random 3′ to 16′ lengths including felt underlayment, nailed in place over prepared subfloor.
Install	S.F.	5.30	1.06		6.36	
Demolish and Install	S.F.	5.30	2.03		7.33	
Remove and Install	S.F.		.85		.85	
Clean	S.F.		.27		.27	
Paint	S.F.	1.14	1.68		2.82	
Minimum Charge	Job		150		150	
Sand and Finish						
Install	S.F.	1.14	1.68		2.82	Includes labor and material to sand and finish (3 passes) with two coats of urethane on a new floor.
Minimum Charge	Job		150		150	

Wood Plank Flooring	Unit	Material	Labor	Equip.	Total	Specification
Walnut						
3" to 7"						
Demolish	S.F.		.97		.97	Includes material and labor to install unfinished T&G walnut flooring, 25/32" thick, random 3' to 16' lengths including felt underlayment, nailed in place over prepared subfloor.
Install	S.F.	9.05	1.06		10.11	
Demolish and Install	S.F.	9.05	2.03		11.08	
Remove and Install	S.F.		.85		.85	
Clean	S.F.		.27		.27	
Paint	S.F.	1.14	1.68		2.82	
Minimum Charge	Job		150		150	
Sand and Finish						
Install	S.F.	1.14	1.68		2.82	Includes labor and material to sand and finish (3 passes) with two coats of urethane on a new floor.
Minimum Charge	Job		150		150	
Pine						
2-1/4"						
Demolish	S.F.		.97		.97	Includes material and labor to install unfinished T&G pine flooring, 25/32" thick, 2-1/4" wide, random 3' to 16' lengths including felt underlayment, nailed in place over prepared subfloor.
Install	S.F.	7.35	1.13		8.48	
Demolish and Install	S.F.	7.35	2.10		9.45	
Remove and Install	S.F.		.91		.91	
Clean	S.F.		.27		.27	
Paint	S.F.	.76	.55		1.31	
Minimum Charge	Job		150		150	
Sand and Finish						
Install	S.F.	1.14	1.68		2.82	Includes labor and material to sand and finish (3 passes) with two coats of urethane on a new floor.
Minimum Charge	Job		150		150	
Prefinished Oak						
2-1/4"						
Demolish	S.F.		.97		.97	Includes material and labor to install prefinished T&G oak flooring, 2-1/4" wide, random 3' to 16' lengths including felt underlayment, nailed in place over prepared subfloor.
Install	S.F.	5.60	1.50		7.10	
Demolish and Install	S.F.	5.60	2.47		8.07	
Remove and Install	S.F.		1.20		1.20	
Clean	S.F.		.27		.27	
Minimum Charge	Job		150		150	
3-1/4"						
Demolish	S.F.		.97		.97	Includes material and labor to install prefinished T&G oak flooring, 25/32" thick, 3" to 7" wide, random 3' to 16' lengths including felt underlayment, nailed in place over prepared subfloor.
Install	S.F.	7.15	1.38		8.53	
Demolish and Install	S.F.	7.15	2.35		9.50	
Remove and Install	S.F.		1.10		1.10	
Clean	S.F.		.27		.27	
Minimum Charge	Job		150		150	
Sand and Finish New						
Install	S.F.	.76	.74		1.50	Includes labor and material to sand (3 passes) and finish, two coats of urethane, new floor.
Minimum Charge	Job		150		150	
Refinish Existing						
Install	S.F.	.76	.55		1.31	Includes labor and material to sand (3 passes) and finish and two coats of urethane existing floor.
Minimum Charge	Job		150		150	
Urethane Coat						
Install	S.F.	.13	.14		.27	Includes labor and material to install two coats of urethane, on existing floor.
Minimum Charge	Job		150		150	
Clean and Wax						
Clean	S.F.	.02	.17		.19	Includes labor and material to wash and wax a hardwood floor.
Minimum Charge	Job		150		150	

Wood Plank Flooring	Unit	Material	Labor	Equip.	Total	Specification
Underlayment Per S.F.						
Demolish	S.F.		.44		.44	Cost includes material and labor to install 1/4" lauan subfloor, standard interior grade, nailed every 6".
Install	S.F.	.50	.40		.90	
Demolish and Install	S.F.	.50	.84		1.34	
Clean	S.F.		.21		.21	
Minimum Charge	Job		150		150	

Wood Parquet Tile Floor	Unit	Material	Labor	Equip.	Total	Specification
Oak						
9" x 9"						
Demolish	S.F.		.74		.74	Includes material and labor to install prefinished oak 9" x 9" parquet block flooring, 5/16" thick, installed in mastic.
Install	S.F.	4.62	2.41		7.03	
Demolish and Install	S.F.	4.62	3.15		7.77	
Remove and Install	S.F.		1.93		1.93	
Clean	S.F.		.27		.27	
Paint	S.F.	1.14	1.68		2.82	
Minimum Charge	Job		150		150	
13" x 13"						
Demolish	S.F.		.74		.74	Includes material and labor to install prefinished oak 13" x 13" parquet block flooring, 5/16" thick, installed in mastic.
Install	S.F.	8.75	1.88		10.63	
Demolish and Install	S.F.	8.75	2.62		11.37	
Remove and Install	S.F.		1.50		1.50	
Clean	S.F.		.27		.27	
Paint	S.F.	1.14	1.68		2.82	
Minimum Charge	Job		150		150	
Cherry						
9" x 9"						
Demolish	S.F.		.74		.74	Includes material and labor to install prefinished cherry 9" x 9" parquet block flooring, 5/16" thick, installed in mastic.
Install	S.F.	8.15	2.41		10.56	
Demolish and Install	S.F.	8.15	3.15		11.30	
Remove and Install	S.F.		2.41		2.41	
Clean	S.F.		.27		.27	
Paint	S.F.	1.14	1.68		2.82	
Minimum Charge	Job		150		150	
13" x 13"						
Demolish	S.F.		.74		.74	Includes material and labor to install prefinished cherry 13" x 13" parquet block flooring, 5/16" thick, installed in mastic.
Install	S.F.	9.30	1.88		11.18	
Demolish and Install	S.F.	9.30	2.62		11.92	
Remove and Install	S.F.		1.88		1.88	
Clean	S.F.		.27		.27	
Paint	S.F.	1.14	1.68		2.82	
Minimum Charge	Job		150		150	
Walnut						
9" x 9"						
Demolish	S.F.		.74		.74	Includes material and labor to install prefinished walnut 9" x 9" parquet flooring in mastic.
Install	S.F.	5	2.41		7.41	
Demolish and Install	S.F.	5	3.15		8.15	
Remove and Install	S.F.		1.93		1.93	
Clean	S.F.		.27		.27	
Paint	S.F.	1.14	1.68		2.82	
Minimum Charge	Job		150		150	

Flooring

Wood Parquet Tile Floor	Unit	Material	Labor	Equip.	Total	Specification
13" x 13"						
Demolish	S.F.		.74		.74	Includes material and labor to install prefinished walnut 13" x 13" parquet block flooring, 5/16" thick, installed in mastic.
Install	S.F.	9.35	1.88		11.23	
Demolish and Install	S.F.	9.35	2.62		11.97	
Remove and Install	S.F.		1.88		1.88	
Clean	S.F.		.27		.27	
Paint	S.F.	1.14	1.68		2.82	
Minimum Charge	Job		150		150	
Acrylic Impregnated Oak 12" x 12"						
Demolish	S.F.		.74		.74	Includes material and labor to install red oak 12" x 12" acrylic impregnated parquet block flooring, 5/16" thick, installed in mastic.
Install	S.F.	7.55	2.15		9.70	
Demolish and Install	S.F.	7.55	2.89		10.44	
Remove and Install	S.F.		2.15		2.15	
Clean	S.F.		.27		.27	
Minimum Charge	Job		150		150	
Cherry 12" x 12"						
Demolish	S.F.		.74		.74	Includes material and labor to install prefinished cherry 12" x 12" acrylic impregnated parquet block flooring, 5/16" thick, installed in mastic.
Install	S.F.	11.40	2.15		13.55	
Demolish and Install	S.F.	11.40	2.89		14.29	
Remove and Install	S.F.		2.15		2.15	
Clean	S.F.		.27		.27	
Minimum Charge	Job		150		150	
Ash 12" x 12"						
Demolish	S.F.		.74		.74	Includes material and labor to install ash 12" x 12" acrylic impregnated parquet block flooring, 5/16" thick in mastic.
Install	S.F.	10.25	2.15		12.40	
Demolish and Install	S.F.	10.25	2.89		13.14	
Remove and Install	S.F.		2.15		2.15	
Clean	S.F.		.27		.27	
Minimum Charge	Job		150		150	
Teak						
Demolish	S.F.		.74		.74	Includes material and labor to install prefinished teak parquet block flooring, 5/16" thick, and installed in mastic.
Install	S.F.	7.20	1.88		9.08	
Demolish and Install	S.F.	7.20	2.62		9.82	
Remove and Install	S.F.		1.88		1.88	
Clean	S.F.		.27		.27	
Paint	S.F.	1.14	1.68		2.82	
Minimum Charge	Job		150		150	
Clean and Wax						
Clean	S.F.	.02	.17		.19	Includes labor and material to wash and wax a hardwood floor.
Minimum Charge	Job		150		150	

Ceramic Tile Flooring	Unit	Material	Labor	Equip.	Total	Specification
Economy Grade						
Demolish	S.F.		.72		.72	Cost includes material and labor to install 4-1/4" x 4-1/4" ceramic tile in mortar bed including grout.
Install	S.F.	4.20	5.40		9.60	
Demolish and Install	S.F.	4.20	6.12		10.32	
Clean	S.F.		.42		.42	
Minimum Charge	Job		138		138	
Average Grade						
Demolish	S.F.		.72		.72	Cost includes material and labor to install 4-1/4" x 4-1/4" ceramic tile in mortar bed including grout.
Install	S.F.	5.60	5.50		11.10	
Demolish and Install	S.F.	5.60	6.22		11.82	
Clean	S.F.		.42		.42	
Minimum Charge	Job		138		138	

Ceramic Tile Flooring	Unit	Material	Labor	Equip.	Total	Specification
Premium Grade						
Demolish	S.F.		.72		.72	Cost includes material and labor to install 4-1/4" x 4-1/4" ceramic tile in mortar bed including grout.
Install	S.F.	8.80	5.50		14.30	
Demolish and Install	S.F.	8.80	6.22		15.02	
Clean	S.F.		.42		.42	
Minimum Charge	Job		138		138	
Re-grout						
Install	S.F.	.14	2.20		2.34	Includes material and labor to regrout tile floors.
Minimum Charge	Job		138		138	

Hard Tile Flooring	Unit	Material	Labor	Equip.	Total	Specification
Thick Set Paver						
Brick						
Demolish	S.F.		1.51		1.51	Cost includes material and labor to install 6" x 12" adobe brick paver with 1/2" mortar joints.
Install	S.F.	.72	12.75		13.47	
Demolish and Install	S.F.	.72	14.26		14.98	
Clean	S.F.		.42		.42	
Minimum Charge	Job		138		138	
Mexican Red						
Demolish	S.F.		1.51		1.51	Cost includes material and labor to install 12" x 12" red Mexican paver tile in mortar bed with grout.
Install	S.F.	1.43	5.75		7.18	
Demolish and Install	S.F.	1.43	7.26		8.69	
Clean	S.F.		.42		.42	
Minimum Charge	Job		138		138	
Saltillo						
Demolish	S.F.		.72		.72	Cost includes material and labor to install 12" x 12"saltillo tile in mortar bed with grout.
Install	S.F.	2.24	5.75		7.99	
Demolish and Install	S.F.	2.24	6.47		8.71	
Clean	S.F.		.42		.42	
Minimum Charge	Job		138		138	
Marble						
Premium Grade						
Demolish	S.F.		.72		.72	Cost includes material and labor to install 3/8" x 12" x 12" marble, in mortar bed with grout.
Install	S.F.	18.90	11.95		30.85	
Demolish and Install	S.F.	18.90	12.67		31.57	
Clean	S.F.		.42		.42	
Minimum Charge	Job		138		138	
Terrazzo						
Gray Cement						
Demolish	S.F.		.96		.96	Cost includes material and labor to install 1-3/4" terrazzo , #1 and #2 chips in gray Portland cement.
Install	S.F.	2.68	3.84	1.53	8.05	
Demolish and Install	S.F.	2.68	4.80	1.53	9.01	
Clean	S.F.		.42		.42	
Minimum Charge	Job		249	99	348	
White Cement						
Demolish	S.F.		.96		.96	Cost includes material and labor to install 1-3/4" terrazzo, #1 and #2 chips in white Portland cement.
Install	S.F.	3.06	3.84	1.53	8.43	
Demolish and Install	S.F.	3.06	4.80	1.53	9.39	
Clean	S.F.		.42		.42	
Minimum Charge	Job		249	99	348	
Non-skid Gray						
Demolish	S.F.		.96		.96	Cost includes material and labor to install 1-3/4" terrazzo, #1 and #2 chips in gray Portland cement with light non-skid abrasive.
Install	S.F.	4.88	8.30	3.30	16.48	
Demolish and Install	S.F.	4.88	9.26	3.30	17.44	
Clean	S.F.		.42		.42	
Minimum Charge	Job		249	99	348	

Hard Tile Flooring

Hard Tile Flooring	Unit	Material	Labor	Equip.	Total	Specification
Non-skid White						
Demolish	S.F.		.96		.96	Cost includes material and labor to install 1-3/4" terrazzo, #1 and #2 chips in white Portland cement with light non-skid abrasive.
Install	S.F.	5.35	8.30	3.30	16.95	
Demolish and Install	S.F.	5.35	9.26	3.30	17.91	
Clean	S.F.		.42		.42	
Minimum Charge	Job		249	99	348	
Gray w / Brass Divider						
Demolish	S.F.		.97		.97	Cost includes material and labor to install 1-3/4" terrazzo, #1 and #2 chips in gray Portland cement with brass strips 2' O.C. each way.
Install	S.F.	6.70	9.05	3.60	19.35	
Demolish and Install	S.F.	6.70	10.02	3.60	20.32	
Clean	S.F.		.42		.42	
Minimum Charge	Job		249	99	348	
Clean						
Clean	S.F.	.12	.28		.40	Includes labor and material to clean terrazzo flooring.
Minimum Charge	Job		249	99	348	
Slate						
Demolish	S.F.		.87		.87	Includes material and labor to install slate tile flooring in thin set with grout.
Install	S.F.	4.55	2.73		7.28	
Demolish and Install	S.F.	4.55	3.60		8.15	
Clean	S.F.		.42		.42	
Minimum Charge	Job		138		138	
Granite						
Demolish	S.F.		.71		.71	Includes material and labor to install granite tile flooring in thin set with grout.
Install	S.F.	18.55	23		41.55	
Demolish and Install	S.F.	18.55	23.71		42.26	
Remove and Install	S.F.		22.93		22.93	
Clean	S.F.		.42		.42	
Minimum Charge	Job		138		138	
Re-grout						
Install	S.F.	.14	2.20		2.34	Includes material and labor to regrout tile floors.
Minimum Charge	Job		138		138	

Sheet Vinyl Per S.F.

Sheet Vinyl Per S.F.	Unit	Material	Labor	Equip.	Total	Specification
Economy Grade						
Demolish	S.F.		.35		.35	Includes material and labor to install resilient sheet vinyl flooring.
Install	S.F.	1.38	.86		2.24	
Demolish and Install	S.F.	1.38	1.21		2.59	
Clean	S.F.		.27		.27	
Minimum Charge	Job		138		138	
Average Grade						
Demolish	S.F.		.35		.35	Includes material and labor to install resilient sheet vinyl flooring.
Install	S.F.	2.23	.86		3.09	
Demolish and Install	S.F.	2.23	1.21		3.44	
Clean	S.F.		.27		.27	
Minimum Charge	Job		138		138	
Premium Grade						
Demolish	S.F.		.35		.35	Includes material and labor to install resilient sheet vinyl flooring.
Install	S.F.	3.51	.86		4.37	
Demolish and Install	S.F.	3.51	1.21		4.72	
Clean	S.F.		.27		.27	
Minimum Charge	Job		138		138	
Clean and Wax						
Clean	S.F.	.02	.14		.16	Includes labor and material to wash and wax a vinyl tile floor.
Minimum Charge	Job		138		138	

Flooring

Sheet Vinyl	Unit	Material	Labor	Equip.	Total	Specification
Economy Grade						
Demolish	S.Y.		3.13		3.13	Includes material and labor to install no wax sheet vinyl flooring.
Install	S.Y.	12.45	7.75		20.20	
Demolish and Install	S.Y.	12.45	10.88		23.33	
Clean	S.Y.		1.90		1.90	
Minimum Charge	Job		138		138	
Average Grade						
Demolish	S.Y.		3.13		3.13	Includes material and labor to install no wax sheet vinyl flooring.
Install	S.Y.	20	7.75		27.75	
Demolish and Install	S.Y.	20	10.88		30.88	
Clean	S.Y.		1.90		1.90	
Minimum Charge	Job		138		138	
Premium Grade						
Demolish	S.Y.		3.13		3.13	Includes material and labor to install no wax sheet vinyl flooring.
Install	S.Y.	31.50	7.75		39.25	
Demolish and Install	S.Y.	31.50	10.88		42.38	
Clean	S.Y.		1.90		1.90	
Minimum Charge	Job		138		138	
Clean and Wax						
Clean	S.F.	.02	.14		.16	Includes labor and material to wash and wax a vinyl tile floor.
Minimum Charge	Job		138		138	

Vinyl Tile	Unit	Material	Labor	Equip.	Total	Specification
Economy Grade						
Demolish	S.F.		.48		.48	Includes material and labor to install no wax 12" x 12" vinyl tile.
Install	S.F.	1.78	.55		2.33	
Demolish and Install	S.F.	1.78	1.03		2.81	
Clean	S.F.		.27		.27	
Minimum Charge	Job		138		138	
Average Grade						
Demolish	S.F.		.48		.48	Includes material and labor to install no wax 12" x 12" vinyl tile.
Install	S.F.	5.05	.55		5.60	
Demolish and Install	S.F.	5.05	1.03		6.08	
Clean	S.F.		.27		.27	
Minimum Charge	Job		138		138	
Premium Grade						
Demolish	S.F.		.48		.48	Includes material and labor to install no wax 12" x 12" vinyl tile.
Install	S.F.	8.45	.55		9	
Demolish and Install	S.F.	8.45	1.03		9.48	
Clean	S.F.		.27		.27	
Minimum Charge	Job		138		138	
Clean and Wax						
Clean	S.F.	.02	.14		.16	Includes labor and material to wash and wax a vinyl tile floor.
Minimum Charge	Job		138		138	

Carpeting Per S.F.	Unit	Material	Labor	Equip.	Total	Specification
Economy Grade						
Demolish	S.F.		.05		.05	Includes material and labor to install carpet including tack strips and hot melt tape on seams.
Install	S.F.	1.05	.54		1.59	
Demolish and Install	S.F.	1.05	.59		1.64	
Clean	S.F.		.38		.38	
Minimum Charge	Job		138		138	

Carpeting Per S.F.	Unit	Material	Labor	Equip.	Total	Specification
Average Grade						
Demolish	S.F.		.05		.05	Includes material and labor to install carpet including tack strips and hot melt tape on seams.
Install	S.F.	1.28	.54		1.82	
Demolish and Install	S.F.	1.28	.59		1.87	
Clean	S.F.		.38		.38	
Minimum Charge	Job		138		138	
Premium Grade						
Demolish	S.F.		.05		.05	Includes material and labor to install carpet including tack strips and hot melt tape on seams.
Install	S.F.	2.28	.54		2.82	
Demolish and Install	S.F.	2.28	.59		2.87	
Clean	S.F.		.38		.38	
Minimum Charge	Job		138		138	
Indoor/Outdoor						
Demolish	S.F.		.05		.05	Includes material and labor to install indoor-outdoor carpet.
Install	S.F.	.98	.65		1.63	
Demolish and Install	S.F.	.98	.70		1.68	
Clean	S.F.		.38		.38	
Minimum Charge	Job		138		138	
100% Wool						
Average Grade						
Demolish	S.F.		.05		.05	Includes material and labor to install carpet including tack strips and hot melt tape on seams.
Install	S.F.	5.60	.55		6.15	
Demolish and Install	S.F.	5.60	.60		6.20	
Clean	S.F.		.38		.38	
Minimum Charge	Job		138		138	
Premium Grade						
Demolish	S.F.		.05		.05	Includes material and labor to install carpet including tack strips and hot melt tape on seams.
Install	S.F.	6.25	.55		6.80	
Demolish and Install	S.F.	6.25	.60		6.85	
Clean	S.F.		.38		.38	
Minimum Charge	Job		138		138	
Luxury Grade						
Demolish	S.F.		.05		.05	Includes material and labor to install carpet including tack strips and hot melt tape on seams.
Install	S.F.	8.85	.55		9.40	
Demolish and Install	S.F.	8.85	.60		9.45	
Clean	S.F.		.38		.38	
Minimum Charge	Job		138		138	
Berber						
Average Grade						
Demolish	S.F.		.05		.05	Includes material and labor to install carpet including tack strips and hot melt tape on seams.
Install	S.F.	2.63	.65		3.28	
Demolish and Install	S.F.	2.63	.70		3.33	
Clean	S.F.		.38		.38	
Minimum Charge	Job		138		138	
Premium Grade						
Demolish	S.F.		.05		.05	Includes material and labor to install carpet including tack strips and hot melt tape on seams.
Install	S.F.	5.65	.65		6.30	
Demolish and Install	S.F.	5.65	.70		6.35	
Clean	S.F.		.38		.38	
Minimum Charge	Job		138		138	

Carpeting	Unit	Material	Labor	Equip.	Total	Specification
Indoor/Outdoor						
Demolish	S.Y.		.81		.81	Includes material and labor to install indoor-outdoor carpet.
Install	S.Y.	8.80	5.85		14.65	
Demolish and Install	S.Y.	8.80	6.66		15.46	
Remove and Install	S.Y.		4.68		4.68	
Clean	S.Y.		3.41		3.41	
Minimum Charge	Job		138		138	
Economy Grade						
Demolish	S.Y.		.81		.81	Includes material and labor to install carpet including tack strips and hot melt tape on seams.
Install	S.Y.	9.40	4.83		14.23	
Demolish and Install	S.Y.	9.40	5.64		15.04	
Remove and Install	S.Y.		3.86		3.86	
Clean	S.Y.		3.41		3.41	
Minimum Charge	Job		138		138	
Average Grade						
Demolish	S.Y.		.81		.81	Includes material and labor to install carpet including tack strips and hot melt tape on seams.
Install	S.Y.	11.45	4.83		16.28	
Demolish and Install	S.Y.	11.45	5.64		17.09	
Remove and Install	S.Y.		3.86		3.86	
Clean	S.Y.		3.41		3.41	
Minimum Charge	Job		138		138	
Premium Grade						
Demolish	S.Y.		.81		.81	Includes material and labor to install carpet including tack strips and hot melt tape on seams.
Install	S.Y.	20.50	4.83		25.33	
Demolish and Install	S.Y.	20.50	5.64		26.14	
Remove and Install	S.Y.		3.86		3.86	
Clean	S.Y.		3.41		3.41	
Minimum Charge	Job		138		138	
100% Wool						
Average Grade						
Demolish	S.Y.		.81		.81	Includes material and labor to install wool carpet including tack strips and hot melt tape on seams.
Install	S.Y.	50.50	4.91		55.41	
Demolish and Install	S.Y.	50.50	5.72		56.22	
Remove and Install	S.Y.		3.93		3.93	
Clean	S.Y.		3.41		3.41	
Minimum Charge	Job		138		138	
Premium Grade						
Demolish	S.Y.		.81		.81	Includes material and labor to install carpet including tack strips and hot melt tape on seams.
Install	S.Y.	56	4.91		60.91	
Demolish and Install	S.Y.	56	5.72		61.72	
Remove and Install	S.Y.		3.93		3.93	
Clean	S.Y.		3.41		3.41	
Minimum Charge	Job		138		138	
Luxury Grade						
Demolish	S.Y.		.81		.81	Includes material and labor to install carpet including tack strips and hot melt tape on seams.
Install	S.Y.	80	4.91		84.91	
Demolish and Install	S.Y.	80	5.72		85.72	
Remove and Install	S.Y.		3.93		3.93	
Clean	S.Y.		3.41		3.41	
Minimum Charge	Job		138		138	
Berber						
Average Grade						
Demolish	S.Y.		.81		.81	Includes material and labor to install berber carpet including tack strips and hot melt tape on seams.
Install	S.Y.	23.50	5.85		29.35	
Demolish and Install	S.Y.	23.50	6.66		30.16	
Remove and Install	S.Y.		4.68		4.68	
Clean	S.Y.		3.41		3.41	
Minimum Charge	Job		138		138	

Carpeting	Unit	Material	Labor	Equip.	Total	Specification
Premium Grade						
Demolish	S.Y.		.81		.81	Includes material and labor to install carpet including tack strips and hot melt tape on seams.
Install	S.Y.	51	5.85		56.85	
Demolish and Install	S.Y.	51	6.66		57.66	
Remove and Install	S.Y.		4.68		4.68	
Clean	S.Y.		3.41		3.41	
Minimum Charge	Job		138		138	

Carpet Pad Per S.F.	Unit	Material	Labor	Equip.	Total	Specification
Urethane						
Demolish	S.F.		.02		.02	Includes material and labor to install urethane carpet pad.
Install	S.F.	.31	.20		.51	
Demolish and Install	S.F.	.31	.22		.53	
Minimum Charge	Job		138		138	
Foam Rubber Slab						
Demolish	S.F.		.02		.02	Includes material and labor to install carpet pad, 3/8" thick foam rubber.
Install	S.F.	.51	.20		.71	
Demolish and Install	S.F.	.51	.22		.73	
Minimum Charge	Job		138		138	
Waffle						
Demolish	S.F.		.02		.02	Includes material and labor to install rubber waffle carpet pad.
Install	S.F.	.33	.02		.35	
Demolish and Install	S.F.	.33	.04		.37	
Minimum Charge	Job		138		138	
Jute						
Demolish	S.F.		.02		.02	Includes material and labor to install jute hair carpet pad.
Install	S.F.	.59	.18		.77	
Demolish and Install	S.F.	.59	.20		.79	
Minimum Charge	Job		138		138	
Rebound						
Demolish	S.F.		.02		.02	Includes material and labor to install rebound carpet pad.
Install	S.F.	.28	.15		.43	
Demolish and Install	S.F.	.28	.17		.45	
Minimum Charge	Job		138		138	

Carpet Pad	Unit	Material	Labor	Equip.	Total	Specification
Urethane						
Demolish	S.Y.		.14		.14	Includes material and labor to install urethane carpet pad.
Install	S.Y.	2.73	1.83		4.56	
Demolish and Install	S.Y.	2.73	1.97		4.70	
Remove and Install	S.Y.		1.47		1.47	
Minimum Charge	Job		138		138	
Foam Rubber Slab						
Demolish	S.Y.		.14		.14	Includes material and labor to install rubber slab carpet pad.
Install	S.Y.	4.55	1.83		6.38	
Demolish and Install	S.Y.	4.55	1.97		6.52	
Remove and Install	S.Y.		1.47		1.47	
Minimum Charge	Job		138		138	
Waffle						
Demolish	S.Y.		.14		.14	Includes material and labor to install rubber waffle carpet pad.
Install	S.Y.	2.93	.20		3.13	
Demolish and Install	S.Y.	2.93	.34		3.27	
Remove and Install	S.Y.		.16		.16	
Minimum Charge	Job		138		138	

Flooring

Carpet Pad	Unit	Material	Labor	Equip.	Total	Specification
Jute						
Demolish	S.Y.		.14		.14	Includes material and labor to install jute hair carpet pad.
Install	S.Y.	5.30	1.62		6.92	
Demolish and Install	S.Y.	5.30	1.76		7.06	
Remove and Install	S.Y.		1.30		1.30	
Minimum Charge	Job		138		138	
Rebound						
Demolish	S.Y.		.14		.14	Includes material and labor to install rebound carpet pad.
Install	S.Y.	2.49	1.37		3.86	
Demolish and Install	S.Y.	2.49	1.51		4	
Remove and Install	S.Y.		1.10		1.10	
Minimum Charge	Job		138		138	

Stair Components	Unit	Material	Labor	Equip.	Total	Specification
Carpeting						
Demolish	Ea.		1.47		1.47	Includes material and labor to install carpet and pad on stairs.
Install	Riser	28.50	7.25		35.75	
Demolish and Install	Riser	28.50	8.72		37.22	
Remove and Install	Riser		5.79		5.79	
Clean	Ea.		3.41		3.41	
Minimum Charge	Job		138		138	

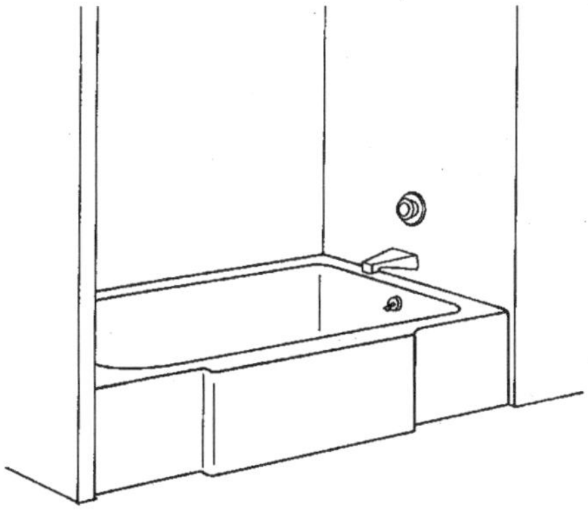

Tub/Shower

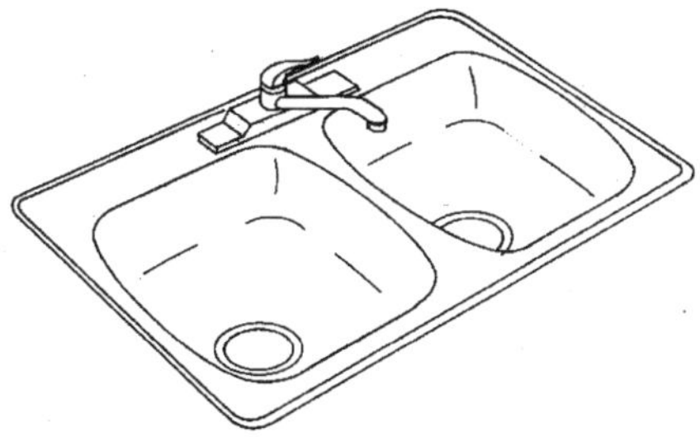

Stainless Steel Double Sink

Water Closet

Commode	Unit	Material	Labor	Equip.	Total	Specification
Demolish	Ea.		32.50		32.50	Includes material and labor to install a toilet with valve, seat and cover.
Install	Ea.	190	110		300	
Demolish and Install	Ea.	190	142.50		332.50	
Remove and Install	Ea.		88.34		88.34	
Clean	Ea.	.28	10.90		11.18	
Minimum Charge	Job		163		163	
Wall w / Flush Valve						
Demolish	Ea.		32.50		32.50	Includes material and labor to install a wall hung toilet with flush valve, seat and cover.
Install	Ea.	380	101		481	
Demolish and Install	Ea.	380	133.50		513.50	
Remove and Install	Ea.		80.73		80.73	
Clean	Ea.	.28	10.90		11.18	
Minimum Charge	Job		163		163	
Designer Floor Mount						
Demolish	Ea.		32.50		32.50	Includes material and labor to install a toilet with valve, seat and cover.
Install	Ea.	545	110		655	
Demolish and Install	Ea.	545	142.50		687.50	
Remove and Install	Ea.		88.34		88.34	
Clean	Ea.	.28	10.90		11.18	
Minimum Charge	Job		163		163	
Urinal						
Demolish	Ea.		32.50		32.50	Includes material and labor to install a wall hung urinal with flush valve.
Install	Ea.	315	195		510	
Demolish and Install	Ea.	315	227.50		542.50	
Remove and Install	Ea.		156.07		156.07	
Clean	Ea.	.28	10.90		11.18	
Minimum Charge	Job		163		163	
Bidet						
Vitreous China						
Demolish	Ea.		32.50		32.50	Includes material and labor to install a vitreous china bidet complete with trim.
Install	Ea.	600	117		717	
Demolish and Install	Ea.	600	149.50		749.50	
Remove and Install	Ea.		93.64		93.64	
Clean	Ea.	.28	10.90		11.18	
Minimum Charge	Job		163		163	
Chrome Fittings						
Install	Ea.	145	40.50		185.50	Includes labor and materials to install chrome fittings for a bidet.
Brass Fittings						
Install	Ea.	160	40.50		200.50	Includes labor and materials to install brass fittings for a bidet.

Commode	Unit	Material	Labor	Equip.	Total	Specification
Seat						
Demolish	Ea.		13.55		13.55	Includes material and labor to install a toilet seat and cover.
Install	Ea.	27.50	13.55		41.05	
Demolish and Install	Ea.	27.50	27.10		54.60	
Remove and Install	Ea.		10.84		10.84	
Clean	Ea.	.13	4.54		4.67	
Minimum Charge	Job		163		163	
Rough-in						
Install	Ea.	152	206		358	Includes labor and materials to install water supply pipe with valves and drain, waste and vent pipe with all couplings, hangers and fasteners necessary.

Sink (assembly)	Unit	Material	Labor	Equip.	Total	Specification
Single						
Porcelain						
Demolish	Ea.		20.50		20.50	Includes material and labor to install a single bowl enamel finished cast iron kitchen sink with faucet and drain.
Install	Ea.	274	105		379	
Demolish and Install	Ea.	274	125.50		399.50	
Remove and Install	Ea.		83.61		83.61	
Clean	Ea.	.18	7.80		7.98	
Minimum Charge	Job		163		163	
Stainless Steel						
Demolish	Ea.		20.50		20.50	Includes material and labor to install a single bowl stainless steel kitchen sink with faucet and drain.
Install	Ea.	360	105		465	
Demolish and Install	Ea.	360	125.50		485.50	
Remove and Install	Ea.		83.61		83.61	
Clean	Ea.	.18	7.80		7.98	
Minimum Charge	Job		163		163	
Double						
Porcelain						
Demolish	Ea.		23		23	Includes material and labor to install a double bowl enamel finished cast iron kitchen sink with faucet and drain.
Install	Ea.	525	122		647	
Demolish and Install	Ea.	525	145		670	
Remove and Install	Ea.		97.55		97.55	
Clean	Ea.	.28	10.90		11.18	
Minimum Charge	Job		163		163	
Stainless Steel						
Demolish	Ea.		23		23	Includes material and labor to install a double bowl stainless steel kitchen sink with faucet and drain.
Install	Ea.	223	122		345	
Demolish and Install	Ea.	223	145		368	
Remove and Install	Ea.		97.55		97.55	
Clean	Ea.	.28	10.90		11.18	
Minimum Charge	Job		163		163	
Bar						
Demolish	Ea.		20.50		20.50	Includes material and labor to install a small single bowl stainless steel bar / vegetable / kitchen sink with faucet and drain.
Install	Ea.	70	81.50		151.50	
Demolish and Install	Ea.	70	102		172	
Remove and Install	Ea.		65.04		65.04	
Clean	Ea.	.18	7.80		7.98	
Minimum Charge	Job		163		163	

Sink (assembly)	Unit	Material	Labor	Equip.	Total	Specification
Floor						
Demolish	Ea.		65		65	Includes material and labor to install an enameled cast iron floor mounted corner service sink with faucet and drain.
Install	Ea.	540	133		673	
Demolish and Install	Ea.	540	198		738	
Remove and Install	Ea.		106.41		106.41	
Clean	Ea.	.28	5.45		5.73	
Minimum Charge	Job		163		163	
Pedestal						
Demolish	Ea.		20.50		20.50	Includes material and labor to install a vitreous china pedestal lavatory with faucet set and pop-up drain.
Install	Ea.	445	91.50		536.50	
Demolish and Install	Ea.	445	112		557	
Remove and Install	Ea.		73.16		73.16	
Clean	Ea.	.18	7.80		7.98	
Minimum Charge	Job		163		163	
Vanity Lavatory						
Demolish	Ea.		20.50		20.50	Includes material and labor to install a vitreous china lavatory with faucet set and pop-up drain.
Install	Ea.	287	108		395	
Demolish and Install	Ea.	287	128.50		415.50	
Remove and Install	Ea.		86.71		86.71	
Clean	Ea.	.18	7.80		7.98	
Minimum Charge	Job		163		163	
Rough-in						
Install	Ea.	96.50	274		370.50	Includes labor and materials to install copper supply pipe with valves and drain, waste and vent pipe with all couplings, hangers and fasteners necessary.
Laundry						
Demolish	Ea.		22		22	Includes material and labor to install a laundry sink (wall mounted or with legs) including faucet and pop up drain.
Install	Ea.	96.50	90		186.50	
Demolish and Install	Ea.	96.50	112		208.50	
Clean	Ea.	.18	7.80		7.98	
Minimum Charge	Job		163		163	

Sink Only	Unit	Material	Labor	Equip.	Total	Specification
Single						
Porcelain						
Demolish	Ea.		20.50		20.50	Includes material and labor to install a single bowl enamel finished cast iron kitchen sink.
Install	Ea.	204	116		320	
Demolish and Install	Ea.	204	136.50		340.50	
Remove and Install	Ea.		92.91		92.91	
Clean	Ea.	.18	7.80		7.98	
Minimum Charge	Job		163		163	
Stainless Steel						
Demolish	Ea.		20.50		20.50	Includes material and labor to install a single bowl stainless steel kitchen sink.
Install	Ea.	285	81.50		366.50	
Demolish and Install	Ea.	285	102		387	
Remove and Install	Ea.		65.04		65.04	
Clean	Ea.	.18	7.80		7.98	
Minimum Charge	Job		163		163	

Sink Only	Unit	Material	Labor	Equip.	Total	Specification
Double						
Porcelain						
Demolish	Ea.		23		23	Includes material and labor to install a double bowl enamel finished cast iron kitchen sink.
Install	Ea.	234	136		370	
Demolish and Install	Ea.	234	159		393	
Remove and Install	Ea.		108.40		108.40	
Clean	Ea.	.28	10.90		11.18	
Minimum Charge	Job		163		163	
Stainless Steel						
Demolish	Ea.		23		23	Includes material and labor to install a double bowl stainless steel kitchen sink.
Install	Ea.	425	105		530	
Demolish and Install	Ea.	425	128		553	
Remove and Install	Ea.		83.92		83.92	
Clean	Ea.	.28	10.90		11.18	
Minimum Charge	Job		163		163	
Bar						
Demolish	Ea.		20.50		20.50	Includes material and labor to install a small single bowl stainless steel bar / vegetable / kitchen sink.
Install	Ea.	36.50	81.50		118	
Demolish and Install	Ea.	36.50	102		138.50	
Remove and Install	Ea.		65.04		65.04	
Clean	Ea.	.18	7.80		7.98	
Minimum Charge	Job		163		163	
Floor						
Demolish	Ea.		65		65	Includes material and labor to install an enameled cast iron floor mounted corner service sink.
Install	Ea.	445	148		593	
Demolish and Install	Ea.	445	213		658	
Remove and Install	Ea.		118.25		118.25	
Clean	Ea.	.28	5.45		5.73	
Minimum Charge	Job		163		163	
Pedestal						
Demolish	Ea.		20.50		20.50	Includes material and labor to install a vitreous china pedestal lavatory.
Install	Ea.	345	102		447	
Demolish and Install	Ea.	345	122.50		467.50	
Remove and Install	Ea.		81.30		81.30	
Clean	Ea.	.18	7.80		7.98	
Minimum Charge	Job		163		163	
Vanity Lavatory						
Demolish	Ea.		20.50		20.50	Includes material and labor to install a vitreous china lavatory.
Install	Ea.	145	102		247	
Demolish and Install	Ea.	145	122.50		267.50	
Remove and Install	Ea.		81.30		81.30	
Clean	Ea.	.18	7.80		7.98	
Minimum Charge	Job		163		163	
Rough-in						
Install	Ea.	84.50	254		338.50	Includes labor and materials to install copper supply pipe with valves and drain, waste and vent pipe with all couplings, hangers and fasteners necessary.

Sink Only	Unit	Material	Labor	Equip.	Total	Specification
Faucet Sink						
Good Grade						
Demolish	Ea.		16.25		16.25	Includes material and labor to install a sink faucet and fittings.
Install	Ea.	56.50	22.50		79	
Demolish and Install	Ea.	56.50	38.75		95.25	
Remove and Install	Ea.		17.95		17.95	
Clean	Ea.	.11	6.80		6.91	
Minimum Charge	Job		163		163	
Better Grade						
Demolish	Ea.		16.25		16.25	Includes material and labor to install a sink faucet and fittings.
Install	Ea.	95	22.50		117.50	
Demolish and Install	Ea.	95	38.75		133.75	
Remove and Install	Ea.		17.95		17.95	
Clean	Ea.	.11	6.80		6.91	
Minimum Charge	Job		163		163	
Premium Grade						
Demolish	Ea.		16.25		16.25	Includes material and labor to install a sink faucet and fittings.
Install	Ea.	186	28		214	
Demolish and Install	Ea.	186	44.25		230.25	
Remove and Install	Ea.		22.39		22.39	
Clean	Ea.	.11	6.80		6.91	
Minimum Charge	Job		163		163	
Drain & Basket						
Demolish	Ea.		10.15		10.15	Includes material and labor to install a sink drain and basket.
Install	Ea.	13.35	20.50		33.85	
Demolish and Install	Ea.	13.35	30.65		44	
Remove and Install	Ea.		16.26		16.26	
Clean	Ea.	.11	6.80		6.91	
Minimum Charge	Job		163		163	

Bathtub	Unit	Material	Labor	Equip.	Total	Specification
Enameled						
Steel						
Demolish	Ea.		58.50		58.50	Includes material and labor to install a formed steel bath tub with spout, mixing valve, shower head and pop-up drain.
Install	Ea.	315	106		421	
Demolish and Install	Ea.	315	164.50		479.50	
Remove and Install	Ea.		85.13		85.13	
Clean	Ea.	.55	18.15		18.70	
Minimum Charge	Job		163		163	
Cast Iron						
Demolish	Ea.		58.50		58.50	Includes material and labor to install a cast iron bath tub with spout, mixing valve, shower head and pop-up drain.
Install	Ea.	415	133		548	
Demolish and Install	Ea.	415	191.50		606.50	
Remove and Install	Ea.		106.41		106.41	
Clean	Ea.	.55	18.15		18.70	
Minimum Charge	Job		163		163	
Fiberglass						
Demolish	Ea.		58.50		58.50	Includes material and labor to install an acrylic soaking tub with spout, mixing valve, shower head and pop-up drain.
Install	Ea.	900	106		1006	
Demolish and Install	Ea.	900	164.50		1064.50	
Remove and Install	Ea.		85.13		85.13	
Clean	Ea.	.55	18.15		18.70	
Minimum Charge	Job		163		163	

Bathtub	Unit	Material	Labor	Equip.	Total	Specification
Institutional						
Demolish	Ea.		58.50		58.50	Includes material and labor to install a hospital / institutional type bathtub with spout, mixing valve, shower head and pop-up drain.
Install	Ea.	975	195		1170	
Demolish and Install	Ea.	975	253.50		1228.50	
Remove and Install	Ea.		156.07		156.07	
Clean	Ea.	.55	18.15		18.70	
Minimum Charge	Job		163		163	
Whirlpool (Acrylic)						
Demolish	Ea.		181		181	Includes material and labor to install a molded fiberglass whirlpool tub.
Install	Ea.	2025	585		2610	
Demolish and Install	Ea.	2025	766		2791	
Remove and Install	Ea.		468.22		468.22	
Clean	Ea.	.55	18.15		18.70	
Minimum Charge	Job		163		163	
Pump / Motor						
Demolish	Ea.		45		45	Includes material and labor to install a whirlpool tub pump / motor.
Install	Ea.	163	136		299	
Demolish and Install	Ea.	163	181		344	
Remove and Install	Ea.		108.40		108.40	
Clean	Ea.	.13	13.65		13.78	
Minimum Charge	Job		163		163	
Heater / Motor						
Demolish	Ea.		45		45	Includes material and labor to install a whirlpool tub heater.
Install	Ea.	217	145		362	
Demolish and Install	Ea.	217	190		407	
Remove and Install	Ea.		115.63		115.63	
Clean	Ea.	.13	13.65		13.78	
Minimum Charge	Job		163		163	
Thermal Cover						
Demolish	Ea.		7.25		7.25	Includes material and labor to install a whirlpool tub thermal cover.
Install	Ea.	146	20.50		166.50	
Demolish and Install	Ea.	146	27.75		173.75	
Remove and Install	Ea.		16.26		16.26	
Clean	Ea.	1.36	13.65		15.01	
Minimum Charge	Job		163		163	
Tub / Shower Combination						
Demolish	Ea.		36		36	Includes material and labor to install a fiberglass module tub with shower surround with spout, mixing valve, shower head and pop-up drain.
Install	Ea.	775	146		921	
Demolish and Install	Ea.	775	182		957	
Remove and Install	Ea.		117.06		117.06	
Clean	Ea.	.68	27.50		28.18	
Minimum Charge	Job		163		163	
Accessories						
Sliding Door						
Demolish	Ea.		15.15		15.15	Includes material and labor to install a 48″ aluminum framed tempered glass shower door.
Install	Ea.	360	69		429	
Demolish and Install	Ea.	360	84.15		444.15	
Remove and Install	Ea.		55.38		55.38	
Clean	Ea.	1.14	9.10		10.24	
Minimum Charge	Job		163		163	
Shower Head						
Demolish	Ea.		4.07		4.07	Includes material and labor to install a water saving shower head.
Install	Ea.	69.50	13.55		83.05	
Demolish and Install	Ea.	69.50	17.62		87.12	
Remove and Install	Ea.		10.84		10.84	
Clean	Ea.	.01	2.73		2.74	
Minimum Charge	Job		163		163	

Finish Mechanical

Bathtub	Unit	Material	Labor	Equip.	Total	Specification
Faucet Set						
Demolish	Ea.		20.50		20.50	Includes material and labor to install a combination spout / diverter for a bathtub.
Install	Ea.	79	40.50		119.50	
Demolish and Install	Ea.	79	61		140	
Remove and Install	Ea.		32.52		32.52	
Clean	Ea.	.11	6.80		6.91	
Minimum Charge	Job		163		163	
Shower Rod						
Demolish	Ea.		2.52		2.52	Includes material and labor to install a chrome plated shower rod.
Install	Ea.	13.15	23		36.15	
Demolish and Install	Ea.	13.15	25.52		38.67	
Remove and Install	Ea.		18.51		18.51	
Clean	Ea.		2.73		2.73	
Minimum Charge	Job		163		163	
Rough-in						
Install	Ea.	138	283		421	Includes labor and materials to install copper supply pipe with valves and drain, waste and vent pipe with all couplings, hangers and fasteners necessary.

Shower	Unit	Material	Labor	Equip.	Total	Specification
Fiberglass						
32″ x 32″						
Demolish	Ea.		122		122	Includes material and labor to install a fiberglass shower stall with door, mixing valve and shower head and drain fitting.
Install	Ea.	385	244		629	
Demolish and Install	Ea.	385	366		751	
Remove and Install	Ea.		195.09		195.09	
Clean	Ea.	.55	18.15		18.70	
Minimum Charge	Job		163		163	
36″ x 36″						
Demolish	Ea.		122		122	Includes material and labor to install a fiberglass shower stall with door, mixing valve and shower head and drain fitting.
Install	Ea.	435	244		679	
Demolish and Install	Ea.	435	366		801	
Remove and Install	Ea.		195.09		195.09	
Clean	Ea.	.55	18.15		18.70	
Minimum Charge	Job		163		163	
35″ x 60″						
Demolish	Ea.		122		122	Includes material and labor to install a handicap fiberglass shower stall with door & seat, mixing valve and shower head and drain fitting.
Install	Ea.	705	293		998	
Demolish and Install	Ea.	705	415		1120	
Remove and Install	Ea.		234.11		234.11	
Clean	Ea.	.55	18.15		18.70	
Minimum Charge	Job		163		163	
Accessories						
Single Door						
Demolish	Ea.		15.15		15.15	Includes material and labor to install a shower door.
Install	Ea.	106	36.50		142.50	
Demolish and Install	Ea.	106	51.65		157.65	
Remove and Install	Ea.		29.23		29.23	
Clean	Ea.	1.14	9.10		10.24	
Minimum Charge	Job		163		163	

Shower	Unit	Material	Labor	Equip.	Total	Specification
Shower Head						
Demolish	Ea.		4.07		4.07	Includes material and labor to install a water saving shower head.
Install	Ea.	69.50	13.55		83.05	
Demolish and Install	Ea.	69.50	17.62		87.12	
Remove and Install	Ea.		10.84		10.84	
Clean	Ea.	.01	2.73		2.74	
Minimum Charge	Job		163		163	
Faucet Set						
Demolish	Ea.		20.50		20.50	Includes material and labor to install a combination spout / diverter for a bathtub.
Install	Ea.	79	40.50		119.50	
Demolish and Install	Ea.	79	61		140	
Remove and Install	Ea.		32.52		32.52	
Clean	Ea.	.11	6.80		6.91	
Minimum Charge	Job		163		163	
Shower Pan						
Demolish	Ea.		15.05		15.05	Includes material and labor to install the base portion of a fiberglass shower unit.
Install	Ea.	177	40.50		217.50	
Demolish and Install	Ea.	177	55.55		232.55	
Remove and Install	Ea.		32.52		32.52	
Clean	Ea.	.55	18.15		18.70	
Minimum Charge	Job		163		163	
Rough-in						
Install	Ea.	91	286		377	Includes labor and materials to install copper supply pipe with valves and drain, waste and vent pipe with all couplings, hangers and fasteners necessary.

Ductwork	Unit	Material	Labor	Equip.	Total	Specification
Return Grill						
Under 10″ Wide						
Demolish	Ea.		7.30		7.30	Includes material and labor to install a 10″ x 10″ air return register.
Install	Ea.	33	18.25		51.25	
Demolish and Install	Ea.	33	25.55		58.55	
Remove and Install	Ea.		14.61		14.61	
Clean	Ea.	.13	6.05		6.18	
Minimum Charge	Job		164		164	
12″ to 20″ Wide						
Demolish	Ea.		7.30		7.30	Includes material and labor to install a 16″ x 16″ air return register.
Install	Ea.	60	19.35		79.35	
Demolish and Install	Ea.	60	26.65		86.65	
Remove and Install	Ea.		15.47		15.47	
Clean	Ea.	.13	6.05		6.18	
Minimum Charge	Job		164		164	
Over 20″ Wide						
Demolish	Ea.		11.05		11.05	Includes material and labor to install a 24″ x 24″ air return register.
Install	Ea.	117	30		147	
Demolish and Install	Ea.	117	41.05		158.05	
Remove and Install	Ea.		23.91		23.91	
Clean	Ea.	.13	6.80		6.93	
Minimum Charge	Job		164		164	

Ductwork	Unit	Material	Labor	Equip.	Total	Specification
Supply Grill						
Baseboard Type						
Demolish	Ea.		5.20		5.20	Includes material and labor to install a 14″ x 6″ baseboard air supply register.
Install	Ea.	16.60	16.45		33.05	
Demolish and Install	Ea.	16.60	21.65		38.25	
Remove and Install	Ea.		13.15		13.15	
Clean	Ea.	.13	6.05		6.18	
Minimum Charge	Job		164		164	
10″ Wide						
Demolish	Ea.		6.10		6.10	Includes material and labor to install a 10″ x 6″ air supply register.
Install	Ea.	25	16.45		41.45	
Demolish and Install	Ea.	25	22.55		47.55	
Remove and Install	Ea.		13.15		13.15	
Clean	Ea.	.13	6.05		6.18	
Minimum Charge	Job		164		164	
12″ to 15″ Wide						
Demolish	Ea.		6.10		6.10	Includes material and labor to install a 14″ x 8″ air supply register.
Install	Ea.	34	19.35		53.35	
Demolish and Install	Ea.	34	25.45		59.45	
Remove and Install	Ea.		15.47		15.47	
Clean	Ea.	.13	6.05		6.18	
Minimum Charge	Job		164		164	
18″ to 24″ Wide						
Demolish	Ea.		7.30		7.30	Includes material and labor to install a 24″ x 8″ air supply register.
Install	Ea.	47.50	25.50		73	
Demolish and Install	Ea.	47.50	32.80		80.30	
Remove and Install	Ea.		20.23		20.23	
Clean	Ea.	.13	6.80		6.93	
Minimum Charge	Job		164		164	
30″ to 36″ Wide						
Demolish	Ea.		9.15		9.15	Includes material and labor to install a 30″ x 8″ air supply register.
Install	Ea.	62	23.50		85.50	
Demolish and Install	Ea.	62	32.65		94.65	
Remove and Install	Ea.		18.79		18.79	
Clean	Ea.	.13	13.65		13.78	
Minimum Charge	Job		164		164	
Diffuser						
12″ Louver						
Demolish	Ea.		7.30		7.30	Includes material and labor to install a 12″ aluminum louvered diffuser.
Install	Ea.	25.50	23.50		49	
Demolish and Install	Ea.	25.50	30.80		56.30	
Remove and Install	Ea.		18.79		18.79	
Clean	Ea.	.13	6.05		6.18	
Minimum Charge	Job		164		164	
14″ to 20″ Louver						
Demolish	Ea.		7.30		7.30	Includes material and labor to install a 14″ to 20″ aluminum louvered diffuser.
Install	Ea.	36	33		69	
Demolish and Install	Ea.	36	40.30		76.30	
Remove and Install	Ea.		26.30		26.30	
Clean	Ea.	.13	6.80		6.93	
Minimum Charge	Job		164		164	
25″ to 32″ Louver						
Demolish	Ea.		11.05		11.05	Includes material and labor to install a 25″ to 32″ aluminum louvered diffuser.
Install	Ea.	40	47		87	
Demolish and Install	Ea.	40	58.05		98.05	
Remove and Install	Ea.		37.58		37.58	
Clean	Ea.	.13	9.10		9.23	
Minimum Charge	Job		164		164	

Ductwork		Unit	Material	Labor	Equip.	Total	Specification
12″ Round							
	Demolish	Ea.		10.15		10.15	Includes material and labor to install a 12″ diameter aluminum diffuser with a butterfly damper.
	Install	Ea.	30.50	27.50		58	
	Demolish and Install	Ea.	30.50	37.65		68.15	
	Remove and Install	Ea.		21.92		21.92	
	Clean	Ea.	.13	6.05		6.18	
	Minimum Charge	Job		164		164	
14″ to 20″ Round							
	Demolish	Ea.		13.55		13.55	Includes material and labor to install a 20″ diameter aluminum diffuser with a butterfly damper.
	Install	Ea.	111	36.50		147.50	
	Demolish and Install	Ea.	111	50.05		161.05	
	Remove and Install	Ea.		29.23		29.23	
	Clean	Ea.	.13	6.80		6.93	
	Minimum Charge	Job		164		164	
Flush-out / Sanitize							
	Install	L.F.	.55	2.73		3.28	Includes labor and material costs to clean ductwork.
	Minimum Charge	Job		164		164	

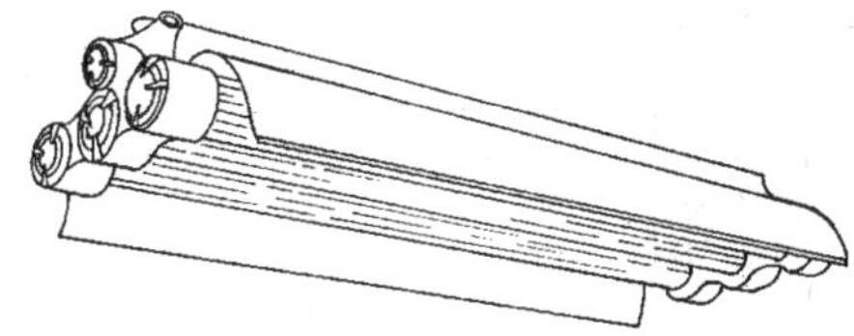

Lighting Fixture

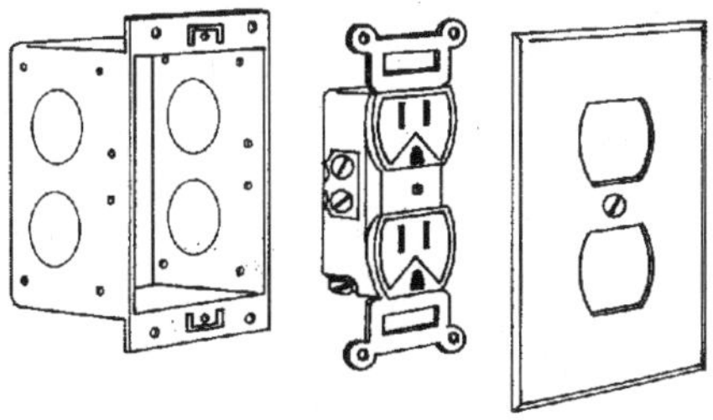

Duplex Receptacle

Electrical Per S.F.	Unit	Material	Labor	Equip.	Total	Specification
Residential						
Light Fixtures						
Install	S.F.	.22	.16		.38	Includes labor and material to replace house light fixtures. Apply S.F. cost to floor area including garages but not basements.

Circuits	Unit	Material	Labor	Equip.	Total	Specification
Outlet with Wiring						
Demolish	Ea.		18.30		18.30	Includes material and labor to install general purpose outlet with cover, up to 20 feet of #12/2 wire.
Install	Ea.	5.90	22.50		28.40	
Demolish and Install	Ea.	5.90	40.80		46.70	
Remove and Install	Ea.		18.10		18.10	
Clean	Ea.		1.36		1.36	
Minimum Charge	Job		165		165	
GFCI Outlet w / Wiring						
Demolish	Ea.		18.30		18.30	Cost includes material and labor to install 120V GFCI receptacle and wall plate, with up to 20 feet of #12/2 wire.
Install	Ea.	36.50	26.50		63	
Demolish and Install	Ea.	36.50	44.80		81.30	
Remove and Install	Ea.		21.39		21.39	
Clean	Ea.		1.36		1.36	
Minimum Charge	Job		165		165	
Exterior Outlet w / Wiring						
Demolish	Ea.		18.30		18.30	Cost includes material and labor to install 110V waterproof external receptacle and cover plate with up to 20′ of #12/2 wire.
Install	Ea.	24	27.50		51.50	
Demolish and Install	Ea.	24	45.80		69.80	
Remove and Install	Ea.		21.95		21.95	
Clean	Ea.		1.36		1.36	
Minimum Charge	Job		165		165	
240V Outlet w / Wiring						
Demolish	Ea.		18.30		18.30	Cost includes material and labor to install 240V receptacle and cover plate with up to 20′ of #12/2 wire.
Install	Ea.	53.50	51.50		105	
Demolish and Install	Ea.	53.50	69.80		123.30	
Remove and Install	Ea.		41.09		41.09	
Clean	Ea.		1.36		1.36	
Minimum Charge	Job		165		165	

Finish Electrical

Circuits	Unit	Material	Labor	Equip.	Total	Specification
TV / Cable Jack w / Wiring						
Demolish	Ea.		18.30		18.30	Includes material and labor to install TV/cable outlet box and cover plate including 20′ of wiring.
Install	Ea.	12.25	20.50		32.75	
Demolish and Install	Ea.	12.25	38.80		51.05	
Remove and Install	Ea.		16.46		16.46	
Clean	Ea.		1.36		1.36	
Minimum Charge	Job		165		165	
Phone Jack with Wiring						
Demolish	Ea.		18.30		18.30	Includes material and labor to install phone jack and cover plate including up to 20′ of telephone cable.
Install	Ea.	7.35	12.65		20	
Demolish and Install	Ea.	7.35	30.95		38.30	
Remove and Install	Ea.		10.13		10.13	
Clean	Ea.		.85		.85	
Minimum Charge	Job		165		165	

Light Switch	Unit	Material	Labor	Equip.	Total	Specification
Single						
Switch						
Demolish	Ea.		3.66		3.66	Includes material and labor to install single switch and wall plate. Wiring is not included.
Install	Ea.	4.75	15.70		20.45	
Demolish and Install	Ea.	4.75	19.36		24.11	
Remove and Install	Ea.		12.54		12.54	
Clean	Ea.		1.36		1.36	
Minimum Charge	Job		165		165	
Plate						
Demolish	Ea.		1.83		1.83	Includes material and labor to install a one gang brown wall switch plate.
Install	Ea.	.33	4.12		4.45	
Demolish and Install	Ea.	.33	5.95		6.28	
Remove and Install	Ea.		3.29		3.29	
Clean	Ea.	.06	.27		.33	
Minimum Charge	Job		165		165	
Double						
Switch						
Demolish	Ea.		7.30		7.30	Includes material and labor to install double switch and wall plate. Wiring is not included.
Install	Ea.	6.25	23.50		29.75	
Demolish and Install	Ea.	6.25	30.80		37.05	
Remove and Install	Ea.		18.81		18.81	
Clean	Ea.	.06	4.36		4.42	
Minimum Charge	Job		165		165	
Plate						
Demolish	Ea.		1.83		1.83	Includes material and labor to install a two gang brown wall switch plate.
Install	Ea.	.85	6.20		7.05	
Demolish and Install	Ea.	.85	8.03		8.88	
Remove and Install	Ea.		4.97		4.97	
Clean	Ea.	.06	.27		.33	
Minimum Charge	Job		165		165	
Triple						
Switch						
Demolish	Ea.		11.10		11.10	Includes material and labor to install triple switch and wall plate. Wiring is not included.
Install	Ea.	20.50	27.50		48	
Demolish and Install	Ea.	20.50	38.60		59.10	
Remove and Install	Ea.		21.95		21.95	
Clean	Ea.	.06	6.55		6.61	
Minimum Charge	Job		165		165	

Finish Electrical

Light Switch	Unit	Material	Labor	Equip.	Total	Specification
Plate						
Demolish	Ea.		1.83		1.83	Includes material and labor to install a three gang brown wall switch plate.
Install	Ea.	1.25	10.30		11.55	
Demolish and Install	Ea.	1.25	12.13		13.38	
Remove and Install	Ea.		8.23		8.23	
Clean	Ea.	.06	.27		.33	
Minimum Charge	Job		165		165	
Dimmer						
Switch						
Demolish	Ea.		3.66		3.66	Includes material and labor to install dimmer switch. Wiring is not included.
Install	Ea.	17.60	19.35		36.95	
Demolish and Install	Ea.	17.60	23.01		40.61	
Remove and Install	Ea.		15.49		15.49	
Clean	Ea.		1.36		1.36	
Minimum Charge	Job		165		165	
Plate						
Demolish	Ea.		1.83		1.83	Includes material and labor to install a one gang brown wall switch plate.
Install	Ea.	.33	4.12		4.45	
Demolish and Install	Ea.	.33	5.95		6.28	
Remove and Install	Ea.		3.29		3.29	
Clean	Ea.	.06	.27		.33	
Minimum Charge	Job		165		165	
With Wiring						
Single						
Demolish	Ea.		18.30		18.30	Includes material and labor to install single switch and plate including up to 20′ of romex cable.
Install	Ea.	7.60	19.25		26.85	
Demolish and Install	Ea.	7.60	37.55		45.15	
Remove and Install	Ea.		15.40		15.40	
Clean	Ea.		1.36		1.36	
Minimum Charge	Job		165		165	
Double						
Demolish	Ea.		18.30		18.30	Includes material and labor to install double switch and plate including up to 20′ of romex cable.
Install	Ea.	18.45	33		51.45	
Demolish and Install	Ea.	18.45	51.30		69.75	
Remove and Install	Ea.		26.34		26.34	
Clean	Ea.	.06	4.36		4.42	
Minimum Charge	Job		165		165	
Triple						
Demolish	Ea.		27.50		27.50	Includes material and labor to install triple switch and plate including up to 20′ of romex cable.
Install	Ea.	26	37		63	
Demolish and Install	Ea.	26	64.50		90.50	
Remove and Install	Ea.		29.62		29.62	
Clean	Ea.	.06	6.55		6.61	
Minimum Charge	Job		165		165	
Dimmer						
Demolish	Ea.		18.30		18.30	Includes material and labor to install dimmer switch and cover plate including up to 20′ of romex cable.
Install	Ea.	20.50	22.50		43	
Demolish and Install	Ea.	20.50	40.80		61.30	
Remove and Install	Ea.		18.10		18.10	
Clean	Ea.		1.36		1.36	
Minimum Charge	Job		165		165	

Finish Electrical

Cover for Outlet / Switch	Unit	Material	Labor	Equip.	Total	Specification
High Grade						
Demolish	Ea.		1.83		1.83	Includes material and labor to install wall plate, stainless steel, 1 gang.
Install	Ea.	1.98	4.12		6.10	
Demolish and Install	Ea.	1.98	5.95		7.93	
Remove and Install	Ea.		3.29		3.29	
Clean	Ea.	.06	.27		.33	
Minimum Charge	Job		165		165	
Deluxe Grade						
Demolish	Ea.		1.83		1.83	Includes material and labor to install wall plate, brushed brass, 1 gang.
Install	Ea.	4.24	4.12		8.36	
Demolish and Install	Ea.	4.24	5.95		10.19	
Remove and Install	Ea.		3.29		3.29	
Clean	Ea.	.06	.27		.33	
Minimum Charge	Job		165		165	

Residential Light Fixture	Unit	Material	Labor	Equip.	Total	Specification
Ceiling						
Good Quality						
Demolish	Ea.		7.60		7.60	Includes material and labor to install standard quality incandescent ceiling light fixture. Wiring is not included.
Install	Ea.	28	8.25		36.25	
Demolish and Install	Ea.	28	15.85		43.85	
Remove and Install	Ea.		6.58		6.58	
Clean	Ea.	.13	9.10		9.23	
Minimum Charge	Job		165		165	
Better Quality						
Demolish	Ea.		7.60		7.60	Includes material and labor to install custom quality incandescent ceiling light fixture. Wiring is not included.
Install	Ea.	83.50	17.35		100.85	
Demolish and Install	Ea.	83.50	24.95		108.45	
Remove and Install	Ea.		13.86		13.86	
Clean	Ea.	.13	9.10		9.23	
Minimum Charge	Job		165		165	
Premium Quality						
Demolish	Ea.		7.60		7.60	Includes material and labor to install deluxe quality incandescent ceiling light fixture. Wiring is not included.
Install	Ea.	248	17.35		265.35	
Demolish and Install	Ea.	248	24.95		272.95	
Remove and Install	Ea.		13.86		13.86	
Clean	Ea.	.13	9.10		9.23	
Minimum Charge	Job		165		165	
Recessed Ceiling						
Demolish	Ea.		15.25		15.25	Includes material and labor to install standard quality recessed eyeball spotlight with housing. Wiring is not included.
Install	Ea.	38.50	10.95		49.45	
Demolish and Install	Ea.	38.50	26.20		64.70	
Remove and Install	Ea.		8.78		8.78	
Clean	Ea.	.02	2.27		2.29	
Minimum Charge	Job		165		165	
Recessed Eyeball						
Demolish	Ea.		15.25		15.25	Includes material and labor to install custom quality recessed eyeball spotlight with housing. Wiring is not included.
Install	Ea.	55	11.75		66.75	
Demolish and Install	Ea.	55	27		82	
Remove and Install	Ea.		9.41		9.41	
Clean	Ea.	.02	2.27		2.29	
Minimum Charge	Job		165		165	

Residential Light Fixture	Unit	Material	Labor	Equip.	Total	Specification
Recessed Shower Type						
Demolish	Ea.		15.25		15.25	Includes material and labor to install deluxe quality recessed eyeball spotlight with housing. Wiring is not included.
Install	Ea.	47.50	10.95		58.45	
Demolish and Install	Ea.	47.50	26.20		73.70	
Remove and Install	Ea.		8.78		8.78	
Clean	Ea.	.02	2.27		2.29	
Minimum Charge	Job		165		165	
Run Wiring						
Install	Ea.	8.20	13.15		21.35	Includes labor and material to install a electric fixture utility box with non-metallic cable.
Minimum Charge	Job		165		165	
Fluorescent						
Good Quality						
Demolish	Ea.		16.65		16.65	Includes material and labor to install standard quality 12″ wide by 48″ long decorative surface mounted fluorescent fixture w/acrylic diffuser.
Install	Ea.	47	47		94	
Demolish and Install	Ea.	47	63.65		110.65	
Remove and Install	Ea.		37.62		37.62	
Clean	Ea.	.02	2.27		2.29	
Minimum Charge	Job		165		165	
Better Quality						
Demolish	Ea.		16.65		16.65	Includes material and labor to install an interior fluorescent lighting fixture.
Install	Ea.	87	53		140	
Demolish and Install	Ea.	87	69.65		156.65	
Remove and Install	Ea.		42.48		42.48	
Clean	Ea.	.02	2.27		2.29	
Minimum Charge	Job		165		165	
Premium Quality						
Demolish	Ea.		16.65		16.65	Includes material and labor to install deluxe quality 24″ wide by 48″ long decorative surface mounted fluorescent fixture w/acrylic diffuser.
Install	Ea.	100	62		162	
Demolish and Install	Ea.	100	78.65		178.65	
Remove and Install	Ea.		49.69		49.69	
Clean	Ea.	.02	2.27		2.29	
Minimum Charge	Job		165		165	
Run Wiring						
Install	Ea.	8.20	13.15		21.35	Includes labor and material to install a electric fixture utility box with non-metallic cable.
Minimum Charge	Job		165		165	
Exterior						
Good Quality						
Demolish	Ea.		14.65		14.65	Includes material and labor to install standard quality incandescent outdoor light fixture. Wiring is not included.
Install	Ea.	34	20.50		54.50	
Demolish and Install	Ea.	34	35.15		69.15	
Remove and Install	Ea.		16.46		16.46	
Clean	Ea.	.13	9.10		9.23	
Minimum Charge	Job		165		165	
Better Quality						
Demolish	Ea.		14.65		14.65	Includes material and labor to install custom quality incandescent outdoor light fixture. Wiring is not included.
Install	Ea.	86	20.50		106.50	
Demolish and Install	Ea.	86	35.15		121.15	
Remove and Install	Ea.		16.46		16.46	
Clean	Ea.	.13	9.10		9.23	
Minimum Charge	Job		165		165	
Premium Quality						
Demolish	Ea.		14.65		14.65	Includes material and labor to install deluxe quality incandescent outdoor light fixture. Wiring is not included.
Install	Ea.	164	82.50		246.50	
Demolish and Install	Ea.	164	97.15		261.15	
Remove and Install	Ea.		65.84		65.84	
Clean	Ea.	.13	9.10		9.23	
Minimum Charge	Job		165		165	

Finish Electrical

Residential Light Fixture		Unit	Material	Labor	Equip.	Total	Specification
Outdoor w / Pole							
	Demolish	Ea.		14.65		14.65	Includes material and labor to install post lantern type incandescent outdoor light fixture. Wiring is not included.
	Install	Ea.	121	82.50		203.50	
	Demolish and Install	Ea.	121	97.15		218.15	
	Remove and Install	Ea.		65.84		65.84	
	Clean	Ea.	.13	9.10		9.23	
	Minimum Charge	Job		165		165	
Run Wiring							
	Install	Ea.	8.20	13.15		21.35	Includes labor and material to install a electric fixture utility box with non-metallic cable.
	Minimum Charge	Job		165		165	
Track Lighting							
	Demolish	L.F.		5.10		5.10	Includes material and labor to install track lighting.
	Install	L.F.	12.20	13.70		25.90	
	Demolish and Install	L.F.	12.20	18.80		31	
	Remove and Install	L.F.		10.97		10.97	
	Clean	L.F.	.13	4.36		4.49	
	Minimum Charge	Job		165		165	

Commercial Light Fixture		Unit	Material	Labor	Equip.	Total	Specification
Fluorescent							
2 Bulb, 2′							
	Demolish	Ea.		14.65		14.65	Includes material and labor to install a 2′ long surface mounted fluorescent light fixture with two 40 watt bulbs.
	Install	Ea.	31	41		72	
	Demolish and Install	Ea.	31	55.65		86.65	
	Remove and Install	Ea.		32.92		32.92	
	Clean	Ea.	.68	15.55		16.23	
	Minimum Charge	Job		165		165	
2 Bulb, 4′							
	Demolish	Ea.		14.65		14.65	Includes material and labor to install a 4′ long surface mounted fluorescent light fixture with two 40 watt bulbs.
	Install	Ea.	58	47		105	
	Demolish and Install	Ea.	58	61.65		119.65	
	Remove and Install	Ea.		37.62		37.62	
	Clean	Ea.	.68	15.55		16.23	
	Minimum Charge	Job		165		165	
3 Bulb, 4′							
	Demolish	Ea.		14.65		14.65	Includes material and labor to install a 4′ long pendant mounted fluorescent light fixture with three 34 watt bulbs.
	Install	Ea.	98	47		145	
	Demolish and Install	Ea.	98	61.65		159.65	
	Remove and Install	Ea.		37.62		37.62	
	Clean	Ea.	.68	15.55		16.23	
	Minimum Charge	Job		165		165	
4 Bulb, 4′							
	Demolish	Ea.		14.65		14.65	Includes material and labor to install a 4′ long pendant mounted fluorescent light fixture with four 34 watt bulbs.
	Install	Ea.	89	50.50		139.50	
	Demolish and Install	Ea.	89	65.15		154.15	
	Remove and Install	Ea.		40.52		40.52	
	Clean	Ea.	.68	15.55		16.23	
	Minimum Charge	Job		165		165	
2 Bulb, 8′							
	Demolish	Ea.		18.30		18.30	Includes material and labor to install a 8′ long surface mounted fluorescent light fixture with two 110 watt bulbs.
	Install	Ea.	89	62		151	
	Demolish and Install	Ea.	89	80.30		169.30	
	Remove and Install	Ea.		49.69		49.69	
	Clean	Ea.	.68	15.55		16.23	
	Minimum Charge	Job		165		165	

Commercial Light Fixture	Unit	Material	Labor	Equip.	Total	Specification
Run Wiring						
Install	Ea.	8.20	13.15		21.35	Includes labor and material to install a electric fixture utility box with non-metallic cable.
Minimum Charge	Job		165		165	
Ceiling, Recessed						
Eyeball Type						
Demolish	Ea.		15.25		15.25	Includes material and labor to install custom quality recessed eyeball spotlight with housing. Wiring is not included.
Install	Ea.	55	11.75		66.75	
Demolish and Install	Ea.	55	27		82	
Remove and Install	Ea.		9.41		9.41	
Clean	Ea.	.02	2.27		2.29	
Minimum Charge	Job		165		165	
Shower Type						
Demolish	Ea.		15.25		15.25	Includes material and labor to install deluxe quality recessed eyeball spotlight with housing. Wiring is not included.
Install	Ea.	47.50	10.95		58.45	
Demolish and Install	Ea.	47.50	26.20		73.70	
Remove and Install	Ea.		8.78		8.78	
Clean	Ea.	.02	2.27		2.29	
Minimum Charge	Job		165		165	
Run Wiring						
Install	Ea.	8.20	13.15		21.35	Includes labor and material to install a electric fixture utility box with non-metallic cable.
Minimum Charge	Job		165		165	
Ceiling Grid Type						
2 Tube						
Demolish	Ea.		21.50		21.50	Cost includes material and labor to install 2 lamp fluorescent light fixture in 2' x 4' ceiling grid system including lens and tubes.
Install	Ea.	55.50	62		117.50	
Demolish and Install	Ea.	55.50	83.50		139	
Remove and Install	Ea.		49.69		49.69	
Clean	Ea.	.68	15.55		16.23	
Minimum Charge	Job		165		165	
4 Tube						
Demolish	Ea.		24.50		24.50	Cost includes material and labor to install 4 lamp fluorescent light fixture in 2' x 4' ceiling grid system including lens and tubes.
Install	Ea.	62	70		132	
Demolish and Install	Ea.	62	94.50		156.50	
Remove and Install	Ea.		56.03		56.03	
Clean	Ea.	.68	15.55		16.23	
Minimum Charge	Job		165		165	
Acrylic Diffuser						
Demolish	Ea.		3.81		3.81	Includes material and labor to install fluorescent light fixture w/acrylic diffuser.
Install	Ea.	38.50	10		48.50	
Demolish and Install	Ea.	38.50	13.81		52.31	
Remove and Install	Ea.		7.98		7.98	
Clean	Ea.	.44	2.73		3.17	
Minimum Charge	Job		165		165	
Louver Diffuser						
Demolish	Ea.		4.57		4.57	Includes material and labor to install fluorescent light fixture w/louver diffuser.
Install	Ea.	42	10		52	
Demolish and Install	Ea.	42	14.57		56.57	
Remove and Install	Ea.		7.98		7.98	
Clean	Ea.	.44	2.73		3.17	
Minimum Charge	Job		165		165	
Run Wiring						
Install	Ea.	8.20	13.15		21.35	Includes labor and material to install a electric fixture utility box with non-metallic cable.
Minimum Charge	Job		165		165	

Commercial Light Fixture	Unit	Material	Labor	Equip.	Total	Specification
Track Lighting						
Demolish	L.F.		5.10		5.10	Includes material and labor to install track lighting.
Install	L.F.	12.20	13.70		25.90	
Demolish and Install	L.F.	12.20	18.80		31	
Remove and Install	L.F.		10.97		10.97	
Clean	L.F.	.13	4.36		4.49	
Minimum Charge	Job		165		165	
Emergency						
Demolish	Ea.		26		26	Includes material and labor to install battery-powered emergency lighting unit with two (2) lights, including wiring.
Install	Ea.	121	82.50		203.50	
Demolish and Install	Ea.	121	108.50		229.50	
Remove and Install	Ea.		65.84		65.84	
Clean	Ea.	.34	15.55		15.89	
Minimum Charge	Job		165		165	
Run Wiring						
Install	Ea.	8.20	13.15		21.35	Includes labor and material to install a electric fixture utility box with non-metallic cable.
Minimum Charge	Job		165		165	

Fan	Unit	Material	Labor	Equip.	Total	Specification
Ceiling Paddle						
Good Quality						
Demolish	Ea.		20.50		20.50	Includes material and labor to install standard ceiling fan.
Install	Ea.	82.50	33		115.50	
Demolish and Install	Ea.	82.50	53.50		136	
Remove and Install	Ea.		26.34		26.34	
Clean	Ea.	.28	13.65		13.93	
Minimum Charge	Job		165		165	
Better Quality						
Demolish	Ea.		20.50		20.50	Includes material and labor to install better ceiling fan.
Install	Ea.	137	33		170	
Demolish and Install	Ea.	137	53.50		190.50	
Remove and Install	Ea.		26.34		26.34	
Clean	Ea.	.28	13.65		13.93	
Minimum Charge	Job		165		165	
Premium Quality						
Demolish	Ea.		20.50		20.50	Includes material and labor to install deluxe ceiling fan.
Install	Ea.	320	41		361	
Demolish and Install	Ea.	320	61.50		381.50	
Remove and Install	Ea.		32.92		32.92	
Clean	Ea.	.28	13.65		13.93	
Minimum Charge	Job		165		165	
Light Kit						
Install	Ea.	29.50	41		70.50	Includes labor and material to install a ceiling fan light kit.
Minimum Charge	Job		165		165	
Bathroom Exhaust						
Ceiling Mounted						
Demolish	Ea.		16.70		16.70	Cost includes material and labor to install 60 CFM ceiling mounted bathroom exhaust fan, up to 10′ of wiring with one wall switch.
Install	Ea.	39.50	33		72.50	
Demolish and Install	Ea.	39.50	49.70		89.20	
Remove and Install	Ea.		26.26		26.26	
Clean	Ea.	.13	5.45		5.58	
Minimum Charge	Job		165		165	

Fan	Unit	Material	Labor	Equip.	Total	Specification
Wall Mounted						
Demolish	Ea.		16.70		16.70	Cost includes material and labor to install 110 CFM wall mounted bathroom exhaust fan, up to 10' of wiring with one wall switch.
Install	Ea.	53.50	33		86.50	
Demolish and Install	Ea.	53.50	49.70		103.20	
Remove and Install	Ea.		26.26		26.26	
Clean	Ea.	.13	5.45		5.58	
Minimum Charge	Job		165		165	
Lighted						
Demolish	Ea.		16.70		16.70	Cost includes material and labor to install 50 CFM ceiling mounted lighted exhaust fan, up to 10' of wiring with one wall switch..
Install	Ea.	37	33		70	
Demolish and Install	Ea.	37	49.70		86.70	
Remove and Install	Ea.		26.26		26.26	
Clean	Ea.	.13	5.45		5.58	
Minimum Charge	Job		165		165	
Blower Heater w / Light						
Demolish	Ea.		16.70		16.70	Includes material and labor to install ceiling blower type ventilator with switch, light and snap-on grill. Includes 1300 watt/120 volt heater.
Install	Ea.	86.50	54.50		141	
Demolish and Install	Ea.	86.50	71.20		157.70	
Remove and Install	Ea.		43.76		43.76	
Clean	Ea.	.13	5.45		5.58	
Minimum Charge	Job		165		165	
Fan Heater w / Light						
Demolish	Ea.		16.70		16.70	Includes material and labor to install circular ceiling fan heater unit, 1500 watt, surface mounted.
Install	Ea.	109	54.50		163.50	
Demolish and Install	Ea.	109	71.20		180.20	
Remove and Install	Ea.		43.76		43.76	
Clean	Ea.	.13	5.45		5.58	
Minimum Charge	Job		165		165	
Whole House Exhaust						
36" Ceiling Mounted						
Install	Ea.	505	82		587	Includes material and labor to install a whole house exhaust fan.
Clean	Ea.	.04	3.92		3.96	
Minimum Charge	Job		165		165	
Kitchen Exhaust						
Standard Model						
Install	Ea.	62.50	22		84.50	Includes labor and materials to install a bathroom or kitchen vent fan.
Clean	Ea.	.04	3.92		3.96	
Minimum Charge	Job		165		165	
Low Noise Model						
Install	Ea.	82.50	22		104.50	Includes labor and materials to install a bathroom or kitchen vent fan, low noise model.
Clean	Ea.	.04	3.92		3.96	
Minimum Charge	Job		165		165	

Low Voltage Systems	Unit	Material	Labor	Equip.	Total	Specification
Smoke Detector						
Pre-wired						
Demolish	Ea.		7.60		7.60	Includes material and labor to install AC type smoke detector including wiring.
Install	Ea.	19.65	62		81.65	
Demolish and Install	Ea.	19.65	69.60		89.25	
Remove and Install	Ea.		49.41		49.41	
Clean	Ea.	.06	6.80		6.86	
Minimum Charge	Job		165		165	

Low Voltage Systems		Unit	Material	Labor	Equip.	Total	Specification
Battery-powered							
	Demolish	Ea.		7.60		7.60	Includes material and labor to install battery-powered type smoke detector.
	Install	Ea.	24	20.50		44.50	
	Demolish and Install	Ea.	24	28.10		52.10	
	Remove and Install	Ea.		16.46		16.46	
	Clean	Ea.	.06	6.80		6.86	
	Minimum Charge	Job		165		165	
Run Wiring							
	Install	Ea.	8.20	13.15		21.35	Includes labor and material to install a electric fixture utility box with non-metallic cable.
	Minimum Charge	Job		165		165	
Door Bell / Chimes							
Good Quality							
	Demolish	Ea.		23		23	Includes material and labor to install surface mounted 2 note door bell. Wiring is not included.
	Install	Ea.	42	82.50		124.50	
	Demolish and Install	Ea.	42	105.50		147.50	
	Remove and Install	Ea.		65.84		65.84	
	Clean	Ea.	.13	6.80		6.93	
	Minimum Charge	Job		165		165	
Better Quality							
	Demolish	Ea.		23		23	Includes material and labor to install standard quality surface mounted 2 note door bell. Wiring is not included.
	Install	Ea.	52.50	82.50		135	
	Demolish and Install	Ea.	52.50	105.50		158	
	Remove and Install	Ea.		65.84		65.84	
	Clean	Ea.	.13	6.80		6.93	
	Minimum Charge	Job		165		165	
Premium Quality							
	Demolish	Ea.		23		23	Includes material and labor to install deluxe quality surface mounted 2 note door bell. Wiring is not included.
	Install	Ea.	65.50	82.50		148	
	Demolish and Install	Ea.	65.50	105.50		171	
	Remove and Install	Ea.		65.84		65.84	
	Clean	Ea.	.13	6.80		6.93	
	Minimum Charge	Job		165		165	
Run Wiring							
	Install	Ea.	8.20	13.15		21.35	Includes labor and material to install a electric fixture utility box with non-metallic cable.
	Minimum Charge	Job		165		165	
Intercom System							
Intercom (master)							
	Demolish	Ea.		61		61	Includes material and labor to install master intercom station with AM/FM music and room monitoring. Wiring is not included.
	Install	Ea.	455	165		620	
	Demolish and Install	Ea.	455	226		681	
	Remove and Install	Ea.		131.68		131.68	
	Clean	Ea.	.13	18.15		18.28	
	Minimum Charge	Job		165		165	
Intercom (remote)							
	Demolish	Ea.		36.50		36.50	Includes material and labor to install remote station for master intercom system. Wiring is not included.
	Install	Ea.	41.50	41		82.50	
	Demolish and Install	Ea.	41.50	77.50		119	
	Remove and Install	Ea.		32.92		32.92	
	Clean	Ea.	.13	6.80		6.93	
	Minimum Charge	Job		165		165	

Low Voltage Systems	Unit	Material	Labor	Equip.	Total	Specification
Security System						
Alarm Panel						
Demolish	Ea.		73		73	Includes material and labor to install a 6 zone burglar alarm panel.
Install	Ea.	165	165		330	
Demolish and Install	Ea.	165	238		403	
Remove and Install	Ea.		131.68		131.68	
Clean	Ea.	.13	6.80		6.93	
Minimum Charge	Job		165		165	
Motion Detector						
Demolish	Ea.		52.50		52.50	Includes material and labor to install a ultrasonic motion detector for a security system.
Install	Ea.	227	143		370	
Demolish and Install	Ea.	227	195.50		422.50	
Remove and Install	Ea.		114.50		114.50	
Clean	Ea.	.13	6.80		6.93	
Minimum Charge	Job		165		165	
Door Switch						
Demolish	Ea.		23		23	Includes material and labor to install a door / window contact for a security system.
Install	Ea.	34.50	62		96.50	
Demolish and Install	Ea.	34.50	85		119.50	
Remove and Install	Ea.		49.69		49.69	
Clean	Ea.	.13	2.18		2.31	
Minimum Charge	Job		165		165	
Window Switch						
Demolish	Ea.		23		23	Includes material and labor to install a door / window contact for a security system.
Install	Ea.	34.50	62		96.50	
Demolish and Install	Ea.	34.50	85		119.50	
Remove and Install	Ea.		49.69		49.69	
Clean	Ea.	.13	2.18		2.31	
Minimum Charge	Job		165		165	
Satellite System						
Dish						
Demolish	Ea.		102		102	Includes material, labor and equipment to install 10′ mesh dish.
Install	Ea.	605	274		879	
Demolish and Install	Ea.	605	376		981	
Remove and Install	Ea.		219.47		219.47	
Clean	Ea.	2.73	27.50		30.23	
Minimum Charge	Job		165		165	
Motor						
Demolish	Ea.		51		51	Includes material and labor to install a television satellite dish motor unit.
Install	Ea.	305	137		442	
Demolish and Install	Ea.	305	188		493	
Remove and Install	Ea.		109.73		109.73	
Clean	Ea.	.13	6.80		6.93	
Minimum Charge	Job		165		165	
Television Antenna						
40′ Pole						
Demolish	Ea.		57		57	Includes material and labor to install stand-alone 1″-2″ metal pole, 10′ long. Cost does not include antenna.
Install	Ea.	22	103		125	
Demolish and Install	Ea.	22	160		182	
Remove and Install	Ea.		82.30		82.30	
Clean	Ea.	.55	27.50		28.05	
Paint	Ea.	6.60	33		39.60	
Minimum Charge	Job		165		165	
Rotor Unit						
Demolish	Ea.		45.50		45.50	Includes material and labor to install a television antenna rotor.
Install	Ea.	105	41		146	
Demolish and Install	Ea.	105	86.50		191.50	
Remove and Install	Ea.		32.92		32.92	
Clean	Ea.	.55	27.50		28.05	
Minimum Charge	Job		165		165	

Low Voltage Systems	Unit	Material	Labor	Equip.	Total	Specification
Single Booster						
Demolish	Ea.		45.50		45.50	Includes material and labor to install standard grade VHF television signal booster unit and all hardware required for connection to TV cable.
Install	Ea.	14.30	41		55.30	
Demolish and Install	Ea.	14.30	86.50		100.80	
Remove and Install	Ea.		32.92		32.92	
Clean	Ea.	.02	3.41		3.43	
Minimum Charge	Job		165		165	

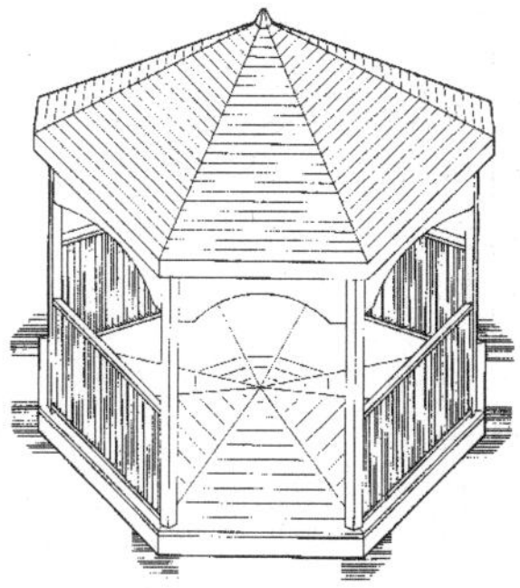

Gazebo

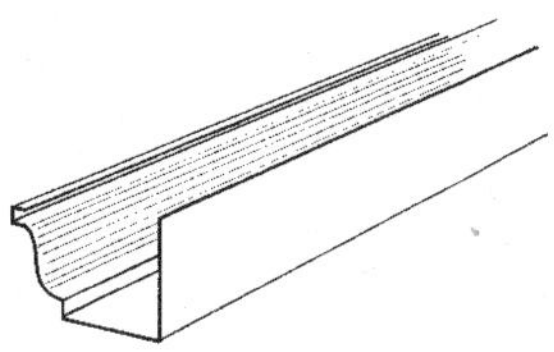

Gutter

Shutter

Window Treatment	Unit	Material	Labor	Equip.	Total	Specification
Drapery						
Good Grade						
Install	L.F.	32.50	4.36		36.86	Includes material and labor to install good quality drapery.
Remove and Install	L.F.		3.49		3.49	
Clean	L.F.	2.31	.11		2.42	
Minimum Charge	Job		109		109	
Premium Grade						
Install	L.F.	56.50	4.36		60.86	Includes material and labor to install premium quality drapery.
Remove and Install	L.F.		3.49		3.49	
Clean	L.F.	2.31	.12		2.43	
Minimum Charge	Job		109		109	
Lining						
Install	L.F.	14.10	4.36		18.46	Includes material and labor to install drapery lining.
Remove and Install	L.F.		3.49		3.49	
Clean	L.F.	1.16	.06		1.22	
Minimum Charge	Job		109		109	
Valance Board						
Demolish	L.F.		1.21		1.21	Includes material and labor to install a drapery valance board.
Install	L.F.	4.22	8.70		12.92	
Demolish and Install	L.F.	4.22	9.91		14.13	
Remove and Install	L.F.		6.98		6.98	
Clean	L.F.	1.16	.22		1.38	
Minimum Charge	Job		109		109	
Rod						
Demolish	L.F.		3.03		3.03	Includes material and labor to install standard traverse drapery hardware.
Install	L.F.	1.97	5.10		7.07	
Demolish and Install	L.F.	1.97	8.13		10.10	
Remove and Install	L.F.		4.08		4.08	
Clean	L.F.	.06	.20		.26	
Paint	L.F.	.06	.51		.57	
Minimum Charge	Job		109		109	
Blinds						
Mini-type						
Demolish	S.F.		.65		.65	Includes material and labor to install horizontal 1" slat blinds.
Install	S.F.	2.79	.51		3.30	
Demolish and Install	S.F.	2.79	1.16		3.95	
Remove and Install	S.F.		.41		.41	
Clean	S.F.	.44	.02		.46	
Minimum Charge	Job		109		109	

Window Treatment	Unit	Material	Labor	Equip.	Total	Specification
Vertical Type						
Demolish	S.F.		.65		.65	Includes material and labor to install vertical 3" - 5" cloth or PVC blinds.
Install	S.F.	7.65	.65		8.30	
Demolish and Install	S.F.	7.65	1.30		8.95	
Remove and Install	S.F.		.52		.52	
Clean	S.F.	.23	.02		.25	
Minimum Charge	Job		109		109	
Vertical Type per L.F.						
Demolish	L.F.		1.01		1.01	Includes material and labor to install vertical 3" - 5" cloth or PVC blinds x 72" long.
Install	L.F.	45.50	2.91		48.41	
Demolish and Install	L.F.	45.50	3.92		49.42	
Remove and Install	L.F.		2.33		2.33	
Clean	L.F.	1.64	.15		1.79	
Minimum Charge	Job		109		109	
Venetian Type						
Demolish	S.F.		.65		.65	Includes material and labor to install horizontal 2" slat blinds.
Install	S.F.	3.27	.51		3.78	
Demolish and Install	S.F.	3.27	1.16		4.43	
Remove and Install	S.F.		.41		.41	
Clean	S.F.	.23	.02		.25	
Minimum Charge	Job		109		109	
Wood Shutter						
27" x 36", 4 Panel						
Demolish	Ea.		21		21	Includes material and labor to install moveable louver, up to 33" x 36", 4 panel shutter.
Install	Pr.	80	17.70		97.70	
Demolish and Install	Pr.	80	38.70		118.70	
Remove and Install	Pr.		14.16		14.16	
Clean	Ea.	.06	6.55		6.61	
Paint	Ea.	1.06	18.85		19.91	
Minimum Charge	Job		150		150	
33" x 36", 4 Panel						
Demolish	Ea.		21		21	Includes material and labor to install moveable louver, up to 33" x 36", 4 panel shutter.
Install	Pr.	80	17.70		97.70	
Demolish and Install	Pr.	80	38.70		118.70	
Remove and Install	Pr.		14.16		14.16	
Clean	Ea.	.06	6.55		6.61	
Paint	Ea.	1.10	20.50		21.60	
Minimum Charge	Job		150		150	
39" x 36", 4 Panel						
Demolish	Ea.		21		21	Includes material and labor to install moveable louver, up to 47" x 36", 4 panel shutter.
Install	Pr.	90	17.70		107.70	
Demolish and Install	Pr.	90	38.70		128.70	
Remove and Install	Pr.		14.16		14.16	
Clean	Ea.	.06	6.55		6.61	
Paint	Ea.	1.10	20.50		21.60	
Minimum Charge	Job		150		150	
47" x 36", 4 Panel						
Demolish	Ea.		21		21	Includes material and labor to install moveable louver, up to 47" x 36", 4 panel shutter.
Install	Pr.	90	17.70		107.70	
Demolish and Install	Pr.	90	38.70		128.70	
Remove and Install	Pr.		14.16		14.16	
Clean	Ea.	.06	6.55		6.61	
Paint	Ea.	1.14	22		23.14	
Minimum Charge	Job		150		150	

Stairs	Unit	Material	Labor	Equip.	Total	Specification
Concrete Steps						
Precast						
Demolish	Ea.		41.50	31	72.50	Includes material and labor to install precast concrete front entrance stairs, 4' wide, with 48" platform and 2 risers.
Install	Flight	310	103	41	454	
Demolish and Install	Flight	310	144.50	72	526.50	
Clean	Ea.		3.03		3.03	
Paint	Ea.	2.22	33		35.22	
Minimum Charge	Job		267		267	
Metal						
Concrete Filled						
Demolish	Ea.		12.10		12.10	Includes material and labor to install steel, cement filled metal pan stairs with picket rail.
Install	Riser	205	51	2.89	258.89	
Demolish and Install	Riser	205	63.10	2.89	270.99	
Clean	Ea.	.09	1.89		1.98	
Paint	Ea.	.33	1.96		2.29	
Minimum Charge	Job		187		187	
Landing						
Demolish	S.F.		3.03		3.03	Includes material and labor to install pre-erected conventional steel pan landing.
Install	S.F.	42	6	.34	48.34	
Demolish and Install	S.F.	42	9.03	.34	51.37	
Clean	S.F.	.09	.39		.48	
Paint	S.F.	.11	.56		.67	
Minimum Charge	Job		187		187	
Railing						
Demolish	L.F.		1.67		1.67	Includes material and labor to install primed steel pipe.
Install	L.F.	7.05	8.70	.49	16.24	
Demolish and Install	L.F.	7.05	10.37	.49	17.91	
Clean	L.F.	.55	1.36		1.91	
Paint	L.F.	.64	4.40		5.04	
Minimum Charge	Job		187		187	

Elevator	Unit	Material	Labor	Equip.	Total	Specification
2 Stop Residential						
Install	Ea.	8875	3400		12275	Includes labor, material and equipment to install 2 stop elevator for residential use.
Clean	Ea.	.63	13.65		14.28	
Minimum Charge	Job		340		340	
2 Stop Commercial						
Install	Ea.	31100	6825		37925	Includes labor, material and equipment to install 2 stop elevator for apartment or commercial building with 2,000 lb/13 passenger capacity.
Clean	Ea.	.63	13.65		14.28	
Minimum Charge	Job		340		340	
3 Stop Commercial						
Install	Ea.	35900	10500		46400	Includes labor, material and equipment to install 3 stop elevator for apartment or commercial building with 2,000 lb/13 passenger capacity.
Clean	Ea.	.63	13.65		14.28	
Minimum Charge	Job		340		340	
4 Stop Commercial						
Install	Ea.	40600	14200		54800	Includes labor, material and equipment to install 4 stop elevator for apartment or commercial building with 2,000 lb/13 passenger capacity.
Clean	Ea.	.63	13.65		14.28	
Minimum Charge	Job		340		340	

Elevator	Unit	Material	Labor	Equip.	Total	Specification
5 Stop Commercial						
Install	Ea.	45400	18000		63400	Includes labor, material and equipment to install 5 stop elevator for apartment or commercial building with 2,000 lb/13 passenger capacity.
Clean	Ea.	.63	13.65		14.28	
Minimum Charge	Job		340		340	

Rangetop	Unit	Material	Labor	Equip.	Total	Specification
4-burner						
Demolish	Ea.		22		22	Cost includes material and labor to install 4-burner electric surface unit.
Install	Ea.	196	55		251	
Demolish and Install	Ea.	196	77		273	
Remove and Install	Ea.		43.89		43.89	
Clean	Ea.	.04	19.60		19.64	
Minimum Charge	Job		54.50		54.50	
6-burner						
Demolish	Ea.		22		22	Includes material and labor to install a 6-burner cooktop unit.
Install	Ea.	450	66		516	
Demolish and Install	Ea.	450	88		538	
Remove and Install	Ea.		52.67		52.67	
Clean	Ea.	.04	19.60		19.64	
Minimum Charge	Job		54.50		54.50	
Premium Brand w / Grill						
Demolish	Ea.		22		22	Includes material and labor to install electric downdraft cooktop with grill.
Install	Ea.	880	66		946	
Demolish and Install	Ea.	880	88		968	
Remove and Install	Ea.		52.67		52.67	
Clean	Ea.	.04	19.60		19.64	
Minimum Charge	Job		54.50		54.50	

Rangehood	Unit	Material	Labor	Equip.	Total	Specification
Vented						
30″						
Demolish	Ea.		20		20	Includes material and labor to install a vented range hood.
Install	Ea.	39.50	66		105.50	
Demolish and Install	Ea.	39.50	86		125.50	
Remove and Install	Ea.		52.67		52.67	
Clean	Ea.	.04	5.60		5.64	
Minimum Charge	Job		54.50		54.50	
36″						
Demolish	Ea.		20		20	Includes material and labor to install a vented range hood.
Install	Ea.	89.50	66		155.50	
Demolish and Install	Ea.	89.50	86		175.50	
Remove and Install	Ea.		52.67		52.67	
Clean	Ea.	.04	5.60		5.64	
Minimum Charge	Job		54.50		54.50	
42″						
Demolish	Ea.		20		20	Includes material and labor to install a vented range hood.
Install	Ea.	245	76.50		321.50	
Demolish and Install	Ea.	245	96.50		341.50	
Remove and Install	Ea.		61.30		61.30	
Clean	Ea.	.04	5.60		5.64	
Minimum Charge	Job		54.50		54.50	

Rangehood	Unit	Material	Labor	Equip.	Total	Specification
Ventless						
30″						
Demolish	Ea.		20		20	Includes material and labor to install a ventless range hood.
Install	Ea.	39.50	41		80.50	
Demolish and Install	Ea.	39.50	61		100.50	
Remove and Install	Ea.		32.92		32.92	
Clean	Ea.	.04	5.60		5.64	
Minimum Charge	Job		54.50		54.50	
36″						
Demolish	Ea.		20		20	Includes material and labor to install a ventless range hood.
Install	Ea.	41.50	66		107.50	
Demolish and Install	Ea.	41.50	86		127.50	
Remove and Install	Ea.		52.67		52.67	
Clean	Ea.	.04	5.60		5.64	
Minimum Charge	Job		54.50		54.50	
42″						
Demolish	Ea.		20		20	Includes material and labor to install a ventless range hood.
Install	Ea.	221	55		276	
Demolish and Install	Ea.	221	75		296	
Remove and Install	Ea.		43.89		43.89	
Clean	Ea.	.04	5.60		5.64	
Minimum Charge	Job		54.50		54.50	

Oven	Unit	Material	Labor	Equip.	Total	Specification
Free Standing						
Demolish	Ea.		22		22	Cost includes material and labor to install 30″ free standing electric range with self-cleaning oven.
Install	Ea.	450	43.50		493.50	
Demolish and Install	Ea.	450	65.50		515.50	
Remove and Install	Ea.		34.88		34.88	
Clean	Ea.	.04	19.60		19.64	
Minimum Charge	Job		54.50		54.50	
Single Wall Oven						
Demolish	Ea.		22		22	Includes material and labor to install a built-in oven.
Install	Ea.	400	82.50		482.50	
Demolish and Install	Ea.	400	104.50		504.50	
Remove and Install	Ea.		65.84		65.84	
Clean	Ea.	.04	19.60		19.64	
Minimum Charge	Job		54.50		54.50	
Double Wall Oven						
Demolish	Ea.		22		22	Includes material and labor to install a conventional double oven.
Install	Ea.	1025	150		1175	
Demolish and Install	Ea.	1025	172		1197	
Remove and Install	Ea.		120.32		120.32	
Clean	Ea.	.04	19.60		19.64	
Minimum Charge	Job		54.50		54.50	
Drop-in Type w / Range						
Demolish	Ea.		22		22	Includes material and labor to install a built-in cooking range with oven.
Install	Ea.	470	55		525	
Demolish and Install	Ea.	470	77		547	
Remove and Install	Ea.		43.89		43.89	
Clean	Ea.	.04	19.60		19.64	
Minimum Charge	Job		54.50		54.50	

Oven	Unit	Material	Labor	Equip.	Total	Specification
Hi/Lo w / Range, Microwave						
Demolish	Ea.		22		22	Includes material and labor to install a double oven, one conventional, one microwave.
Install	Ea.	1600	300		1900	
Demolish and Install	Ea.	1600	322		1922	
Remove and Install	Ea.		240.64		240.64	
Clean	Ea.	.04	19.60		19.64	
Minimum Charge	Job		54.50		54.50	
Hi/Lo w/ Range / Double Oven						
Demolish	Ea.		22		22	Includes material and labor to install a conventional double oven.
Install	Ea.	1025	150		1175	
Demolish and Install	Ea.	1025	172		1197	
Remove and Install	Ea.		120.32		120.32	
Clean	Ea.	.04	19.60		19.64	
Minimum Charge	Job		54.50		54.50	
Countertop Microwave						
Demolish	Ea.		7.55		7.55	Includes material and labor to install a 1.0 cubic foot microwave oven.
Install	Ea.	104	27.50		131.50	
Demolish and Install	Ea.	104	35.05		139.05	
Remove and Install	Ea.		21.80		21.80	
Clean	Ea.	.04	9.80		9.84	
Minimum Charge	Job		54.50		54.50	
Cabinet / Wall Mounted Microwave						
Demolish	Ea.		15.15		15.15	Includes material and labor to install a space saver microwave oven.
Install	Ea.	550	82.50		632.50	
Demolish and Install	Ea.	550	97.65		647.65	
Remove and Install	Ea.		65.84		65.84	
Clean	Ea.	.04	9.80		9.84	
Minimum Charge	Job		54.50		54.50	

Dryer	Unit	Material	Labor	Equip.	Total	Specification
Basic Grade						
Demolish	Ea.		10.10		10.10	Includes material and labor to install electric or gas dryer. Vent kit not included.
Install	Ea.	282	174		456	
Demolish and Install	Ea.	282	184.10		466.10	
Remove and Install	Ea.		139.43		139.43	
Clean	Ea.	.04	19.60		19.64	
Minimum Charge	Job		54.50		54.50	
Good Grade						
Demolish	Ea.		10.10		10.10	Includes material and labor to install electric or gas dryer. Vent kit not included.
Install	Ea.	350	174		524	
Demolish and Install	Ea.	350	184.10		534.10	
Remove and Install	Ea.		139.43		139.43	
Clean	Ea.	.04	19.60		19.64	
Minimum Charge	Job		54.50		54.50	
Premium Grade						
Demolish	Ea.		10.10		10.10	Includes material and labor to install electric or gas dryer. Vent kit not included.
Install	Ea.	860	261		1121	
Demolish and Install	Ea.	860	271.10		1131.10	
Remove and Install	Ea.		209.15		209.15	
Clean	Ea.	.04	19.60		19.64	
Minimum Charge	Job		54.50		54.50	

Dryer

Dryer	Unit	Material	Labor	Equip.	Total	Specification
Washer / Dryer Combination Unit						
Demolish	Ea.		10.10		10.10	Includes material and labor to install electric washer/dryer combination unit, 27" wide, stackable or unitized.
Install	Ea.	900	43.50		943.50	
Demolish and Install	Ea.	900	53.60		953.60	
Remove and Install	Ea.		34.88		34.88	
Clean	Ea.	.04	19.60		19.64	
Minimum Charge	Job		54.50		54.50	

Washer

Washer	Unit	Material	Labor	Equip.	Total	Specification
Basic Grade						
Demolish	Ea.		10.10		10.10	Includes material and labor to install washing machine. New hoses not included.
Install	Ea.	298	108		406	
Demolish and Install	Ea.	298	118.10		416.10	
Remove and Install	Ea.		86.72		86.72	
Clean	Ea.	.04	19.60		19.64	
Minimum Charge	Job		54.50		54.50	
Good Grade						
Demolish	Ea.		10.10		10.10	Includes material and labor to install washing machine. New hoses not included.
Install	Ea.	490	163		653	
Demolish and Install	Ea.	490	173.10		663.10	
Remove and Install	Ea.		130.08		130.08	
Clean	Ea.	.04	19.60		19.64	
Minimum Charge	Job		54.50		54.50	
Premium Grade						
Demolish	Ea.		10.10		10.10	Includes material and labor to install washing machine with digital computer readouts and large capacity. New hoses not included.
Install	Ea.	1200	325		1525	
Demolish and Install	Ea.	1200	335.10		1535.10	
Remove and Install	Ea.		260.16		260.16	
Clean	Ea.	.04	19.60		19.64	
Minimum Charge	Job		54.50		54.50	
Washer / Dryer Combination Unit						
Demolish	Ea.		10.10		10.10	Includes material and labor to install electric washer/dryer combination unit, 27" wide, stackable or unitized.
Install	Ea.	900	43.50		943.50	
Demolish and Install	Ea.	900	53.60		953.60	
Remove and Install	Ea.		34.88		34.88	
Clean	Ea.	.04	19.60		19.64	
Minimum Charge	Job		54.50		54.50	

Refrigerator

Refrigerator	Unit	Material	Labor	Equip.	Total	Specification
12 Cubic Foot						
Demolish	Ea.		30.50		30.50	Includes material and labor to install a refrigerator.
Install	Ea.	475	43.50		518.50	
Demolish and Install	Ea.	475	74		549	
Remove and Install	Ea.		34.88		34.88	
Clean	Ea.	.04	19.60		19.64	
Minimum Charge	Job		54.50		54.50	
16 Cubic Foot						
Demolish	Ea.		30.50		30.50	Includes material and labor to install a refrigerator.
Install	Ea.	485	48.50		533.50	
Demolish and Install	Ea.	485	79		564	
Remove and Install	Ea.		38.76		38.76	
Clean	Ea.	.04	19.60		19.64	
Minimum Charge	Job		54.50		54.50	

Refrigerator	Unit	Material	Labor	Equip.	Total	Specification
18 Cubic Foot						
Demolish	Ea.		30.50		30.50	Includes material and labor to install a refrigerator.
Install	Ea.	580	54.50		634.50	
Demolish and Install	Ea.	580	85		665	
Remove and Install	Ea.		43.60		43.60	
Clean	Ea.	.04	19.60		19.64	
Minimum Charge	Job		54.50		54.50	
21 Cubic Foot						
Demolish	Ea.		30.50		30.50	Includes material and labor to install a refrigerator.
Install	Ea.	760	62.50		822.50	
Demolish and Install	Ea.	760	93		853	
Remove and Install	Ea.		49.83		49.83	
Clean	Ea.	.04	19.60		19.64	
Minimum Charge	Job		54.50		54.50	
24 Cubic Foot						
Demolish	Ea.		30.50		30.50	Includes material and labor to install a refrigerator.
Install	Ea.	1000	62.50		1062.50	
Demolish and Install	Ea.	1000	93		1093	
Remove and Install	Ea.		49.83		49.83	
Clean	Ea.	.04	19.60		19.64	
Minimum Charge	Job		54.50		54.50	
27 Cubic Foot						
Demolish	Ea.		30.50		30.50	Includes material and labor to install a refrigerator.
Install	Ea.	1650	43.50		1693.50	
Demolish and Install	Ea.	1650	74		1724	
Remove and Install	Ea.		34.88		34.88	
Clean	Ea.	.04	19.60		19.64	
Minimum Charge	Job		54.50		54.50	
Ice Maker						
Install	Ea.	66	40.50		106.50	Includes labor and material to install an automatic ice maker in a refrigerator.
Minimum Charge	Job		54.50		54.50	

Freezer	Unit	Material	Labor	Equip.	Total	Specification
15 Cubic Foot						
Demolish	Ea.		30.50		30.50	Cost includes material and labor to install 15 cubic foot freezer.
Install	Ea.	415	43.50		458.50	
Demolish and Install	Ea.	415	74		489	
Remove and Install	Ea.		34.88		34.88	
Clean	Ea.	.04	19.60		19.64	
Minimum Charge	Job		54.50		54.50	
18 Cubic Foot						
Demolish	Ea.		30.50		30.50	Cost includes material and labor to install 18 cubic foot freezer.
Install	Ea.	550	43.50		593.50	
Demolish and Install	Ea.	550	74		624	
Remove and Install	Ea.		34.88		34.88	
Clean	Ea.	.04	19.60		19.64	
Minimum Charge	Job		54.50		54.50	
21 Cubic Foot						
Demolish	Ea.		30.50		30.50	Includes material and labor to install a 15 to 23 cubic foot deep freezer.
Install	Ea.	515	43.50		558.50	
Demolish and Install	Ea.	515	74		589	
Remove and Install	Ea.		34.88		34.88	
Clean	Ea.	.04	19.60		19.64	
Minimum Charge	Job		54.50		54.50	

Freezer

Freezer	Unit	Material	Labor	Equip.	Total	Specification
24 Cubic Foot						
Demolish	Ea.		30.50		30.50	Includes material and labor to install a 24 cubic foot deep freezer.
Install	Ea.	580	87		667	
Demolish and Install	Ea.	580	117.50		697.50	
Remove and Install	Ea.		69.76		69.76	
Clean	Ea.	.04	19.60		19.64	
Minimum Charge	Job		54.50		54.50	
Chest-type						
Demolish	Ea.		30.50		30.50	Cost includes material and labor to install 15 cubic foot freezer.
Install	Ea.	415	43.50		458.50	
Demolish and Install	Ea.	415	74		489	
Clean	Ea.	.04	19.60		19.64	
Minimum Charge	Job		54.50		54.50	

Dishwasher

Dishwasher	Unit	Material	Labor	Equip.	Total	Specification
4 Cycle						
Demolish	Ea.		24		24	Includes material and labor to install good grade automatic dishwasher.
Install	Ea.	293	102		395	
Demolish and Install	Ea.	293	126		419	
Remove and Install	Ea.		81.50		81.50	
Clean	Ea.	.04	19.60		19.64	
Minimum Charge	Job		54.50		54.50	

Garbage Disposal

Garbage Disposal	Unit	Material	Labor	Equip.	Total	Specification
In Sink Unit						
Demolish	Ea.		22.50		22.50	Cost includes material and labor to install 1/2 HP custom disposal unit.
Install	Ea.	107	41		148	
Demolish and Install	Ea.	107	63.50		170.50	
Remove and Install	Ea.		32.60		32.60	
Clean	Ea.	.02	5.60		5.62	
Minimum Charge	Job		54.50		54.50	

Trash Compactor

Trash Compactor	Unit	Material	Labor	Equip.	Total	Specification
4 to 1 Compaction						
Demolish	Ea.		23		23	Includes material and labor to install a trash compactor.
Install	Ea.	580	60		640	
Demolish and Install	Ea.	580	83		663	
Remove and Install	Ea.		48.13		48.13	
Clean	Ea.	.04	13.65		13.69	
Minimum Charge	Job		54.50		54.50	

Kitchenette Unit

Kitchenette Unit	Unit	Material	Labor	Equip.	Total	Specification
Range, Refrigerator and Sink						
Demolish	Ea.		91.50		91.50	Includes material and labor to install range/refrigerator/sink unit, 72" wide, base unit only.
Install	Ea.	3000	340		3340	
Demolish and Install	Ea.	3000	431.50		3431.50	
Remove and Install	Ea.		271.67		271.67	
Clean	Ea.	.68	9.80		10.48	
Minimum Charge	Job		54.50		54.50	

Bath Accessories	Unit	Material	Labor	Equip.	Total	Specification
Towel Bar						
Demolish	Ea.		5.05		5.05	Includes material and labor to install towel bar up to 24" long.
Install	Ea.	28	12.55		40.55	
Demolish and Install	Ea.	28	17.60		45.60	
Remove and Install	Ea.		10.03		10.03	
Clean	Ea.		2.73		2.73	
Minimum Charge	Job		150		150	
Toothbrush Holder						
Demolish	Ea.		5.05		5.05	Includes material and labor to install a surface mounted tumbler and toothbrush holder.
Install	Ea.	23	15.05		38.05	
Demolish and Install	Ea.	23	20.10		43.10	
Remove and Install	Ea.		12.03		12.03	
Clean	Ea.		2.73		2.73	
Minimum Charge	Job		150		150	
Grab Rail						
Demolish	Ea.		7.55		7.55	Includes material and labor to install a 24" long grab bar. Blocking is not included.
Install	Ea.	49.50	13.10		62.60	
Demolish and Install	Ea.	49.50	20.65		70.15	
Remove and Install	Ea.		10.46		10.46	
Clean	Ea.	.28	4.54		4.82	
Minimum Charge	Job		150		150	
Soap Dish						
Demolish	Ea.		5.05		5.05	Includes material and labor to install a surface mounted soap dish.
Install	Ea.	19.80	15.05		34.85	
Demolish and Install	Ea.	19.80	20.10		39.90	
Remove and Install	Ea.		12.03		12.03	
Clean	Ea.		2.73		2.73	
Minimum Charge	Job		150		150	
Toilet Paper Roller						
Demolish	Ea.		5.05		5.05	Includes material and labor to install a surface mounted toilet tissue dispenser.
Install	Ea.	11.90	10.05		21.95	
Demolish and Install	Ea.	11.90	15.10		27	
Remove and Install	Ea.		8.02		8.02	
Clean	Ea.	.28	6.80		7.08	
Minimum Charge	Job		150		150	
Dispenser						
Soap						
Demolish	Ea.		6.75		6.75	Includes material and labor to install a surface mounted soap dispenser.
Install	Ea.	44	15.05		59.05	
Demolish and Install	Ea.	44	21.80		65.80	
Remove and Install	Ea.		12.03		12.03	
Clean	Ea.	.28	6.80		7.08	
Minimum Charge	Job		150		150	
Towel						
Demolish	Ea.		7.55		7.55	Includes material and labor to install a surface mounted towel dispenser.
Install	Ea.	39.50	18.80		58.30	
Demolish and Install	Ea.	39.50	26.35		65.85	
Remove and Install	Ea.		15.04		15.04	
Clean	Ea.	.28	9.10		9.38	
Minimum Charge	Job		150		150	
Seat Cover						
Demolish	Ea.		5.05		5.05	Includes material and labor to install a surface mounted toilet seat cover dispenser.
Install	Ea.	25.50	20		45.50	
Demolish and Install	Ea.	25.50	25.05		50.55	
Remove and Install	Ea.		16.04		16.04	
Clean	Ea.	.28	6.80		7.08	
Minimum Charge	Job		150		150	

Bath Accessories	Unit	Material	Labor	Equip.	Total	Specification
Sanitary Napkin						
Demolish	Ea.		12.75		12.75	Includes material and labor to install a surface mounted sanitary napkin dispenser.
Install	Ea.	21	31.50		52.50	
Demolish and Install	Ea.	21	44.25		65.25	
Remove and Install	Ea.		25.33		25.33	
Clean	Ea.	.13	9.10		9.23	
Minimum Charge	Job		150		150	
Partition						
Urinal						
Demolish	Ea.		15.15		15.15	Includes material and labor to install a urinal screen.
Install	Ea.	177	75		252	
Demolish and Install	Ea.	177	90.15		267.15	
Remove and Install	Ea.		60.16		60.16	
Clean	Ea.	.41	6.05		6.46	
Minimum Charge	Job		150		150	
Toilet						
Demolish	Ea.		45		45	Includes material and labor to install baked enamel standard size partition, floor or ceiling mounted with one wall and one door.
Install	Ea.	430	120		550	
Demolish and Install	Ea.	430	165		595	
Remove and Install	Ea.		96.26		96.26	
Clean	Ea.	2.73	13.65		16.38	
Minimum Charge	Job		150		150	
Handicap						
Demolish	Ea.		45		45	Includes material and labor to install baked enamel handicap partition, floor or ceiling mounted with one wall and one door.
Install	Ea.	730	120		850	
Demolish and Install	Ea.	730	165		895	
Remove and Install	Ea.		96.26		96.26	
Clean	Ea.	2.73	13.65		16.38	
Minimum Charge	Job		150		150	

Fire Prevention / Protection	Unit	Material	Labor	Equip.	Total	Specification
Fire Extinguisher						
5# Carbon Dioxide						
Install	Ea.	111			111	Includes labor and material to install one wall mounted 5 lb. carbon dioxide factory charged unit, complete with hose, horn and wall mounting bracket.
10# Carbon Dioxide						
Install	Ea.	165			165	Includes labor and material to install one wall mounted 10 lb. carbon dioxide factory charged unit, complete with hose, horn and wall mounting bracket.
15# Carbon Dioxide						
Install	Ea.	189			189	Includes labor and material to install one wall mounted 15 lb. carbon dioxide factory charged unit, complete with hose, horn and wall mounting bracket.
5# Dry Chemical						
Install	Ea.	44			44	Includes labor and material to install one wall mounted 5 lb. dry chemical factory charged unit, complete with hose, horn and wall mounting bracket.

Fire Prevention / Protection	Unit	Material	Labor	Equip.	Total	Specification
20# Dry Chemical						
Install	Ea.	99			99	Includes labor and material to install one wall mounted 20 lb. dry chemical factory charged unit, complete with hose, horn and wall mounting bracket.

Gutters	Unit	Material	Labor	Equip.	Total	Specification
Galvanized						
Demolish	L.F.		1.01		1.01	Cost includes material and labor to install 5″ galvanized steel gutter with enamel finish, fittings, hangers, and corners.
Install	L.F.	1.19	2.74		3.93	
Demolish and Install	L.F.	1.19	3.75		4.94	
Remove and Install	L.F.		2.19		2.19	
Clean	L.F.		.29		.29	
Paint	L.F.	.18	.81		.99	
Minimum Charge	Job		164		164	
Aluminum						
Demolish	L.F.		1.01		1.01	Cost includes material and labor to install 5″ aluminum gutter with enamel finish, fittings, hangers, and corners.
Install	L.F.	1.34	2.74		4.08	
Demolish and Install	L.F.	1.34	3.75		5.09	
Remove and Install	L.F.		2.19		2.19	
Clean	L.F.		.29		.29	
Paint	L.F.	.18	.81		.99	
Minimum Charge	Job		164		164	
Copper						
Demolish	L.F.		1.01		1.01	Cost includes material and labor to install 5″ half-round copper gutter with fittings, hangers, and corners.
Install	L.F.	4.08	2.74		6.82	
Demolish and Install	L.F.	4.08	3.75		7.83	
Remove and Install	L.F.		2.19		2.19	
Clean	L.F.		.29		.29	
Minimum Charge	Job		164		164	
Plastic						
Demolish	L.F.		1.01		1.01	Cost includes material and labor to install 4″ white vinyl gutter fittings, hangers, and corners.
Install	L.F.	.94	2.73		3.67	
Demolish and Install	L.F.	.94	3.74		4.68	
Remove and Install	L.F.		2.19		2.19	
Clean	L.F.		.29		.29	
Paint	L.F.	.18	.81		.99	
Minimum Charge	Job		164		164	

Downspouts	Unit	Material	Labor	Equip.	Total	Specification
Galvanized						
Demolish	L.F.		.69		.69	Cost includes material and labor to install 4″ diameter downspout with enamel finish.
Install	L.F.	1.19	2.27		3.46	
Demolish and Install	L.F.	1.19	2.96		4.15	
Remove and Install	L.F.		1.81		1.81	
Clean	L.F.		.29		.29	
Paint	L.F.	.18	.81		.99	
Minimum Charge	Job		164		164	

Downspouts	Unit	Material	Labor	Equip.	Total	Specification
Aluminum						
Demolish	L.F.		.69		.69	Includes material and labor to install aluminum enameled downspout.
Install	L.F.	1.75	2.35		4.10	
Demolish and Install	L.F.	1.75	3.04		4.79	
Remove and Install	L.F.		1.88		1.88	
Clean	L.F.		.29		.29	
Paint	L.F.	.18	.81		.99	
Minimum Charge	Job		164		164	
Copper						
Demolish	L.F.		.69		.69	Includes material and labor to install round copper downspout.
Install	L.F.	4.96	1.73		6.69	
Demolish and Install	L.F.	4.96	2.42		7.38	
Remove and Install	L.F.		1.38		1.38	
Clean	L.F.		.29		.29	
Paint	L.F.	.18	.81		.99	
Minimum Charge	Job		164		164	
Plastic						
Demolish	L.F.		.69		.69	Cost includes material and labor to install 2″ x 3″ vinyl rectangular downspouts.
Install	L.F.	.79	1.57		2.36	
Demolish and Install	L.F.	.79	2.26		3.05	
Remove and Install	L.F.		1.25		1.25	
Clean	L.F.		.29		.29	
Paint	L.F.	.18	.81		.99	
Minimum Charge	Job		164		164	

Shutters	Unit	Material	Labor	Equip.	Total	Specification
Exterior						
25″ High						
Demolish	Ea.		7.55		7.55	Includes material and labor to install pair of exterior shutters 25″ x 16″, unpainted, installed as fixed shutters on concrete or wood frame.
Install	Pr.	87	30		117	
Demolish and Install	Pr.	87	37.55		124.55	
Remove and Install	Pr.		30.08		30.08	
Clean	Ea.	.06	6.55		6.61	
Paint	Ea.	1.40	14.20		15.60	
Minimum Charge	Job		150		150	
39″ High						
Demolish	Ea.		7.55		7.55	Includes material and labor to install pair of shutters 39″ x 16″, unpainted, installed as fixed shutters on concrete or wood frame.
Install	Pr.	93.50	30		123.50	
Demolish and Install	Pr.	93.50	37.55		131.05	
Remove and Install	Pr.		30.08		30.08	
Clean	Ea.	.06	6.55		6.61	
Paint	Ea.	2.17	19.15		21.32	
Minimum Charge	Job		150		150	
51″ High						
Demolish	Ea.		7.55		7.55	Includes material and labor to install pair of shutters 51″ x 16″, unpainted, installed as fixed shutters on concrete or wood frame.
Install	Pr.	104	30		134	
Demolish and Install	Pr.	104	37.55		141.55	
Remove and Install	Pr.		30.08		30.08	
Clean	Ea.	.06	6.55		6.61	
Paint	Ea.	2.83	19.85		22.68	
Minimum Charge	Job		150		150	

Shutters	Unit	Material	Labor	Equip.	Total	Specification
59" High						
Demolish	Ea.		7.55		7.55	Includes material and labor to install pair of shutters 59" x 16", unpainted, installed as fixed shutters on concrete or wood frame.
Install	Pr.	146	30		176	
Demolish and Install	Pr.	146	37.55		183.55	
Remove and Install	Pr.		30.08		30.08	
Clean	Ea.	.06	6.55		6.61	
Paint	Ea.	3.26	22		25.26	
Minimum Charge	Job		150		150	
67" High						
Demolish	Ea.		7.55		7.55	Includes material and labor to install pair of shutters 67" x 16", unpainted, installed as fixed shutters on concrete or wood frame.
Install	Pr.	147	33.50		180.50	
Demolish and Install	Pr.	147	41.05		188.05	
Remove and Install	Pr.		33.42		33.42	
Clean	Ea.	.06	6.55		6.61	
Paint	Ea.	3.73	24.50		28.23	
Minimum Charge	Job		150		150	
Interior Wood						
23" x 24", 2 Panel						
Demolish	Ea.		7.55		7.55	Cost includes material and labor to install 23" x 24", 4 panel interior moveable pine shutter, 1-1/4" louver, knobs and hooks and hinges.
Install	Ea.	81	17.70		98.70	
Demolish and Install	Ea.	81	25.25		106.25	
Remove and Install	Ea.		14.16		14.16	
Clean	Ea.	.06	6.55		6.61	
Paint	Ea.	1.06	18.85		19.91	
Minimum Charge	Job		150		150	
27" x 24", 4 Panel						
Demolish	Ea.		7.55		7.55	Cost includes material and labor to install 27" x 24", 4 panel interior moveable pine shutter, 1-1/4" louver, knobs and hooks and hinges.
Install	Ea.	94	17.70		111.70	
Demolish and Install	Ea.	94	25.25		119.25	
Remove and Install	Ea.		14.16		14.16	
Clean	Ea.	.06	6.55		6.61	
Paint	Ea.	1.06	18.85		19.91	
Minimum Charge	Job		150		150	
31" x 24", 4 Panel						
Demolish	Ea.		7.55		7.55	Cost includes material and labor to install 31" x 24", 4 panel interior moveable pine shutter, 1-1/4" louver, knobs and hooks and hinges.
Install	Ea.	108	17.70		125.70	
Demolish and Install	Ea.	108	25.25		133.25	
Remove and Install	Ea.		14.16		14.16	
Clean	Ea.	.06	6.55		6.61	
Paint	Ea.	1.06	18.85		19.91	
Minimum Charge	Job		150		150	
35" x 24", 4 Panel						
Demolish	Ea.		7.55		7.55	Cost includes material and labor to install 35" x 24", 4 panel interior moveable pine shutter, 1-1/4" louver, knobs and hooks and hinges.
Install	Ea.	108	17.70		125.70	
Demolish and Install	Ea.	108	25.25		133.25	
Remove and Install	Ea.		14.16		14.16	
Clean	Ea.	.06	6.55		6.61	
Paint	Ea.	1.06	18.85		19.91	
Minimum Charge	Job		150		150	
39" x 24", 4 Panel						
Demolish	Ea.		7.55		7.55	Cost includes material and labor to install 39" x 24", 4 panel interior moveable pine shutter, 1-1/4" louver, knobs and hooks and hinges.
Install	Ea.	124	17.70		141.70	
Demolish and Install	Ea.	124	25.25		149.25	
Remove and Install	Ea.		14.16		14.16	
Clean	Ea.	.06	6.55		6.61	
Paint	Ea.	1.06	18.85		19.91	
Minimum Charge	Job		150		150	

Shutters	Unit	Material	Labor	Equip.	Total	Specification
47" x 24", 4 Panel						
Demolish	Ea.		7.55		7.55	Cost includes material and labor to install 47" x 24", 4 panel interior moveable pine shutter, 1-1/4" louver, knobs and hooks and hinges.
Install	Ea.	138	17.70		155.70	
Demolish and Install	Ea.	138	25.25		163.25	
Remove and Install	Ea.		14.16		14.16	
Clean	Ea.	.06	6.55		6.61	
Paint	Ea.	1.10	20.50		21.60	
Minimum Charge	Job		150		150	

Awning	Unit	Material	Labor	Equip.	Total	Specification
Standard Aluminum						
36" Wide x 30" Deep						
Demolish	Ea.		7.55		7.55	Includes material and labor to install one aluminum awning with weather-resistant finish.
Install	Ea.	171	25		196	
Demolish and Install	Ea.	171	32.55		203.55	
Remove and Install	Ea.		20.05		20.05	
Clean	Ea.		9.10		9.10	
Minimum Charge	Job		150		150	
48" Wide x 30" Deep						
Demolish	Ea.		7.55		7.55	Includes material and labor to install one aluminum awning with weather-resistant finish.
Install	Ea.	193	30		223	
Demolish and Install	Ea.	193	37.55		230.55	
Remove and Install	Ea.		24.06		24.06	
Clean	Ea.		9.10		9.10	
Minimum Charge	Job		150		150	
60" Wide x 30" Deep						
Demolish	Ea.		15.15		15.15	Includes material and labor to install one aluminum awning with weather-resistant finish.
Install	Ea.	215	33.50		248.50	
Demolish and Install	Ea.	215	48.65		263.65	
Remove and Install	Ea.		26.74		26.74	
Clean	Ea.		9.10		9.10	
Minimum Charge	Job		150		150	
72" Wide x 30" Deep						
Demolish	Ea.		15.15		15.15	Includes material and labor to install one aluminum awning with weather-resistant finish.
Install	Ea.	237	37.50		274.50	
Demolish and Install	Ea.	237	52.65		289.65	
Remove and Install	Ea.		30.08		30.08	
Clean	Ea.		9.10		9.10	
Minimum Charge	Job		150		150	
Roll-up Type						
Demolish	S.F.		.55		.55	Includes material and labor to install an aluminum roll-up awning.
Install	S.F.	6.40	3.01		9.41	
Demolish and Install	S.F.	6.40	3.56		9.96	
Remove and Install	S.F.		2.41		2.41	
Clean	S.F.	.02	.45		.47	
Minimum Charge	Job		150		150	
Canvas						
30" Wide						
Demolish	Ea.		4.54		4.54	Includes material and labor to install a canvas window awning.
Install	Ea.	182	25		207	
Demolish and Install	Ea.	182	29.54		211.54	
Remove and Install	Ea.		20.05		20.05	
Clean	Ea.		2.73		2.73	
Minimum Charge	Job		150		150	

Awning	Unit	Material	Labor	Equip.	Total	Specification
36″ Wide						
Demolish	Ea.		4.54		4.54	Includes material and labor to install a canvas window awning.
Install	Ea.	193	30		223	
Demolish and Install	Ea.	193	34.54		227.54	
Remove and Install	Ea.		24.06		24.06	
Clean	Ea.		2.99		2.99	
Minimum Charge	Job		150		150	
42″ Wide						
Demolish	Ea.		4.54		4.54	Includes material and labor to install a canvas window awning.
Install	Ea.	204	37.50		241.50	
Demolish and Install	Ea.	204	42.04		246.04	
Remove and Install	Ea.		30.08		30.08	
Clean	Ea.		3.54		3.54	
Minimum Charge	Job		150		150	
48″ Wide						
Demolish	Ea.		4.54		4.54	Includes material and labor to install a canvas window awning.
Install	Ea.	215	43		258	
Demolish and Install	Ea.	215	47.54		262.54	
Remove and Install	Ea.		34.38		34.38	
Clean	Ea.		4.11		4.11	
Minimum Charge	Job		150		150	
Cloth Canopy Cover						
8′ x 10′						
Demolish	Ea.		252		252	Includes material and labor to install one 8′ x 10′ cloth canopy patio cover with front bar and tension support rafters and 9″ valance.
Install	Ea.	305	150		455	
Demolish and Install	Ea.	305	402		707	
Remove and Install	Ea.		120.32		120.32	
Clean	Ea.	2.19	36.50		38.69	
Minimum Charge	Job		300		300	
8′ x 15′						
Demolish	Ea.		252		252	Includes material and labor to install one 8′ x 15′ cloth canopy patio cover with front bar and tension support rafters and 9″ valance.
Install	Ea.	415	150		565	
Demolish and Install	Ea.	415	402		817	
Remove and Install	Ea.		120.32		120.32	
Clean	Ea.	2.19	54.50		56.69	
Minimum Charge	Job		300		300	

Vent	Unit	Material	Labor	Equip.	Total	Specification
Small Roof Turbine						
Demolish	Ea.		7.55		7.55	Includes material and labor to install spinner ventilator, wind driven, galvanized, 4″ neck diam, 180 CFM.
Install	Ea.	57	29.50		86.50	
Demolish and Install	Ea.	57	37.05		94.05	
Remove and Install	Ea.		23.67		23.67	
Clean	Ea.		4.84		4.84	
Minimum Charge	Job		163		163	
Large Roof Turbine						
Demolish	Ea.		6.05		6.05	Includes material and labor to install spinner ventilator, wind driven, galvanized, 8″ neck diam, 360 CFM.
Install	Ea.	66.50	42.50		109	
Demolish and Install	Ea.	66.50	48.55		115.05	
Remove and Install	Ea.		33.81		33.81	
Clean	Ea.		5.45		5.45	
Minimum Charge	Job		163		163	

Vent	Unit	Material	Labor	Equip.	Total	Specification
Roof Ventilator						
Demolish	Ea.		7.55		7.55	Includes material and labor to install a roof ventilator.
Install	Ea.	163	37		200	
Demolish and Install	Ea.	163	44.55		207.55	
Remove and Install	Ea.		29.58		29.58	
Clean	Ea.	.55	5.45		6	
Paint	Ea.	1.32	6.60		7.92	
Minimum Charge	Job		163		163	
Gable Louvered						
Demolish	Ea.		15.15		15.15	Includes material and labor to install a vinyl gable end vent.
Install	Ea.	31	10.05		41.05	
Demolish and Install	Ea.	31	25.20		56.20	
Remove and Install	Ea.		8.02		8.02	
Clean	Ea.		2.91		2.91	
Paint	Ea.	.88	4.06		4.94	
Minimum Charge	Job		150		150	
Ridge Vent						
Demolish	L.F.		.78		.78	Includes material and labor to install a mill finish aluminum ridge vent strip.
Install	L.F.	2.35	2.12		4.47	
Demolish and Install	L.F.	2.35	2.90		5.25	
Minimum Charge	Job		150		150	
Foundation						
Demolish	Ea.		5.35		5.35	Includes material and labor to install galvanized foundation block vent.
Install	Ea.	19.80	10.15		29.95	
Demolish and Install	Ea.	19.80	15.50		35.30	
Remove and Install	Ea.		8.11		8.11	
Clean	Ea.	.13	5.45		5.58	
Minimum Charge	Job		150		150	
Frieze						
Demolish	L.F.		.54		.54	Includes material and labor to install pine frieze board.
Install	L.F.	.85	.68		1.53	
Demolish and Install	L.F.	.85	1.22		2.07	
Remove and Install	L.F.		.55		.55	
Clean	L.F.	.03	.18		.21	
Minimum Charge	Job		150		150	
Water Heater Cap						
Demolish	Ea.		11.30		11.30	Includes material and labor to install a galvanized steel water heater cap.
Install	Ea.	23	41		64	
Demolish and Install	Ea.	23	52.30		75.30	
Remove and Install	Ea.		32.88		32.88	
Clean	Ea.	.96	13.65		14.61	

Cupolas	Unit	Material	Labor	Equip.	Total	Specification
22″ x 22″						
Demolish	Ea.		145		145	Includes material and labor to install a cedar cupola with aluminum roof covering.
Install	Ea.	267	75		342	
Demolish and Install	Ea.	267	220		487	
Remove and Install	Ea.		60.16		60.16	
Minimum Charge	Job		150		150	
29″ x 29″						
Demolish	Ea.		145		145	Includes material and labor to install a cedar cupola with aluminum roof covering.
Install	Ea.	345	75		420	
Demolish and Install	Ea.	345	220		565	
Remove and Install	Ea.		60.16		60.16	
Minimum Charge	Job		150		150	

Cupolas	Unit	Material	Labor	Equip.	Total	Specification
35″ x 35″						
Demolish	Ea.		145		145	Includes material and labor to install a cedar cupola with aluminum roof covering.
Install	Ea.	650	100		750	
Demolish and Install	Ea.	650	245		895	
Remove and Install	Ea.		80.21		80.21	
Minimum Charge	Job		150		150	
47″ x 47″						
Demolish	Ea.		145		145	Includes material and labor to install a cedar cupola with aluminum roof covering.
Install	Ea.	1050	150		1200	
Demolish and Install	Ea.	1050	295		1345	
Remove and Install	Ea.		120.32		120.32	
Minimum Charge	Job		150		150	
Weather Vane						
Install	Ea.	44	37.50		81.50	Includes minimum labor and equipment to install residential type weathervane.
Minimum Charge	Job		150		150	

Aluminum Carport	Unit	Material	Labor	Equip.	Total	Specification
Natural Finish						
Demolish	S.F.		1.51		1.51	Includes material, labor and equipment to install an aluminum carport. Foundations are not included.
Install	S.F.	19.35	5.05	.76	25.16	
Demolish and Install	S.F.	19.35	6.56	.76	26.67	
Remove and Install	S.F.		4.06	.61	4.67	
Clean	S.F.	.02	.17		.19	
Minimum Charge	Job		300		300	
Enamel Finish						
Demolish	S.F.		1.51		1.51	Includes material, labor and equipment to install an aluminum carport. Foundations are not included.
Install	S.F.	19.35	5.05	.76	25.16	
Demolish and Install	S.F.	19.35	6.56	.76	26.67	
Remove and Install	S.F.		4.06	.61	4.67	
Clean	S.F.	.02	.17		.19	
Minimum Charge	Job		300		300	
20′ x 20′ Complete						
Demolish	Ea.		151		151	Includes material, labor and equipment to install an aluminum carport. Foundations are not included.
Install	Ea.	6075	505	76	6656	
Demolish and Install	Ea.	6075	656	76	6807	
Remove and Install	Ea.		405.79	60.77	466.56	
Clean	Ea.	.14	54.50		54.64	
Minimum Charge	Job		300		300	
24′ x 24′ Complete						
Demolish	Ea.		151		151	Includes material, labor and equipment to install an aluminum carport. Foundations are not included.
Install	Ea.	8250	675	101	9026	
Demolish and Install	Ea.	8250	826	101	9177	
Remove and Install	Ea.		541.06	81.02	622.08	
Clean	Ea.	.14	109		109.14	
Minimum Charge	Job		300		300	
Metal Support Posts						
Demolish	Ea.		2.02		2.02	Includes material and labor to install metal support posts for a carport.
Install	Ea.	13.50	37.50		51	
Demolish and Install	Ea.	13.50	39.52		53.02	
Remove and Install	Ea.		30.08		30.08	
Clean	Ea.	.14	2.27		2.41	
Paint	Ea.	.12	5.30		5.42	
Minimum Charge	Job		300		300	

Storage Shed	Unit	Material	Labor	Equip.	Total	Specification
Aluminum						
Pre-fab (8′ x 10′)						
Demolish	Ea.		44		44	Includes material and labor to install an aluminum storage shed.
Install	Ea.	273	435		708	
Demolish and Install	Ea.	273	479		752	
Clean	Ea.	.91	27.50		28.41	
Minimum Charge	Ea.		375		375	
Pre-fab (10′ x 12′)						
Demolish	Ea.		84		84	Includes material and labor to install an aluminum storage shed.
Install	Ea.	460	435		895	
Demolish and Install	Ea.	460	519		979	
Clean	Ea.	.91	30.50		31.41	
Minimum Charge	Ea.		375		375	
Wood						
Pre-fab (8′ x 10′)						
Demolish	Ea.		44		44	Includes material and labor to install a pre-fabricated wood storage shed.
Install	Ea.	760	435		1195	
Demolish and Install	Ea.	760	479		1239	
Clean	Ea.	.91	27.50		28.41	
Minimum Charge	Ea.		375		375	
Pre-fab (10′ x 12′)						
Demolish	Ea.		84		84	Includes material and labor to install a pre-fabricated wood storage shed.
Install	Ea.	990	435		1425	
Demolish and Install	Ea.	990	519		1509	
Clean	Ea.	.91	30.50		31.41	
Minimum Charge	Ea.		375		375	

Gazebo	Unit	Material	Labor	Equip.	Total	Specification
Good Grade						
Demolish	S.F.		2.42		2.42	Includes material and labor to install a gazebo.
Install	S.F.	18.90	22.50		41.40	
Demolish and Install	S.F.	18.90	24.92		43.82	
Remove and Install	S.F.		18.05		18.05	
Clean	S.F.	.11	1.45		1.56	
Minimum Charge	Job		150		150	
Custom Grade						
Demolish	S.F.		2.42		2.42	Includes material and labor to install a gazebo.
Install	S.F.	30.50	22.50		53	
Demolish and Install	S.F.	30.50	24.92		55.42	
Remove and Install	S.F.		18.05		18.05	
Clean	S.F.	.11	1.45		1.56	
Minimum Charge	Job		150		150	
Bench Seating						
Demolish	L.F.		4.04		4.04	Includes material and labor to install pressure treated bench seating.
Install	L.F.	2.15	15.05		17.20	
Demolish and Install	L.F.	2.15	19.09		21.24	
Remove and Install	L.F.		12.03		12.03	
Clean	L.F.	.11	1.09		1.20	
Paint	L.F.	.48	2.11		2.59	
Minimum Charge	Job		150		150	
Re-screen						
Install	S.F.	.52	1.20		1.72	Includes labor and material to re-screen wood frame.
Minimum Charge	Job		150		150	

Pool	Unit	Material	Labor	Equip.	Total	Specification
Screen Enclosure						
Re-screen						
Install	SF Wall		3.74		3.74	Includes material and labor to install a screen enclosure for a pool.
Minimum Charge	Job		150		150	
Screen Door						
Demolish	Ea.		5.95		5.95	Includes material and labor to install wood screen door with aluminum cloth screen. Frame and hardware not included.
Install	Ea.	134	50		184	
Demolish and Install	Ea.	134	55.95		189.95	
Remove and Install	Ea.		40.11		40.11	
Clean	Ea.	.10	7.85		7.95	
Paint	Ea.	2.94	10.55		13.49	
Minimum Charge	Job		150		150	
Heater / Motor						
Demolish	Ea.		28.50		28.50	Includes material and labor to install swimming pool heater, not incl. base or pad, elec., 12 kW, 4,800 gal pool, incl. pump.
Install	Ea.	1850	365		2215	
Demolish and Install	Ea.	1850	393.50		2243.50	
Remove and Install	Ea.		293.91		293.91	
Clean	Ea.	2.73	27.50		30.23	
Minimum Charge	Job		164		164	
Filter System						
Demolish	Ea.		48.50		48.50	Includes material, labor and equipment to install sand filter system tank including hook-up to existing equipment. Pump and motor not included.
Install	Total	545	260		805	
Demolish and Install	Total	545	308.50		853.50	
Remove and Install	Total		208.13		208.13	
Minimum Charge	Job		163		163	
Pump / Motor						
Demolish	Ea.		28.50		28.50	Includes material and labor to install a 22 GPM pump.
Install	Ea.	535	65		600	
Demolish and Install	Ea.	535	93.50		628.50	
Remove and Install	Ea.		52.03		52.03	
Clean	Ea.	2.73	27.50		30.23	
Minimum Charge	Job		164		164	
Pump Water						
Install	Day		298	74	372	Includes labor and equipment to pump water from a swimming pool.
Minimum Charge	Job		137		137	
Replaster						
Install	SF Surf	.91	7.25		8.16	Includes labor and material to replaster gunite pool with a surface area of 500 to 600 S.F.
Minimum Charge	Job		137		137	
Solar Panels						
Demolish	S.F.		1.20		1.20	Includes material and labor to install solar panel.
Install	S.F.	18.75	2.14		20.89	
Demolish and Install	S.F.	18.75	3.34		22.09	
Remove and Install	S.F.		1.71		1.71	
Minimum Charge	Ea.		375		375	

Construction Clean-up	Unit	Material	Labor	Equip.	Total	Specification
Final						
Clean	S.F.		.19		.19	Includes labor for final site/job clean-up including detail cleaning.
Minimum Charge	Job		109		109	

Location Factors

Costs shown in *Means cost data publications* are based on National Averages for materials and installation. To adjust these costs to a specific location, simply multiply the base cost by the factor for that city. The data is arranged alphabetically by state and postal zip code numbers. For a city not listed, use the factor for a nearby city with similar economic characteristics.

STATE/ZIP	CITY	Residential	Commercial
ALABAMA			
350-352	Birmingham	.87	.88
354	Tuscaloosa	.80	.78
355	Jasper	.76	.77
356	Decatur	.78	.79
357-358	Huntsville	.85	.86
359	Gadsden	.79	.80
360-361	Montgomery	.82	.80
362	Anniston	.72	.73
363	Dothan	.77	.76
364	Evergreen	.78	.76
365-366	Mobile	.80	.81
367	Selma	.78	.76
368	Phenix City	.81	.79
369	Butler	.77	.75
ALASKA			
995-996	Anchorage	1.27	1.26
997	Fairbanks	1.26	1.25
998	Juneau	1.26	1.25
999	Ketchikan	1.33	1.32
ARIZONA			
850,853	Phoenix	.91	.88
852	Mesa/Tempe	.89	.86
855	Globe	.87	.84
856-857	Tucson	.89	.86
859	Show Low	.88	.84
860	Flagstaff	.91	.87
863	Prescott	.88	.84
864	Kingman	.88	.84
865	Chambers	.86	.83
ARKANSAS			
716	Pine Bluff	.80	.80
717	Camden	.69	.69
718	Texarkana	.73	.72
719	Hot Springs	.68	.68
720-722	Little Rock	.82	.82
723	West Memphis	.77	.77
724	Jonesboro	.77	.77
725	Batesville	.74	.74
726	Harrison	.75	.75
727	Fayetteville	.68	.65
728	Russellville	.76	.73
729	Fort Smith	.82	.79
CALIFORNIA			
900-902	Los Angeles	1.07	1.07
903-905	Inglewood	1.04	1.04
906-908	Long Beach	1.05	1.05
910-912	Pasadena	1.05	1.05
913-916	Van Nuys	1.07	1.07
917-918	Alhambra	1.07	1.07
919-921	San Diego	1.09	1.05
922	Palm Springs	1.08	1.04
923-924	San Bernardino	1.07	1.03
925	Riverside	1.11	1.07
926-927	Santa Ana	1.08	1.05
928	Anaheim	1.10	1.08
930	Oxnard	1.12	1.07
931	Santa Barbara	1.10	1.07
932-933	Bakersfield	1.11	1.06
934	San Luis Obispo	1.12	1.07
935	Mojave	1.08	1.04
936-938	Fresno	1.12	1.08
939	Salinas	1.11	1.11
940-941	San Francisco	1.21	1.24
942,956-958	Sacramento	1.12	1.11
943	Palo Alto	1.14	1.17
944	San Mateo	1.15	1.18
945	Vallejo	1.10	1.13
946	Oakland	1.15	1.18
947	Berkeley	1.15	1.18
948	Richmond	1.14	1.17
949	San Rafael	1.25	1.19
950	Santa Cruz	1.15	1.13
951	San Jose	1.22	1.20
952	Stockton	1.12	1.08
953	Modesto	1.12	1.08
CALIFORNIA (CONT'D)			
954	Santa Rosa	1.13	1.16
955	Eureka	1.08	1.07
959	Marysville	1.09	1.08
960	Redding	1.11	1.10
961	Susanville	1.11	1.10
COLORADO			
800-802	Denver	1.00	.96
803	Boulder	.88	.85
804	Golden	.99	.95
805	Fort Collins	.97	.91
806	Greeley	.90	.85
807	Fort Morgan	.97	.91
808-809	Colorado Springs	.95	.93
810	Pueblo	.94	.92
811	Alamosa	.89	.87
812	Salida	.89	.87
813	Durango	.89	.87
814	Montrose	.86	.85
815	Grand Junction	.89	.85
816	Glenwood Springs	.98	.93
CONNECTICUT			
060	New Britain	1.05	1.06
061	Hartford	1.04	1.05
062	Willimantic	1.04	1.05
063	New London	1.05	1.04
064	Meriden	1.05	1.06
065	New Haven	1.06	1.07
066	Bridgeport	1.03	1.06
067	Waterbury	1.06	1.06
068	Norwalk	1.03	1.07
069	Stamford	1.05	1.09
D.C.			
200-205	Washington	.94	.96
DELAWARE			
197	Newark	1.01	1.02
198	Wilmington	1.01	1.02
199	Dover	1.01	1.02
FLORIDA			
320,322	Jacksonville	.83	.82
321	Daytona Beach	.86	.85
323	Tallahassee	.74	.76
324	Panama City	.69	.71
325	Pensacola	.83	.81
326,344	Gainesville	.83	.80
327-328,347	Orlando	.85	.83
329	Melbourne	.89	.88
330-332,340	Miami	.83	.85
333	Fort Lauderdale	.82	.84
334,349	West Palm Beach	.85	.82
335-336,346	Tampa	.78	.80
337	St. Petersburg	.79	.81
338	Lakeland	.78	.80
339,341	Fort Myers	.78	.78
342	Sarasota	.77	.79
GEORGIA			
300-303,399	Atlanta	.85	.90
304	Statesboro	.71	.73
305	Gainesville	.74	.79
306	Athens	.76	.81
307	Dalton	.67	.66
308-309	Augusta	.75	.76
310-312	Macon	.80	.80
313-314	Savannah	.79	.80
315	Waycross	.73	.73
316	Valdosta	.75	.75
317	Albany	.76	.78
318-319	Columbus	.77	.77
HAWAII			
967	Hilo	1.26	1.22
968	Honolulu	1.27	1.23

Location Factors

STATE/ZIP	CITY	Residential	Commercial
STATES & POSS.			
969	Guam	1.34	1.30
IDAHO			
832	Pocatello	.93	.92
833	Twin Falls	.78	.77
834	Idaho Falls	.82	.81
835	Lewiston	1.08	1.00
836-837	Boise	.94	.93
838	Coeur d'Alene	.94	.87
ILLINOIS			
600-603	North Suburban	1.11	1.10
604	Joliet	1.11	1.10
605	South Suburban	1.11	1.10
606	Chicago	1.15	1.14
609	Kankakee	1.03	1.03
610-611	Rockford	1.06	1.05
612	Rock Island	1.07	.98
613	La Salle	1.08	1.01
614	Galesburg	1.08	1.01
615-616	Peoria	1.09	1.02
617	Bloomington	1.05	1.01
618-619	Champaign	1.05	1.02
620-622	East St. Louis	1.00	1.00
623	Quincy	1.00	.98
624	Effingham	1.02	.99
625	Decatur	1.01	.98
626-627	Springfield	1.03	.99
628	Centralia	.98	.98
629	Carbondale	.97	.97
INDIANA			
460	Anderson	.94	.92
461-462	Indianapolis	.97	.95
463-464	Gary	1.04	1.02
465-466	South Bend	.94	.92
467-468	Fort Wayne	.91	.92
469	Kokomo	.92	.91
470	Lawrenceburg	.92	.89
471	New Albany	.92	.88
472	Columbus	.94	.91
473	Muncie	.92	.91
474	Bloomington	.95	.93
475	Washington	.92	.92
476-477	Evansville	.93	.93
478	Terre Haute	.95	.94
479	Lafayette	.91	.91
IOWA			
500-503,509	Des Moines	.96	.92
504	Mason City	.88	.82
505	Fort Dodge	.86	.80
506-507	Waterloo	.87	.82
508	Creston	.88	.83
510-511	Sioux City	.94	.88
512	Sibley	.79	.77
513	Spencer	.79	.77
514	Carroll	.83	.78
515	Council Bluffs	.95	.89
516	Shenandoah	.81	.76
520	Dubuque	1.00	.89
521	Decorah	.87	.78
522-524	Cedar Rapids	1.00	.92
525	Ottumwa	.94	.86
526	Burlington	.92	.86
527-528	Davenport	.99	.97
KANSAS			
660-662	Kansas City	.99	.97
664-666	Topeka	.85	.84
667	Fort Scott	.86	.84
668	Emporia	.82	.81
669	Belleville	.85	.79
670-672	Wichita	.89	.86
673	Independence	.81	.78
674	Salina	.84	.80
675	Hutchinson	.78	.75
676	Hays	.83	.79
677	Colby	.83	.79
678	Dodge City	.83	.80
679	Liberal	.77	.74
KENTUCKY			
400-402	Louisville	.94	.91
403-405	Lexington	.86	.83
KENTUCKY (CONT'D)			
406	Frankfort	.90	.84
407-409	Corbin	.76	.71
410	Covington	.98	.95
411-412	Ashland	.96	.97
413-414	Campton	.75	.72
415-416	Pikeville	.81	.82
417-418	Hazard	.75	.72
420	Paducah	.96	.92
421-422	Bowling Green	.94	.90
423	Owensboro	.89	.87
424	Henderson	.94	.92
425-426	Somerset	.75	.71
427	Elizabethtown	.93	.89
LOUISIANA			
700-701	New Orleans	.88	.87
703	Thibodaux	.85	.85
704	Hammond	.83	.82
705	Lafayette	.83	.80
706	Lake Charles	.82	.82
707-708	Baton Rouge	.81	.80
710-711	Shreveport	.80	.80
712	Monroe	.78	.78
713-714	Alexandria	.77	.77
MAINE			
039	Kittery	.84	.86
040-041	Portland	.89	.91
042	Lewiston	.90	.91
043	Augusta	.86	.86
044	Bangor	.91	.91
045	Bath	.87	.87
046	Machias	.85	.85
047	Houlton	.87	.87
048	Rockland	.84	.84
049	Waterville	.84	.83
MARYLAND			
206	Waldorf	.87	.87
207-208	College Park	.90	.90
209	Silver Spring	.89	.89
210-212	Baltimore	.91	.91
214	Annapolis	.88	.89
215	Cumberland	.86	.87
216	Easton	.72	.72
217	Hagerstown	.89	.88
218	Salisbury	.75	.76
219	Elkton	.81	.82
MASSACHUSETTS			
010-011	Springfield	1.02	1.01
012	Pittsfield	.98	.98
013	Greenfield	1.00	.98
014	Fitchburg	1.06	1.02
015-016	Worcester	1.10	1.06
017	Framingham	1.04	1.05
018	Lowell	1.07	1.07
019	Lawrence	1.07	1.07
020-022, 024	Boston	1.14	1.15
023	Brockton	1.04	1.06
025	Buzzards Bay	1.01	1.03
026	Hyannis	1.03	1.04
027	New Bedford	1.05	1.06
MICHIGAN			
480,483	Royal Oak	1.03	1.02
481	Ann Arbor	1.04	1.03
482	Detroit	1.09	1.08
484-485	Flint	.98	.99
486	Saginaw	.96	.97
487	Bay City	.96	.97
488-489	Lansing	1.02	.99
490	Battle Creek	1.01	.95
491	Kalamazoo	1.00	.93
492	Jackson	.99	.96
493,495	Grand Rapids	.89	.86
494	Muskegon	.95	.92
496	Traverse City	.90	.86
497	Gaylord	.88	.89
498-499	Iron Mountain	.98	.95
MINNESOTA			
550-551	Saint Paul	1.14	1.12
553-555	Minneapolis	1.17	1.13

Location Factors

STATE/ZIP	CITY	Residential	Commercial
556-558	Duluth	1.02	1.03
559	Rochester	1.05	1.02
560	Mankato	1.00	.99
561	Windom	.91	.90
562	Willmar	.94	.93
563	St. Cloud	1.13	1.05
564	Brainerd	1.07	.99
565	Detroit Lakes	.91	.98
566	Bemidji	.92	.99
567	Thief River Falls	.87	.94
MISSISSIPPI			
386	Clarksdale	.68	.65
387	Greenville	.79	.75
388	Tupelo	.69	.70
389	Greenwood	.70	.67
390-392	Jackson	.79	.75
393	Meridian	.75	.74
394	Laurel	.71	.67
395	Biloxi	.83	.79
396	McComb	.79	.77
397	Columbus	.69	.70
MISSOURI			
630-631	St. Louis	.99	1.02
633	Bowling Green	.92	.94
634	Hannibal	.98	.92
635	Kirksville	.85	.89
636	Flat River	.94	.96
637	Cape Girardeau	.92	.95
638	Sikeston	.90	.92
639	Poplar Bluff	.90	.92
640-641	Kansas City	1.05	1.02
644-645	St. Joseph	.92	.96
646	Chillicothe	.81	.85
647	Harrisonville	1.01	.99
648	Joplin	.83	.85
650-651	Jefferson City	.97	.91
652	Columbia	.99	.93
653	Sedalia	.99	.92
654-655	Rolla	.94	.88
656-658	Springfield	.86	.88
MONTANA			
590-591	Billings	.92	.90
592	Wolf Point	.90	.88
593	Miles City	.90	.88
594	Great Falls	.91	.90
595	Havre	.89	.88
596	Helena	.90	.89
597	Butte	.90	.89
598	Missoula	.88	.87
599	Kalispell	.87	.86
NEBRASKA			
680-681	Omaha	.92	.91
683-685	Lincoln	.90	.85
686	Columbus	.77	.76
687	Norfolk	.84	.83
688	Grand Island	.88	.84
689	Hastings	.84	.80
690	Mccook	.78	.74
691	North Platte	.85	.81
692	Valentine	.76	.72
693	Alliance	.73	.70
NEVADA			
889-891	Las Vegas	1.05	1.04
893	Ely	.93	.94
894-895	Reno	.94	.99
897	Carson City	.96	.99
898	Elko	.91	.93
NEW HAMPSHIRE			
030	Nashua	.92	.93
031	Manchester	.92	.93
032-033	Concord	.91	.92
034	Keene	.76	.77
035	Littleton	.80	.81
036	Charleston	.75	.75
037	Claremont	.74	.75
038	Portsmouth	.90	.89

STATE/ZIP	CITY	Residential	Commercial
NEW JERSEY			
070-071	Newark	1.11	1.10
072	Elizabeth	1.08	1.06
073	Jersey City	1.10	1.09
074-075	Paterson	1.09	1.09
076	Hackensack	1.08	1.08
077	Long Branch	1.08	1.06
078	Dover	1.10	1.08
079	Summit	1.07	1.05
080,083	Vineland	1.09	1.06
081	Camden	1.09	1.06
082,084	Atlantic City	1.09	1.06
085-086	Trenton	1.09	1.08
087	Point Pleasant	1.08	1.06
088-089	New Brunswick	1.10	1.08
NEW MEXICO			
870-872	Albuquerque	.88	.90
873	Gallup	.88	.90
874	Farmington	.88	.90
875	Santa Fe	.87	.89
877	Las Vegas	.87	.89
878	Socorro	.87	.89
879	Truth/Consequences	.86	.86
880	Las Cruces	.84	.84
881	Clovis	.89	.89
882	Roswell	.90	.90
883	Carrizozo	.90	.90
884	Tucumcari	.89	.89
NEW YORK			
100-102	New York	1.33	1.33
103	Staten Island	1.30	1.30
104	Bronx	1.28	1.28
105	Mount Vernon	1.19	1.19
106	White Plains	1.19	1.19
107	Yonkers	1.21	1.21
108	New Rochelle	1.19	1.19
109	Suffern	1.14	1.14
110	Queens	1.29	1.29
111	Long Island City	1.30	1.30
112	Brooklyn	1.30	1.30
113	Flushing	1.31	1.31
114	Jamaica	1.29	1.29
115,117,118	Hicksville	1.24	1.24
116	Far Rockaway	1.31	1.31
119	Riverhead	1.26	1.26
120-122	Albany	.96	.96
123	Schenectady	.96	.96
124	Kingston	1.11	1.09
125-126	Poughkeepsie	1.13	1.11
127	Monticello	1.10	1.08
128	Glens Falls	.93	.91
129	Plattsburgh	.94	.92
130-132	Syracuse	.97	.95
133-135	Utica	.89	.92
136	Watertown	.90	.93
137-139	Binghamton	.93	.93
140-142	Buffalo	1.06	1.02
143	Niagara Falls	1.06	1.02
144-146	Rochester	.98	.99
147	Jamestown	.97	.93
148-149	Elmira	.94	.92
NORTH CAROLINA			
270,272-274	Greensboro	.75	.76
271	Winston-Salem	.74	.75
275-276	Raleigh	.76	.76
277	Durham	.75	.76
278	Rocky Mount	.68	.68
279	Elizabeth City	.69	.69
280	Gastonia	.73	.74
281-282	Charlotte	.73	.74
283	Fayetteville	.75	.75
284	Wilmington	.73	.74
285	Kinston	.67	.67
286	Hickory	.65	.66
287-288	Asheville	.72	.74
289	Murphy	.66	.67
NORTH DAKOTA			
580-581	Fargo	.80	.85
582	Grand Forks	.76	.81
583	Devils Lake	.78	.83
584	Jamestown	.75	.80
585	Bismarck	.81	.85

Location Factors

STATE/ZIP	CITY	Residential	Commercial
NORTH DAKOTA (CONT'D)			
586	Dickinson	.81	.84
587	Minot	.83	.87
588	Williston	.79	.83
OHIO			
430-432	Columbus	.98	.96
433	Marion	.93	.94
434-436	Toledo	1.02	1.01
437-438	Zanesville	.93	.92
439	Steubenville	.97	.97
440	Lorain	1.03	.97
441	Cleveland	1.09	1.03
442-443	Akron	1.00	.99
444-445	Youngstown	1.00	.97
446-447	Canton	.96	.95
448-449	Mansfield	.96	.94
450	Hamilton	1.00	.94
451-452	Cincinnati	1.00	.94
453-454	Dayton	.93	.92
455	Springfield	.95	.93
456	Chillicothe	1.02	.96
457	Athens	.91	.90
458	Lima	.95	.94
OKLAHOMA			
730-731	Oklahoma City	.81	.83
734	Ardmore	.82	.81
735	Lawton	.84	.83
736	Clinton	.79	.81
737	Enid	.82	.81
738	Woodward	.81	.81
739	Guymon	.68	.67
740-741	Tulsa	.84	.81
743	Miami	.85	.82
744	Muskogee	.74	.72
745	Mcalester	.75	.76
746	Ponca City	.81	.80
747	Durant	.78	.80
748	Shawnee	.78	.80
749	Poteau	.84	.80
OREGON			
970-972	Portland	1.07	1.05
973	Salem	1.05	1.04
974	Eugene	1.05	1.04
975	Medford	1.05	1.04
976	Klamath Falls	1.05	1.04
977	Bend	1.05	1.04
978	Pendleton	1.02	1.00
979	Vale	.98	.96
PENNSYLVANIA			
150-152	Pittsburgh	1.02	1.00
153	Washington	1.00	.98
154	Uniontown	.99	.97
155	Bedford	1.01	.94
156	Greensburg	1.00	.98
157	Indiana	1.03	.96
158	Dubois	1.02	.95
159	Johnstown	1.03	.96
160	Butler	1.00	.97
161	New Castle	1.00	.97
162	Kittanning	1.01	.98
163	Oil City	.89	.94
164-165	Erie	.97	.96
166	Altoona	1.03	.95
167	Bradford	.97	.96
168	State College	.95	.95
169	Wellsboro	.92	.93
170-171	Harrisburg	.96	.95
172	Chambersburg	.94	.93
173-174	York	.96	.94
175-176	Lancaster	.94	.92
177	Williamsport	.89	.88
178	Sunbury	.94	.93
179	Pottsville	.93	.92
180	Lehigh Valley	1.02	1.01
181	Allentown	1.02	1.01
182	Hazleton	.96	.96
183	Stroudsburg	.99	.98
184-185	Scranton	.94	.97
186-187	Wilkes-Barre	.92	.95
188	Montrose	.92	.95
189	Doylestown	.94	1.05

STATE/ZIP	CITY	Residential	Commercial
PENNSYLVANIA (CONT'D)			
190-191	Philadelphia	1.13	1.11
193	Westchester	1.07	1.06
194	Norristown	1.09	1.07
195-196	Reading	.95	.96
PUERTO RICO			
009	San Juan	.85	.85
RHODE ISLAND			
028	Newport	1.00	1.02
029	Providence	1.00	1.02
SOUTH CAROLINA			
290-292	Columbia	.71	.74
293	Spartanburg	.70	.72
294	Charleston	.72	.74
295	Florence	.70	.72
296	Greenville	.69	.72
297	Rock Hill	.63	.66
298	Aiken	.81	.85
299	Beaufort	.67	.69
SOUTH DAKOTA			
570-571	Sioux Falls	.87	.81
572	Watertown	.83	.78
573	Mitchell	.83	.77
574	Aberdeen	.86	.80
575	Pierre	.85	.79
576	Mobridge	.84	.77
577	Rapid City	.84	.78
TENNESSEE			
370-372	Nashville	.86	.86
373-374	Chattanooga	.81	.80
375,380-381	Memphis	.86	.86
376	Johnson City	.81	.80
377-379	Knoxville	.79	.79
382	Mckenzie	.76	.76
383	Jackson	.71	.78
384	Columbia	.78	.78
385	Cookeville	.76	.76
TEXAS			
750	Mckinney	.87	.80
751	Waxahackie	.81	.81
752-753	Dallas	.89	.85
754	Greenville	.77	.72
755	Texarkana	.86	.76
756	Longview	.83	.73
757	Tyler	.90	.79
758	Palestine	.71	.71
759	Lufkin	.74	.74
760-761	Fort Worth	.83	.82
762	Denton	.86	.77
763	Wichita Falls	.79	.79
764	Eastland	.72	.71
765	Temple	.76	.75
766-767	Waco	.80	.79
768	Brownwood	.71	.70
769	San Angelo	.78	.74
770-772	Houston	.87	.88
773	Huntsville	.72	.72
774	Wharton	.74	.75
775	Galveston	.85	.86
776-777	Beaumont	.80	.82
778	Bryan	.80	.81
779	Victoria	.76	.76
780	Laredo	.75	.76
781-782	San Antonio	.80	.81
783-784	Corpus Christi	.78	.77
785	Mc Allen	.76	.74
786-787	Austin	.77	.80
788	Del Rio	.66	.66
789	Giddings	.71	.70
790-791	Amarillo	.80	.80
792	Childress	.74	.77
793-794	Lubbock	.77	.78
795-796	Abilene	.78	.78
797	Midland	.76	.77
798-799,885	El Paso	.78	.77
UTAH			
840-841	Salt Lake City	.91	.90
842,844	Ogden	.90	.88

Location Factors

STATE/ZIP	CITY	Residential	Commercial
UTAH (CONT'D)			
843	Logan	.91	.89
845	Price	.80	.79
846-847	Provo	.90	.89
VERMONT			
050	White River Jct.	.77	.76
051	Bellows Falls	.77	.76
052	Bennington	.77	.76
053	Brattleboro	.77	.77
054	Burlington	.85	.86
056	Montpelier	.83	.84
057	Rutland	.86	.85
058	St. Johnsbury	.74	.75
059	Guildhall	.73	.74
VIRGINIA			
220-221	Fairfax	.89	.90
222	Arlington	.89	.90
223	Alexandria	.91	.92
224-225	Fredericksburg	.83	.83
226	Winchester	.77	.78
227	Culpeper	.78	.79
228	Harrisonburg	.74	.75
229	Charlottesville	.83	.81
230-232	Richmond	.86	.84
233-235	Norfolk	.82	.82
236	Newport News	.82	.81
237	Portsmouth	.81	.81
238	Petersburg	.86	.84
239	Farmville	.73	.72
240-241	Roanoke	.76	.75
242	Bristol	.79	.74
243	Pulaski	.72	.71
244	Staunton	.75	.74
245	Lynchburg	.80	.76
246	Grundy	.71	.71
WASHINGTON			
980-981,987	Seattle	.98	1.03
982	Everett	.95	1.01
983-984	Tacoma	1.04	1.02
985	Olympia	1.04	1.02
986	Vancouver	1.09	1.03
988	Wenatchee	.92	.96
989	Yakima	1.00	.98
990-992	Spokane	.98	.97
993	Richland	.99	.98
994	Clarkston	.97	.96
WEST VIRGINIA			
247-248	Bluefield	.89	.89
249	Lewisburg	.91	.91
250-253	Charleston	.95	.95
254	Martinsburg	.83	.83
255-257	Huntington	.93	.95
258-259	Beckley	.93	.93
260	Wheeling	.93	.95
261	Parkersburg	.92	.94
262	Buckhannon	.97	.94
263-264	Clarksburg	.97	.94
265	Morgantown	.97	.94
266	Gassaway	.94	.94
267	Romney	.91	.91
268	Petersburg	.94	.92
WISCONSIN			
530,532	Milwaukee	1.02	1.01
531	Kenosha	1.01	1.00
534	Racine	1.05	1.00
535	Beloit	1.01	.98
537	Madison	1.00	.98
538	Lancaster	.96	.94
539	Portage	.98	.96
540	New Richmond	1.04	.96
541-543	Green Bay	1.00	.97
544	Wausau	.99	.94
545	Rhinelander	1.00	.96
546	La Crosse	.98	.95
547	Eau Claire	1.04	.96
548	Superior	1.04	.98
549	Oshkosh	.98	.95

STATE/ZIP	CITY	Residential	Commercial
WYOMING			
820	Cheyenne	.84	.80
821	Yellowstone Nat. Pk.	.79	.76
822	Wheatland	.81	.77
823	Rawlins	.80	.76
824	Worland	.77	.75
825	Riverton	.80	.77
826	Casper	.85	.81
827	Newcastle	.78	.74
828	Sheridan	.82	.79
829-831	Rock Springs	.82	.77
CANADIAN FACTORS (reflect Canadian currency)			
ALBERTA			
	Calgary	1.00	.97
	Edmonton	1.00	.97
	Fort McMurray	.99	.96
	Lethbridge	.99	.96
	Lloydminster	.99	.96
	Medicine Hat	.99	.96
	Red Deer	.99	.96
BRITISH COLUMBIA			
	Kamloops	1.02	1.03
	Prince George	1.03	1.04
	Vancouver	1.05	1.06
	Victoria	1.04	1.05
MANITOBA			
	Brandon	.97	.96
	Portage la Prairie	.96	.96
	Winnipeg	.97	.96
NEW BRUNSWICK			
	Bathurst	.93	.91
	Dalhousie	.93	.91
	Fredericton	.96	.94
	Moncton	.93	.91
	Newcastle	.93	.91
	Saint John	.96	.94
NEWFOUNDLAND			
	Corner Brook	.95	.94
	St. John's	.95	.94
NORTHWEST TERRITORIES			
	Yellowknife	.92	.91
NOVA SCOTIA			
	Dartmouth	.96	.95
	Halifax	.96	.95
	New Glasgow	.96	.95
	Sydney	.94	.93
	Yarmouth	.96	.95
ONTARIO			
	Barrie	1.08	1.06
	Brantford	1.09	1.07
	Cornwall	1.07	1.05
	Hamilton	1.11	1.07
	Kingston	1.08	1.06
	Kitchener	1.04	1.03
	London	1.08	1.06
	North Bay	1.07	1.05
	Oshawa	1.08	1.06
	Ottawa	1.08	1.06
	Owen Sound	1.08	1.06
	Peterborough	1.07	1.05
	Sarnia	1.10	1.08
	St. Catharines	1.04	1.02
	Sudbury	1.04	1.02
	Thunder Bay	1.04	1.03
	Toronto	1.11	1.10
	Windsor	1.05	1.03
PRINCE EDWARD ISLAND			
	Charlottetown	.92	.90
	Summerside	.92	.90

Location Factors

STATE/ZIP	CITY	Residential	Commercial
QUEBEC			
	Cap-de-la-Madeleine	1.02	1.01
	Charlesbourg	1.02	1.01
	Chicoutimi	1.01	1.00
	Gatineau	1.01	1.00
	Laval	1.02	1.01
	Montreal	1.07	1.01
	Quebec	1.09	1.02
	Sherbrooke	1.02	1.01
	Trois Rivieres	1.02	1.01
SASKATCHEWAN			
	Moose Jaw	.91	.91
	Prince Albert	.91	.91
	Regina	.92	.92
	Saskatoon	.91	.91
YUKON			
	Whitehorse	.92	.91

Abbreviations

A	Area
ASTM	American Society for Testing and Materials
B.F.	Board feet
Carp.	Carpenter
C.F.	Cubic feet
CWJ	Composite wood joist
C.Y.	Cubic yard
Ea.	Each
Equip.	Equipment
Exp.	Exposure
Ext.	Exterior
F	Fahrenheit
Ft.	Foot, feet
Gal.	Gallon
Hr.	Hour
in.	Inch, inches
Inst.	Installation
Int.	Interior
Lb.	Pound
L.F.	Linear feet
LVL	Laminated veneer lumber
Mat.	Material
Max.	Maximum
MBF	Thousand board feet
MBM	Thousand feet board measure
MSF	Thousand square feet
Min.	Minimum
O.C.	On center
O&P	Overhead and profit
OWJ	Open web wood joist
Oz.	Ounce
Pr.	Pair
Quan.	Quantity
S.F.	Square foot
Sq.	Square, 100 square feet
S.Y.	Square yard
V.L.F.	Vertical linear feet
′	Foot, feet
″	Inch, inches
°	Degrees

A

A/C junction box 154, 155, 161
Accessories 129, 177, 254, 255
  components 84
  extras 75
Accordion type 196
Acrylic impregnated 240
  whirlpool 254
Add for additional story 71
  for steep pitch 70
Adhesive tile 187
Adobe 138
Air compressor 4
  handler 159
  vent system negative 13
Airless sprayer 13
Allowance to repair 164
Aluminum .. 64, 67, 95, 97, 131, 282, 289
  standard 285
  valley 68
Antenna television 269
Architectural 70
Asbestos ceiling encapsulate 12
Asphalt 83
Average grade 240, 245
  room 6
Awning sash 106

B

Back bar 226
Backfill 17
  trenching 17
Bar 250, 252
  back 226
  joist 45
  liquor 226
  towel 280
Barge rafter 63
Barrier vapor ... 26, 46, 166, 168, 170
Base moulding vinyl 201
  oak 202
  pad install 127
Baseboard electric 156
Basic grade 276, 277
Basket drain 253
Bathroom exhaust 266
Batt unfaced 165, 167, 169
Bay 109
Beam collar 52
Beams glue-laminated 42
Bell / chimes door 268
Bench seating 58, 59, 289
Berber 244, 245
Better grade 98, 182
  grade wood 210, 216, 221, 229
Bidet 249
Bifold door 194
Black corrugated 76
  flat 77
  weathering 73
Blend color 75
Blinds 271
Block bridge 31
  butcher 228
  glass 179
  w / footing 138
  w / o footing 139
Blown 167, 169, 188
Blue corrugated 77
  flat 77
  glazed 75
Board ledger 53
  particle 33, 205
  plank 34
  rigid 166, 168-170
Boards ledger 32, 49
Bobcat 4
Boiler 154
Bookcase built-in 227
  custom 205
Border wallcovering 185
Bow casement 114
Bowstring truss 56
Box fire 128
Bracing cross 21
Brand w / grill premium 274
Breast (face) brick 128
Brick common 127
  deluxe 138
  standard 138
  structural 124
  veneer 124
Bridge block 31
Bright red corrugated 76
  red flat 77
Bucket concrete 5
Built-in bookcase 227
  desk 227
Bullet proof 2″ 117
Butcher block 228

C

Cabinet medicine 215
  wall mounted microwave 276
Canopy cover cloth 286
Cant strips 81
Canvas 285
Cap metal 127
  vent 146
Carpenter 2
Carpeted 202
Carpeting 247
  per s.f. 243
Cased opening double 203
  opening single 203
Casement 113
  bow 114
Casing per l.f. window 204
  trim 87, 89, 92, 192, 196, 199
Caulk renail 202
Cedar 57, 58, 62, 65
  wood shingles 71
Ceiling 262
  asbestos seal 13
  framing 47
  grid type 265
  paddle 266
  plywood 59
  recessed 265
  texture 235
Ceramic tile 202, 228
Chain link 140
Chair rail 202
Chemical toilet 3
Cherry 239
Chest-type 279
Chloride (PVC) polyvinyl 81
Clad vinyl 96
Cleaning maintenance 126
Clear 1/4″ plate 116
  glass 118
  plate glass 118
Clips hurricane 36, 49
Closet shelf and rod 205
Cloth canopy cover 286
Clothing rod 206
CMU reinforced 123
  unreinforced 123
Coat seal 137
  thin 173, 175
  urethane 238
Collar beam 52
Color blend 75
Commercial grade 90
Common brick 127
Compactor trash 279
Complete panel 163
  system 185
Components 145
  accessories 84
Compressor 157
  air 4
Concealed 161
Concrete 29
  bucket 5
  mixer 4
  natural 137
  pad 161
  precast 20
  pump 5
  steps 273
Condensing unit 158
Container storage 3
Contents store 6
Continuous 24
Cooler evaporative 159
Coping wall cap 68
Copper 68, 80, 282
Cork 179
Corner 203
Corrugated 77, 79
  black 76
  blue 77
  bright red 76
  gray 76
  green 76
  red / brown 76
Cotta red terra 73
Countertop microwave 276
Cove 203
Cover protect floors 6, 233, 234
  protect walls 6, 233, 234
Cross bracing 21
Cross-bracing 23
Crown 203
Cultured marble 228
  marble top 232
Cupolas 287
Curved decorative 119
  hardwood 206
Custom bookcase 205
  grade 289
Cycle 4 279

D

Debris hauling 7
Decontamination unit 13
Decorative curved 119
  type 119
Deluxe brick 138
  grade 262
Designer 184
  floor mount 249
Desk built-in 227
Detector smoke 267
Diffuser 257
Dimmer switch only 261
Dish soap 280
Dispenser 280
Disposal garbage 279
Dome 120
Door 38
  bell / chimes 268
  bifold 194
  garage 98
  only hc 191
  only (sc) 91
  only steel 88
Double 204, 250, 252
  cased opening 203
  hung 110, 111
  hung two lite 104
  switch only 260
  wall oven 275
Douglas fir 65
Drain basket 253
Drapery 271
Drip edge 68
Drop-in type w / range 275
Dryer combination unit washer .. 277
Drywaller 2
Dump fees 7
  truck 4

E

Economy grade 240, 245
Edge drip 68
Electric 147
  baseboard 156
  forced air 153
  opener 100
Electrician 2
Elevator 273
Emergency 266
  repairs 83
Enamel finish 288
Enameled 253
Encapsulate asbestos ceiling 12
Enclosure screen 290
End gable 54
  wall 68
Engineered lumber joist 31
  lumber truss / joist 55
Entrance 119
Entry 85
  w / lites 85
EPDM type 82
Epoxy inject 25
Eq strapping 33, 36, 49
Etched 117
Evaporative cooler 159
Exhaust bathroom 266
  kitchen 267
Existing refinish 238
Exposed 161
Exterior 263, 283
  sheathing 45
  siding 131
  type 89
Extinguisher fire 281
Extra large room 6

F

Face brick breast 128
Faced batt foil 165, 167, 168
  batt kraft 165, 166, 168
Fascia 62
Faucet sink 253
Fees dump 7
Felt underlay 34
Fencing 3
Fiberglass 170, 253, 255

3-tab ... 69
flexible ... 149
sheet ... 79
Filter system ... 290
Finish enamel ... 288
natural ... 288
new sand ... 238
plywood ... 46
Finished ... 175
Fir douglas ... 65
Fire box ... 128
extinguisher ... 281
Fireblock ... 35
Fixed ... 121
100% ... 122
Flat black ... 77
blue ... 77
bright red ... 77
green ... 77
natural gray ... 77
red / brown ... 77
Flatbed truck ... 3
Flexible fiberglass ... 149
Floor ... 251, 252
asbestos seal ... 13
framing system ... 29
gas-fired ... 153
mount designer ... 249
mounted w / tank ... 249
Florentine obscure ... 117
Fluorescent ... 263, 264
Flush valve wall ... 249
Flush-out sanitize ... 258
Fly rafter ... 53
Foam rubber slab ... 246
silicone-urethane ... 83
Foil ... 185
faced batt ... 165, 167, 168
Footings ... 17
Forced air electric ... 153
air gas ... 153
Forklift ... 4
Foundation ... 287
post wood ... 22
Framed wood ... 96
Framing ceiling ... 47
roof ... 49
system floor ... 29
wall ... 34
Free standing ... 275
Freezer ... 278
French style ... 86
Frieze ... 287
Full ... 197
Furnace unit ... 152
Furring ... 126
strips ... 177, 181, 188

G

Gable end ... 54
louvered ... 287
Galvanized ... 67
spiral ... 149
square or rectangular ... 148
Garage ... 216
door ... 98
storage ... 227
Garbage disposal ... 279
Gas forced air ... 153
suspension ... 151
Gas-fired ... 147
floor ... 153
Gate ... 142
General laborer ... 2
Generator ... 5
Glass ... 120, 205
block ... 179
clear ... 118
louvered ... 107
Glazed ... 73
blue ... 75
red ... 75
white ... 75
Glue-laminated beams ... 42
Good grade ... 98, 182, 276, 277, 289
grade laminated 209, 216, 220, 229
Grab rail ... 280
Grade average ... 240, 245
basic ... 276, 277
better ... 98, 182
commercial ... 90
custom ... 289
deluxe ... 262
economy ... 240, 245
good ... 98, 182, 276, 277, 289
high ... 262
laminated good . 209, 216, 220, 229
premium .. 98, 183, 241, 245, 276, 277
wood better ... 210, 216, 221, 229
wood premium . 212, 216, 222, 230
Granite ... 242
Grasscloth string ... 184
Gravel stop ... 68
Gray corrugated ... 76
Green corrugated ... 76
flat ... 77
unfading ... 72
weathering ... 73
Grey / black unfading ... 72
Grid t-bar ... 186
type ceiling ... 265
Grill return ... 256
supply ... 257
Grm membrane ... 81
Grout pressure ... 26
Gunite ... 25
Gutters ... 282

H

H section steel ... 22
Half ... 197
Hand (fine) ... 137
Handler air ... 159
Hardboard ... 180
Hardware ... 44, 101, 110, 192
truss ... 55
Hardwood curved ... 206
oak ... 205
premium ... 213, 216, 224, 231
spiral ... 206
straight ... 206
Hauling debris ... 7
Hc door only ... 191
Headers ... 37
Hearth raised ... 128
Heat pump ... 155
Heater cap water ... 287
motor ... 290
space ... 6
Heavy weight ... 120
Hem-fir std & better ... 63
Hepa vacuum cleaner ... 13
High grade ... 262
Hi-lo range double oven ... 276
range microwave ... 276
Holder toothbrush ... 280
Hole in gypsum wallboard patch 177
Hot water baseboard ... 151
water system solar ... 148
House exhaust whole ... 267
Hung double ... 110, 111
two lite double ... 104
two lite single ... 102
Hurricane clips ... 36, 49

I

Ice maker ... 278
Impregnated acrylic ... 240
walk ... 84
Indoor-outdoor ... 244, 245
Inject epoxy ... 25
Install base pad ... 127
from roof up ... 127
new (w / flue) ... 127
Institutional ... 254
Insulated ... 98, 116
Insulating wrap ... 148
Intercom system ... 268
Interior ... 173, 174
wood ... 284
Iron wrought ... 142
Island ... 225
I-type steel ... 28

J

Jack relevel ... 7
roof ... 70
with wiring phone ... 260
Jackhammer ... 4
Jamb ... 87, 193, 196, 197, 199
metal ... 101
Joist bar ... 45
Junction box A/C ... 154, 155, 161
Jute ... 246, 247

K

Kitchen exhaust ... 267
Kitchenette unit ... 279
Kraft faced batt ... 165, 166, 168

L

Laborer general ... 2
Laminated ... 227
safety ... 117
top ... 232
Large ... 116
load ... 7
roof turbine ... 286
room ... 6
room extra ... 6
Lath ... 174, 175
Laundry ... 251
Lavatory vanity ... 251, 252
Ledger board ... 53
boards ... 32, 49
strips ... 32, 48
Let-ins ... 36
Light stand ... 13
Lighting track ... 264, 266
Lightweight ... 25
Link chain ... 140
Liquor bar ... 226
Load large ... 7
medium ... 7
small ... 7
Lookout rafter ... 53
Louvered ... 195
gable ... 287
glass ... 107
Lumber joist engineered ... 31
stock ... 61, 62
truss / joist engineered ... 55

M

Machine (fine) ... 137
Maintenance cleaning ... 126
Maker ice ... 278
Maple ... 237
Marble ... 179, 204, 241
cultured ... 228
top cultured ... 232
Marlite ... 180
Mason ... 2
Masonite pre-hung ... 190
Masthead ... 164
Medicine cabinet ... 215
Medium load ... 7
Membrane grm ... 81
Metal ... 273
cap ... 127
jamb ... 101
pipe ... 21
steel roll up ... 99
stud framing ... 44
support posts ... 288
Microwave countertop ... 276
Mirror ... 179
wall ... 117
Mixer concrete ... 4
Motor heater ... 290
pump ... 290
Moulding ... 181
shingle ... 70
Mounted w / tank floor ... 249
window ... 160
Mural ... 183

N

Natural concrete ... 137
finish ... 288
gray flat ... 77
stone ... 128
Negative air vent system ... 13
Neoprene ... 81
New (w / flue) install ... 127

O

Oak ... 237, 239
base ... 202
hardwood ... 205
prefinished ... 238
Obscure florentine ... 117
Office trailer ... 3
One piece ... 199
piece oak ... 200
Only (SC) door ... 91
Opener electric ... 100
Opening rough-in ... 40
Operable ... 121
or fink truss w ... 54
OSB ... 46, 56
Outriggers ... 53
Oven single ... 217

P

Pad concrete 161
Paddle ceiling 266
Painter 2
Pair of wood 198
Panel complete 163
Panels solar 290
  tile 186
Paper roller toilet 280
Particle board 33, 205
Partition 281
Patch hole in gypsum wallboard 177
Paving with 1" cap top 137
Pedestal 251, 252
Perlite 171
Phone jack with wiring 260
Picture 203
  window wood 115
Piece oak one 200
  one 199
Pine 204, 238
Pipe metal 21
Plank 57
  board 34
Plastic 179, 282
Plate clear 1/4" 116
  glass clear 118
  glass polished 118
  per l.f. sill 33
Plates 37
Plumber 2
Plywood 33, 56, 62, 64, 180, 205
  ceiling 59
  finish 46
Pole power 3
Polished plate glass 118
Polystyrene 171
Polyvinyl chloride (PVC) 81
Post 142
Power pole 3
Precast concrete 20
Prefabricated 128
Prefinished 201
  oak 238
Pre-hung masonite 190
  steel door 100
  steel-clad 88
  wood 189, 193
Premium brand w / grill 274
  grade 98, 183, 241, 245, 276, 277
  grade wood 212, 216, 222, 230
  hardwood 213, 216, 224, 231
Prep (for flooring) 34
  surface 182
Pressure grout 26
Privacy wall 58, 59
Proof 2" bullet 117
Protect floors cover 6, 233, 234
  walls cover 6, 233, 234
Pump concrete 5
  heat 155
  motor 290
  sump 146
  water 290
Purlins 52
Purple unfading 72
  variegated 72

Q

Quarry tile 202, 228
Quarter round 202

R

Radiant wall type 157
Rafter 50
  barge 63
  fly 53
  lookout 53
Rail chair 202
  grab 280
Railing 3, 58, 59
Raised hearth 128
Range double oven hi-lo 276
  microwave hi-lo 276
Rebound 246, 247
Recessed ceiling 265
Red / brown corrugated 76
  / brown flat 77
  glazed 75
  unfading 72
  unglazed 75
Redwood 57, 58, 62, 65
Refinish existing 238
  strip 216, 219, 227, 232
Refit shave 97, 193, 196, 198, 199
Refrigerator 277
Re-grout 178, 180, 241, 242
Regular weight 120
Reinforced 24
  cmu 123
Relevel jack 7
Renail caulk 202
Repairs 174, 175
  emergency 83
Replace slate tiles 72
Replaster 290
Re-screen 59, 289
Residential 163, 259
Return grill 256
Ribbed 78, 80
Ridge vent 70, 287
Ridgeboard 51
Rigid board 166, 168-170
Rod clothing 206
Rolled roofing 69
Roof framing 49
  jack 70
  turbine large 286
  turbine small 286
  ventilator 287
Roofer 2
Roofing rolled 69
Room average 6
  large 6
  small 6
Rough-in 250-252, 255, 256
  opening 40
Round quarter 202
Rubber slab foam 246
Run wiring 266

S

Safety laminated 117
Sand finish new 238
Sandblaster 5
Sanitize flush-out 258
Sash awning 106
  sliding 105
Satellite system 269
Scaffolding 7
SCPE type 82
Screen enclosure 290
  sliding 97
Seal ceiling asbestos 13
  coat 137
  floor asbestos 13
  wall asbestos 13
Seat 250
Seating bench 58, 59, 289
Security system 269
Self-contained 157
Service water 3
Set of mirrored 198
  paver thick 241
set thick 178
  thin 178
Set trim 110, 115
Shave refit 97, 193, 196, 198, 199
Sheathing exterior 45
Sheet fiberglass 79
  vinyl 243
  vinyl per S.F. 242
Shelf and rod closet 205
Shim support piers 6
Shingle moulding 70
Shoe 201, 202
Shower combination tub 254
Shutter wood 272
Side wall 68
Siding exterior 131
Silicone-urethane foam 83
Sill plate per L.F. 33
Single 203, 250, 251
  cased opening 203
  hung two lite 102
  oven 217
  switch only 260
  wall oven 275
Sink faucet 253
Size standard 116
Skip-type 57
Slate 242
  tiles replace 72
Sliding sash 105
  screen 97
Slumpstone 125
  veneer 128
Small 116
  load 7
  roof turbine 286
  room 6
Smoke detector 267
Soap dish 280
Softener water 148
Solar hot water system 148
  panels 290
Solid surface 228, 232
  vinyl 183
  wood beam 27
Space heater 6
Spiral galvanized 149
  hardwood 206
Sprayer airless 13
Springs 100
Square or rectangular galvanized 148
Stain 196, 198, 199, 219, 227
Stained 117
Stainless steel 81
Stairs 47, 57, 58
Stand light 13
Standard aluminum 285
  brick 138
  size 116
Standing free 275
Stapled tile 187
Std & better hem-fir 63
Steel door only 88
  door pre-hung 100
  H section 22
  I-type 28
  roll up metal 99
  stainless 81
Steel-clad pre-hung 88
Stem wall 24
Steps concrete 273
Stock lumber 61, 62
Stone natural 128
Stop gravel 68
Storage container 3
  garage 227
Store contents 6
Storm window 110
Straight hardwood 206
Strapping eq 33, 36, 49
String grasscloth 184
Strip refinish 216, 219, 227, 232
Strips cant 81
  furring 177, 181, 188
  ledger 32, 48
Structural brick 124
Stucco 126
Stud 36
  framing metal 44
Style french 86
Sump pump 146
Supply grill 257
Support piers shim 6
  posts metal 288
  wire 166
Surface prep 182
  solid 228, 232
Suspension gas 151
Switch only dimmer 261
  only double 260
  only single 260
  only triple 260
System complete 185
  filter 290
  intercom 268
  satellite 269
  security 269

T

T-bar grid 186
Teak 240
Telephone 3
Television antenna 269
Tempered 117
Terra cotta red 73
Terrazzo 241
Texture 177
  ceiling 235
  walls 234
Thermostat 155, 157, 161
Thick set 178
  set paver 241
Thin coat 173, 175
  set 178
Tile adhesive 187
  ceramic 202, 228
  panels 186
  quarry 202, 228
  stapled 187
  vinyl 243
Tinted 116
Toilet chemical 3
  paper roller 280
Toothbrush holder 280
Top laminated 232
  paving with 1" cap 137
Towel bar 280
Track lighting 264, 266
Trailer office 3
Trash compactor 279
Treated 57, 58

wood 12″ butt 19
Trenching backfill 17
Trim casing . 87, 89, 92, 192, 196, 199
set 110, 115
Triple 204
switch only 260
Truck dump 4
flatbed 3
Truss bowstring 56
hardware 55
Tub shower combination 254
TV / cable jack w / wiring 260
Two-story 7
Type accordion 196
decorative 119
EPDM 82
exterior 89
sCPE 82
w / range drop-in 275

U

Underlay felt 34
Underlayment 69
per S.F. 239
Unfaced batt 165, 167, 169
Unfading green 72
grey / black 72
purple 72
red 72
Unfinished 176
Unglazed red 75
white 75
Unit condensing 158
decontamination 13
furnace 152
kitchenette 279
Unreinforced 25
CMU 123
Untreated wood 12″ butt 19
Urethane 172, 246
coat 238
Urinal 249
Utility 218

V

Vacuum cleaner HEPA 13
Valley aluminum 68
Vane weather 288
Vanity lavatory 251, 252
Vapor barrier ... 26, 46, 166, 168, 170
Variegated purple 72
Veneer brick 124
slumpstone 128
Vent cap 146
ridge 70, 287
Vented 274
50% 122
Ventilator roof 287
Ventless 275
Vinyl 62, 64
base moulding 201
clad 96
sheet 243
solid 183
tile 243
Vinyl-coated 183

W

W or fink truss 54
Waffle 246
Walk impregnated 84
Wall asbestos seal 13
cap coping 68
end 68
flush valve 249
framing 34
mirror 117
mounted microwave cabinet .. 276
oven double 275
oven single 275
privacy 58, 59
side 68
stem 24
type radiant 157
Wallcovering border 185
Walls texture 234
Walnut 238, 239
Washer dryer combination unit .. 277
Water baseboard hot 151
heater cap 287
pump 290
service 3
softener 148
Weather vane 288
Weathering black 73
green 73
Weight heavy 120
regular 120
Welder 5
Whirlpool acrylic 254
White glazed 75
unglazed 75
Whole house exhaust 267
Window 41
casing per L.F. 204
mounted 160
storm 110
Wire support 166
Wired 117
Wiring run 266
with 261
With wiring 261
Wood 97, 204, 289
12″ butt treated 19
12″ butt untreated 19
beam solid 27
foundation post 22
framed 96
interior 284
picture window 115
pre-hung 189, 193
shingles cedar 71
shutter 272
Wool 100% 244, 245
Wrap insulating 148
Wrought iron 142

Notes

Notes

Notes

Notes

Notes

Notes

Notes

Notes

Notes

Reed Contruction Data

Reed Construction Data, a leading worldwide provider of total construction information solutions, is comprised of three main product groups designed specifically to help construction professionals advance their businesses with timely, accurate and actionable project, product, and cost data. Reed Construction Data is a division of Reed Business Information, a member of the Reed Elsevier plc group of companies.

The *Project, Product, and Cost & Estimating* divisions offer a variety of innovative products and services designed for the full spectrum of design, construction, and manufacturing professionals. Through it's *International* companies, Reed Construction Data's reputation for quality construction market data is growing worldwide.

Cost Information
R.S. Means, the undisputed market leader and authority on construction costs, publishes current cost and estimating information in annual cost books and on the CostWorks CD-ROM. R.S. Means furnishes the construction industry with a rich library of complementary reference books and a series of professional seminars that are designed to sharpen professional skills and maximize the effective use of cost estimating and management tools. R.S. Means also provides construction cost consulting for Owners, Manufacturers, Designers, and Contractors.

Project Data
Reed Construction Data provides complete, accurate and relevant project information through all stages of construction. Customers are supplied industry data through leads, project reports, contact lists, plans and specifications surveys, market penetration analyses and sales evaluation reports. Any of these products can pinpoint a county, look at a state, or cover the country. Data is delivered via paper, e-mail, CD-ROM or the Internet.

Building Product Information
The First Source suite of products is the only integrated building product information system offered to the commercial construction industry for comparing and specifying building products. These print and online resources include *CMD First Source,* CSI's SPEC-DATA™, CSI's MANU-SPEC™, CADBlocks®, and Manufacturer Catalogs. Written by industry professionals and organized using CSI's MasterFormat™, construction professionals use this information to make better design decisions.

CMD FirstSource.com combines Reed Construction Data's project, product and cost data with news and information from Reed Business Information's *Building Design & Construction* and *Consulting-Specifying Engineer,* this industry-focused site offers easy and unlimited access to vital information for all construction professionals.

International
BIMSA/Mexico provides construction project news, product information, cost-data, seminars and consulting services to construction professionals in Mexico. Its subsidiary, PRISMA, provides job costing software.

Byggfakta Scandinavia AB, founded in 1936, is the parent company for the leaders of customized construction market data for Denmark, Estonia, Finland, Norway and Sweden. Each company fully covers the local construction market and provides information across several platforms including subscription, ad-hoc basis, electronically and on paper.

Reed Construction Data Canada serves the Canadian construction market with reliable and comprehensive project and product information services that cover all facets of construction. Core services include: *BuildSource, BuildSpec, BuildSelect,* product selection and specification tools available in print and on the Internet; Building Reports, a national construction project lead service; CanaData, statistical and forecasting information; *Daily Commercial News,* a construction newspaper reporting on news and projects in Ontario; and *Journal of Commerce,* reporting news in British Columbia and Alberta.

Cordell Building Information Services, with its complete range of project and cost and estimating services, is Australia's specialist in the construction information industry. Cordell provides in-depth and historical information on all aspects of construction projects and estimation, including several customized reports, construction and sales leads, and detailed cost information among others.

For more information, please visit our Web site at www.cmdg.com.

Reed Construction Data Corporate Office
30 Technology Parkway South
Norcross, GA 30092-2912
(800) 322-6996
(800) 895-8661 (fax)
info@cmdg.com
www.cmdg.com

Reed Construction Data

Contractor's Pricing Guides

For more information visit Means Web Site at www.rsmeans.com

Means Mechanical Estimating Methods
New 3rd Edition

This guide assists you in making a review of plans, specs, and bid packages with suggestions for takeoff procedures, listings, substitutions and pre-bid scheduling. Includes suggestions for budgeting labor and equipment usage. Compares materials and construction methods to allow you to select the best option.

$64.95 per copy
Over 350 pages, illustrated, hardcover
Catalog No. 67294A

Contractor's Pricing Guide: Residential Square Foot Costs 2003

Now available in one concise volume, all you need to know to plan and budget the cost of new homes. If you are looking for a quick reference, the model home section contains costs for over 250 different sizes and types of residences, with hundreds of easily applied modifications. If you need even more detail, the Assemblies Section lets you build your own costs or modify the model costs further. Hundreds of graphics are provided, along with forms and procedures to help you get it right.

$39.95 per copy
Over 250 pages, illustrated, 8-1/2 x 11
Catalog No. 60323 ISBN 0-87629-688-6

Contractor's Pricing Guide: Residential Detailed Costs 2003

Every aspect of residential construction, from overhead costs to residential lighting and wiring, is in here. All the detail you need to accurately estimate the costs of your work with or without markups–labor-hours, typical crews and equipment are included as well. When you need a detailed estimate, this publication has all the costs to help you come up with a complete, on the money, price you can rely on to win profitable work.

$38.95 per copy
Over 300 pages, with charts and tables, 8-1/2 x 11
Catalog No. 60333 ISBN 0-87629-686-X

Contractor's Pricing Guide: Residential Repair & Remodeling Costs 2003

This book provides total unit price costs for every aspect of the most common repair & remodeling projects. Organized in the order of construction by component and activity, it includes demolition and installation, cleaning, painting, and more.

With simplified estimating methods; clear, concise descriptions; and technical specifications for each component, the book is a valuable tool for contractors who want to speed up their estimating time, while making sure their costs are on target.

$38.95 per copy
Over 250 pages, illustrated, 8-1/2 x 11
Catalog No. 60343 ISBN 0-87629-687-8

Means Repair & Remodeling Estimating Methods
New 4th Edition

By Edward B. Wetherill and R.S. Means Engineering Staff

This updated edition focuses on the unique problems of estimating renovations in existing structures—using the latest cost resources and construction methods. The book helps you determine the true costs of remodeling, and includes:

Part I–The Estimating Process
Part II–Estimating by CSI Division
Part III–Two Complete Sample Estimates–Unit Price & Assemblies
Part IV–Disaster Reconstruction

$69.95 per copy
Over 450 pages, illustrated, hardcover
Catalog No. 67265B ISBN 0-87629-661-4

Means Landscape Estimating Methods New 4th Edition

By Sylvia H. Fee

Professional Methods for Estimating and Bidding Landscaping Projects and Grounds Maintenance Contracts

- Easy-to-understand text. Clearly explains the estimating process and how to use *Means Site Work & Landscape Cost Data*.
- Sample forms and worksheets to save you time and avoid errors.
- Tips on best techniques for saving money and winning jobs.
- **Updated cost and estimate examples** help you control your equipment costs and bid landscape maintenance projects.

$62.95 per copy
Over 300 pages, illustrated, hardcover
Catalog No. 67295B ISBN 0-87629-633-0

For more information visit Means Web Site at www.rsmeans.com

Annual Cost Guides

Means Building Construction Cost Data 2003

Offers you unchallenged unit price reliability in an easy-to-use arrangement. Whether used for complete, finished estimates or for periodic checks, it supplies more cost facts better and faster than any comparable source. Over 23,000 unit prices for 2003. The City Cost Indexes now cover over 930 areas, for indexing to any project location in North America.

$104.95 per copy
Over 700 pages, softcover
Catalog No. 60013 ISBN 0-87629-665-7

Means Open Shop Building Construction Cost Data 2003

The open-shop version of the *Means Building Construction Cost Data*. More than 22,000 reliable unit cost entries based on open shop trade labor rates. Eliminates time-consuming searches for these prices. The first book with open shop labor rates and crews. Labor information is itemized by labor-hours, crew, hourly/daily output, equipment, overhead and profit. For contractors, owners and facility managers.

$104.95 per copy
Over 680 pages, softcover
Catalog No. 60153 ISBN 0-87629-668-1

Means Plumbing Cost Data 2003

Comprehensive unit prices and assemblies for plumbing, irrigation systems, commercial and residential fire protection, point-of-use water heaters, and the latest approved materials. This publication and its companion, *Means Mechanical Cost Data*, provide full-range cost estimating coverage for all the mechanical trades.

$104.95 per copy
Over 570 pages, softcover
Catalog No. 60213 ISBN 0-87629-681-9

Means Residential Cost Data 2003

Speeds you through residential construction pricing with more than 100 illustrated complete house square-foot costs. Alternate assemblies cost selections are located on adjoining pages, so that you can develop tailor-made estimates in minutes. Complete data for detailed unit cost estimates is also provided.

$92.95 per copy
Over 600 pages, softcover
Catalog No. 60173 ISBN 0-87629-683-5

Means Electrical Cost Data 2003

Pricing information for every part of electrical cost planning: unit and systems costs with design tables; engineering guides and illustrated estimating procedures; complete labor-hour, materials, and equipment costs for better scheduling and procurement. With the latest products and construction methods used in electrical work. More than 15,000 unit and systems costs, clear specifications and drawings.

$104.95 per copy
Over 480 pages, softcover
Catalog No. 60033 ISBN 0-87629-672-X

Means Repair & Remodeling Cost Data 2003

Commercial/Residential

You can use this valuable tool to estimate commercial and residential renovation and remodeling. By using the specialized costs in this manual, you'll find it's not necessary to force fit prices for new construction into remodeling cost planning. Provides comprehensive unit costs, building systems costs, extensive labor data and estimating assistance for every kind of building improvement.

$92.95 per copy
Over 660 pages, softcover
Catalog No. 60043 ISBN 0-87629-682-7

Means Site Work & Landscape Cost Data 2003

Hard-to-find costs are presented in an easy-to-use format for every type of site work and landscape construction. Costs are organized, described, and laid out for earthwork, utilities, roads and bridges, as well as grading, planting, lawns, trees, irrigation systems, and site improvements.

$104.95 per copy
Over 630 pages, softcover
Catalog No. 60283 ISBN 0-87629-684-3

Means Light Commercial Cost Data 2003

Specifically addresses the light commercial market, which is an increasingly specialized niche in the industry. Aids you, the owner/designer/contractor, in preparing all types of estimates, from budgets to detailed bids. Includes new advances in methods and materials. Assemblies section allows you to evaluate alternatives in early stages of design/planning.

$92.95 per copy
Over 672 pages, softcover
Catalog No. 60183 ISBN 0-87629-679-7

For more information visit Means Web Site at www.rsmeans.com

Books for Builders

Builder's Essentials: Plan Reading & Material Takeoff

A complete course in reading and interpreting building plans—and performing quantity takeoffs to professional standards.

This book shows and explains, in clear language and with over 160 illustrations, typical working drawings encountered by contractors in residential and light commercial construction. The author describes not only how all common features are represented, but how to translate that information into a material list. Organized by CSI division, each chapter uses plans, details and tables, and a summary checklist.

$35.95 per copy
Over 420 pages, illustrated, softcover
Catalog No. 67307 ISBN 0-87629-348-8

Builder's Essentials: Best Business Practices for Builders & Remodelers: *An Easy-to-Use Checklist System*

By Thomas N. Frisby

A comprehensive guide covering all aspects of running a construction business, with more than 40 user–friendly checklists. This book provides expert guidance on: increasing your revenue and keeping more of your profit; planning for long-term growth; keeping good employees and managing subcontractors.

$29.95 per copy
Over 220 pages, softcover
Catalog No. 67329 ISBN 0-87629-619-3

Means Estimating Handbook

This comprehensive reference is for use in the field and the office. It covers a full spectrum of technical data required for estimating, with information on sizing, productivity, equipment requirements, codes, design standards and engineering factors. It will help you evaluate architectural plans and specifications; prepare accurate quantity takeoffs; perform value engineering; compare design alternatives; prepare estimates from conceptual to detail; evaluate change orders.

$99.95 per copy
Over 900 pages, hardcover
Catalog No. 67276 ISBN 0-87629-177-9

Builder's Essentials: Framing & Rough Carpentry, 2nd Edition

A complete, illustrated do-it-yourself course on framing and rough carpentry. The book covers walls, floors, stairs, windows, doors, and roofs, as well as nailing patterns and procedures. Additional sections are devoted to equipment and material handling, standards, codes, and safety requirements.

The "framer-friendly" approach includes easy-to-follow, step-by-step instructions. This practical guide will benefit both the carpenter's apprentice and the experienced carpenter, and sets a uniform standard for framing crews.
Also available in Spanish!

$24.95 per copy
Over 125 pages, illustrated, softcover
Catalog No. 67298A ISBN 0-87629-617-7
Spanish 67298AS ISBN 0-87629-654-1

Interior Home Improvement Costs, New 8th Edition

Updated, estimates for 66 interior projects, including:

- Attic/Basement Conversions
- Kitchen/Bath Remodeling
- Stairs, Doors, Walls/Ceilings
- Fireplaces
- Home Offices/In-law Apartments

$19.95 per copy
Over 230 pages, illustrated, softcover
Catalog No. 67308D ISBN 0-87629-656-8

Exterior Home Improvement Costs, New 8th Edition

Updated, quick estimates for 64 projects, including:

- Room Additions/Garages
- Roofing/Siding/Painting
- Windows/Doors
- Landscaping/Patios
- Porches/Decks

$19.95 per copy
Over 250 pages, illustrated, softcover
Catalog No. 67309D ISBN 0-87629-657-6

Practical Pricing Guides for Homeowners and Contractors

These updated resources on the cost and complexity of the nation's most popular home improvement projects include estimates of materials quantities, total project costs, and labor hours. With costs localized to over 900 zip code locations.

For more information
visit Means Web Site
at www.rsmeans.com

Books for Builders

From Model Codes to the IBC:
A *Transitional Guide*
By Rolf Jensen & Associates, Inc.

A time–saving resource for Architects, Engineers, Building Officials and Authorities Having Jurisdiction (AHJs), Contractors, Manufacturers, Building Owners, and Facility Managers.

Provides comprehensive, user-friendly guidance on making the transition to the International Building Code® from the model codes you're familiar with. Includes side-by-side code comparison of the IBC to the UBC, NBC, SBC, and NFPA 101®. Also features professional code commentary, quick-find indexes, and a Web site with regular code updates.

Also contains illustrations, abbreviations key, and an extensive resource section.

$114.95 per copy
880 pages, softcover
Catalog No. 67328

Historic Preservation
Project Planning & Estimating
by Swanke Hayden Connell Architects

Managing Historic Restoration, Rehabilitation, and Preservation Building Projects and Determining and Controlling Their Costs

The authors explain:

- How to determine whether a structure qualifies as historic
- Where to obtain funding and other assistance
- How to evaluate and repair more than 75 historic building materials
- How to properly research, document, and manage the project to meet code, agency, and other special requirements
- How to approach the upgrade of major building systems

$99.95 per copy
Over 675 pages, softcover
Catalog No. 67323

Builder's Essentials: Advanced Framing Methods
By Scot Simpson

A highly illustrated, "framer-friendly" approach to advanced framing elements. Provides expert, but easy-to-interpret, instruction for laying out and framing complex walls, roofs, and stairs, and special requirements for earthquake and hurricane protection. Also helps bring framers up to date on the latest building code changes, and provides tips on the lead framer's role and responsibilities, how to prepare for a job, and how to get the crew started.

$24.95 per copy
250 pages, illustrated, softcover
Catalog No. 67330

Unit Price Estimating Methods
New 3rd Edition

This new edition includes up-to-date cost data and estimating examples, updated to reflect changes to the CSI numbering system and new features of Means cost data. It describes the most productive, universally accepted ways to estimate, and uses checklists and forms to illustrate shortcuts and timesavers. A model estimate demonstrates procedures. A new chapter explores computer estimating alternatives.

$59.95 per copy
Over 350 pages, illustrated, hardcover
Catalog No. 67303A

Means Spanish/English Construction Dictionary
By R.S. Means, The International Conference of Building Officials (ICBO), and Rolf Jensen & Associates (RJA)

Designed to facilitate communication among Spanish- and English-speaking construction personnel—improving performance and job-site safety. Features the most common words and phrases used in the construction industry, with easy-to-follow pronunciations. Includes extensive building systems and tools illustrations.

$22.95 per copy
250 pages, illustrated, softcover
Catalog No. 67327

Means Illustrated Construction Dictionary, 3rd Edition

Long regarded as the Industry's finest, the Means Illustrated Construction Dictionary is now even better. With the addition of over 1,000 new terms and hundreds of new illustrations, it is the clear choice for the most comprehensive and current information.

The companion CD-ROM that comes with this new edition adds many extra features: larger graphics, expanded definitions, and links to both CSI MasterFormat numbers and product information.

Contains 19,000 construction words, terms, phrases, symbols, weights, measures, and equivalents. 1,000 new entries; 1,200 helpful illustrations, easy-to-use format with thumbtabs.

$99.95 per copy
Over 790 pages, illustrated, hardcover
Catalog No. 67292A

MeansData™

CONSTRUCTION COSTS FOR SOFTWARE APPLICATIONS

Your construction estimating software is only as good as your cost data.

A proven construction cost database is a mandatory part of any estimating package. The following list of software providers can offer you MeansData™ as an added feature for their estimating systems. See the table below for what types of products and services they offer (match their numbers). Visit them online at **www.rsmeans.com/demo/** for more information and free demos. Or call their numbers listed below.

1. **3D International**
   713-871-7000
   venegas@3di.com
2. **4Clicks-Solutions, LLC**
   719-574-7721
   mbrown@4clicks-solutions.com
3. **Aepco, Inc.**
   301-670-4642
   blueworks@aepco.com
4. **Applied Flow Technology**
   800-589-4943
   info@aft.com
5. **ArenaSoft Estimating**
   888-370-8806
   info@arenasoft.com
6. **ARES Corporation**
   925-299-6700
   sales@arescorporatio.com
7. **Benchmark Inc.**
   319-393-9100
   sales@benchmark-inc.com
8. **Beneco Enterprises, Inc.**
   801-565-1122
   inquire@beneco.com
9. **BSD - Building Systems Design, Inc.**
   888-273-7638
   bsd@bsdsoftlink.com
10. **Cadafis, Inc.**
    614-444-9851
    cadafis@rrohio.com
11. **Caelix, Inc.**
    602-787-8600
    kwebster@caelix.com
12. **CMS - Computerized Micro Solutions**
    800-255-7407
    cms@proest.com
13. **Conac Software**
    800-663-2338
    sales@conac.com
14. **Estimating Systems, Inc.**
    800-967-8572
    pulsar@capecod.net
15. **eCaliper Estimating Software, A Marathon Systems Service Business**
    877-469-3700
    sales@ecaliper.com
16. **G2 Estimator A Divi. of Valli Info.Systems, Inc.**
    800-627-3283
    info@g2estimator.com
17. **G/C Emuni**
    514-633-8339
    rpa@gcei.com
18. **Geac Computer Corp., LTD.**
    800-554-9865
    info@geac.com
19. **HardDollar**
    800-637-7496
    sales@harddollar.com
20. **LUQS International**
    888-682-5573
    info@luqs.com
21. **MAESTRO Estimator Schwaab Technology Solutions. Inc.**
    281-578-3039
    Stefan@schwaabtech.com.com
22. **Magellan K-12**
    936-447-1744
    sam.wilson@magellan-K12.com
23. **MC^2 - Management Computer Controls**
    800-225-5622
    vkeys@mc2-ice.com
24. **Maximus Asset Solutions**
    800-659-9001
    assetsolutions@maximus.com
25. **Prism Computer Corp.**
    800-774-7622
    famis@prismcc.com
26. **Quest Solutions, Inc.**
    800-452-2342
    info@questsolutions.com
27. **Sanders Software, Inc.**
    800-280-9760
    hsander@vallnet.com
28. **Timberline Software Corp.** 800-628-6583
    pruduct.info@timberline.com
29. **TMA SYSTEMS, Inc.**
    800-862-1130
    sales@tmasys.com
30. **US Cost, Inc.**
    800-372-4003
    sales@uscost.com
31. **Vanderweil Facility Advisors**
    617-451-5100
    info@VFA.com
32. **Vertigraph, Inc.**
    800-989-4243
    info-request@vertigraph.com
33. **Wendleware**
    714-895-7222
    sales@corecon.com
34. **WinEstimator, Inc.**
    800-950-2374
    sales@winest.com

TYPE:	1	2	3	4	5	6	7	8	9	10	11	12	13	14	15	16	17	18	19	20	21	22	23	24	25	26	27	28	29	30	31	32	33	34
BID									•				•				•		•				•			•		•				•		•
Estimating		•			•	•		•	•	•	•	•	•	•	•	•	•		•	•	•		•			•		•		•		•	•	•
DOC/JOC/SABER		•						•						•										•				•						•
IDIQ		•						•						•										•										•
Asset Mgmt										•								•				•		•	•						•			•
Facility Mgmt	•		•							•								•			•	•		•	•				•		•			
Project Mgmt	•					•		•			•		•		•				•		•	•						•					•	
TAKE-OFF					•						•				•	•			•	•			•			•		•		•		•	•	•
EARTHWK																			•	•			•			•						•	•	
Pipe flow				•																														
HVAC/Plumbing																				•							•							
Roofing							•			•																								
Design	•							•																				•						•
Other Offers/Links:																																		
Accounting/HR								•					•						•									•						
Scheduling															•	•			•		•							•		•		•		•
CAD													•		•				•			•						•						•
PDA										•																		•		•				•
Lt Versions																												•		•				•
Consulting	•	•								•					•		•				•			•	•			•		•	•			•
Training		•						•	•	•				•	•	•		•			•			•	•	•		•	•	•	•	•	•	•

Reseller applications now being accepted. Call Carol Polio Ext. 5107.

FOR MORE INFORMATION

CALL 1-800-448-8182, EXT. 5107 OR FAX 1-800-632-6732

For more information
visit Means Web Site
at www.rsmeans.com

New Titles

Residential & Light Commercial Construction Standards, 2nd Edition

By R.S. Means and Contributing Authors

New, updated second edition of this unique collection of industry standards that define quality construction. For contractors, subcontractors, owners, developers, architects, engineers, attorneys, and insurance personnel, this book provides authoritative requirements and recommendations compiled from the nation's leading professional associations, industry publications, and building code organizations. This one-stop reference is enhanced by helpful commentary from respected practitioners, including identification of items most frequently targeted for construction defect claims. The new second edition provides the latest building code requirements.

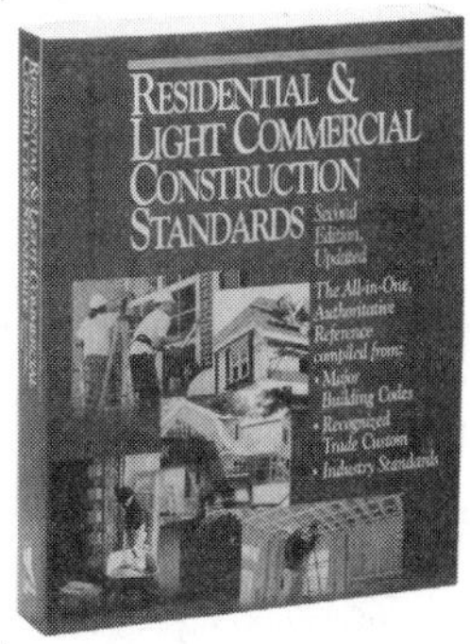

$59.95 per copy
600 pages, illustrated, softcover
Catalog No. 67322A

Green Building: Project Planning & Cost Estimating

By R.S. Means and Contributing Authors

Written by a team of leading experts in sustainable design, this new book is a complete guide to planning and estimating green building projects, a growing trend in building design and construction – commercial, industrial, institutional and residential. It explains:

- All the different criteria for "green-ness"
- What criteria your building needs to meet to get a LEED, Energy Star, or other recognized rating for green buildings
- How the project team works differently on a green versus a traditional building project
- How to select and specify green products
- How to evaluate the cost and value of green products versus conventional ones — not only for their first (installation) cost, but their cost over time (in maintenance and operation).

Features an extensive Green Building Cost Data section, which details the available products, how they are specified, and how much they cost.

$89.95 per copy
350 pages, illustrated, hardcover
Catalog No. 67338

Coming Soon!
Building Security: Strategies & Costs

By R.S. Means and David Owen

- Assessing Risks Unique to Your Facility
- Developing a Rational Response Plan
- Incorporating Protection into Design & Space Planning
- Cost Data for Security Systems & Building Materials

For commercial, industrial, and institutional buildings, this new book provides guidance to owners, facility managers, architects, and interior designers on how to identify security risks—both the likelihood of an event (including terrorist, criminal, or natural disaster), and the monetary risk (the price of a potential loss). Provides guidance on:

- How to develop a plan to minimize risk, and create a plan of action to be used in case of an event.
- How to design buildings and plan interior space in a security-conscious way to minimize risk.
- What building systems, materials, and equipment are available to promote security.

Features a cost data section for building materials and systems, as well as security devices and equipment, including line items with labor, equipment, and materials costs.

$89.95 per copy
Over 250 pages, illustrated, hardcover
Catalog No. 67339

ORDER TOLL FREE 1-800-334-3509
OR FAX 1-800-632-6732.

2003 Order Form

Qty.	Book No.	COST ESTIMATING BOOKS	Unit Price	Total
	60063	Assemblies Cost Data 2003	$173.95	
	60013	Building Construction Cost Data 2003	104.95	
	61013	Building Const. Cost Data–Looseleaf Ed. 2003	124.95	
	63013	Building Const. Cost Data–Metric Version 2003	104.95	
	60223	Building Const. Cost Data–Western Ed. 2003	104.95	
	60113	Concrete & Masonry Cost Data 2003	97.95	
	60143	Construction Cost Indexes 2003	228.95	
	60143A	Construction Cost Index–January 2003	57.25	
	60143B	Construction Cost Index–April 2003	57.25	
	60143C	Construction Cost Index–July 2003	57.25	
	60143D	Construction Cost Index–October 2003	57.25	
	60343	Contr. Pricing Guide: Resid. R & R Costs 2003	38.95	
	60333	Contr. Pricing Guide: Resid. Detailed 2003	38.95	
	60323	Contr. Pricing Guide: Resid. Sq. Ft. 2003	39.95	
	64023	ECHOS Assemblies Cost Book 2003	172.95	
	64013	ECHOS Unit Cost Book 2003	115.95	
	54003	ECHOS (Combo set of both books)	241.95	
	60233	Electrical Change Order Cost Data 2003	104.95	
	60033	Electrical Cost Data 2003	104.95	
	60203	Facilities Construction Cost Data 2003	253.95	
	60303	Facilities Maintenance & Repair Cost Data 2003	230.95	
	60163	Heavy Construction Cost Data 2003	104.95	
	63163	Heavy Const. Cost Data–Metric Version 2003	104.95	
	60093	Interior Cost Data 2003	104.95	
	60123	Labor Rates for the Const. Industry 2003	230.95	
	60183	Light Commercial Cost Data 2003	92.95	
	60023	Mechanical Cost Data 2003	104.95	
	60153	Open Shop Building Const. Cost Data 2003	104.95	
	60213	Plumbing Cost Data 2003	104.95	
	60043	Repair and Remodeling Cost Data 2003	92.95	
	60173	Residential Cost Data 2003	92.95	
	60283	Site Work & Landscape Cost Data 2003	104.95	
	60053	Square Foot Costs 2003	115.95	
		REFERENCE BOOKS		
	67147A	ADA in Practice	59.98	
	67310	ADA Pricing Guide	59.98	
	67273	Basics for Builders: How to Survive and Prosper	34.95	
	67330	Bldrs Essentials: Adv. Framing Methods	24.95	
	67329	Bldrs Essentials: Best Bus. Practices for Bldrs	29.95	
	67298A	Bldrs Essentials: Framing/Carpentry 2nd Ed.	24.95	
	67298AS	Bldrs Essentials: Framing/Carpentry Spanish	24.95	
	67307	Bldrs Essentials: Plan Reading & Takeoff	35.95	
	67261A	Bldg. Prof. Guide to Contract Documents 3rd Ed.	64.95	
	67312	Building Spec Homes Profitably	29.95	
	67146	Concrete Repair & Maintenance Illustrated	69.95	
	67278	Construction Delays	29.48	
	67314	Cost Planning & Est. for Facil. Maint.	89.95	
	67317A	Cyberplaces: The Internet Guide 2nd Ed.	59.95	
	67230A	Electrical Estimating Methods 2nd Ed.	64.95	

Qty.	Book No.	REFERENCE BOOKS (Cont.)	Unit Price	Total
	64777	Environmental Remediation Est. Methods	$ 99.95	
	67160	Estimating for Contractors	35.95	
	67276	Estimating Handbook	99.95	
	67249	Facilities Maintenance Management	86.95	
	67246	Facilities Maintenance Standards	79.95	
	67318	Facilities Operations & Engineering Reference	109.95	
	67301	Facilities Planning & Relocation	89.95	
	67231	Forms for Building Const. Professional	47.48	
	67328	From Model Codes to IBC: Transitional Guide	114.95	
	67260	Fundamentals of the Construction Process	34.98	
	67323	Historic Preservation: Proj. Planning & Est.	99.95	
	67308D	Home Improvement Costs–Int. Projects 8th Ed.	19.95	
	67309D	Home Improvement Costs–Ext. Projects 8th Ed.	19.95	
	67324A	How to Est. w/Means Data & CostWorks, 2nd Ed.	59.95	
	67304	How to Estimate with Metric Units	9.98	
	67306	HVAC: Design Criteria, Options, Select. 2nd Ed.	84.95	
	67281	HVAC Systems Evaluation	84.95	
	67282	Illustrated Construction Dictionary, Condensed	59.95	
	67292A	Illustrated Construction Dictionary, w/CD-ROM	99.95	
	67295B	Landscape Estimating 4th Ed.	62.95	
	67299	Maintenance Management Audit	32.48	
	67302	Managing Construction Purchasing	19.98	
	67294A	Mechanical Estimating 3rd Ed.	64.95	
	67245A	Planning and Managing Interior Projects 2nd Ed.	69.95	
	67283A	Plumbing Estimating Methods 2nd Ed.	59.95	
	67326	Preventive Maint. Guidelines for School Facil.	149.95	
	67337	Preventive Maint. for Higher Education Facilities	149.95	
	67236A	Productivity Standards for Constr.–3rd Ed.	69.98	
	67247A	Project Scheduling & Management for Constr.	64.95	
	67265B	Repair & Remodeling Estimating 4th Ed.	69.95	
	67322A	Resi. & Light Commercial Const. Stds.2nd Ed.	59.95	
	67254	Risk Management for Building Professionals	15.98	
	67327	Spanish/English Construction Dictionary	22.95	
	67145B	Sq. Ft. & Assem. Estimating Methods 3rd Ed.	69.95	
	67287	Successful Estimating Methods	32.48	
	67313	Successful Interior Projects	24.98	
	67233	Superintending for Contractors	35.95	
	67321	Total Productive Facilities Management	39.98	
	67284	Understanding Building Automation Systems	29.98	
	67303A	Unit Price Estimating Methods 3rd Ed.	59.95	
	67319	Value Engineering: Practical Applications	79.95	
		MA residents add 5% state sales tax		
		Shipping & Handling**		
		Total (U.S. Funds)*		

Prices are subject to change and are for U.S. delivery only. *Canadian customers may call for current prices. **Shipping & handling charges: Add 7% of total order for check and credit card payments. Add 9% of total order for invoiced orders.

Send Order To: **ADDV-1001**

Name (Please Print) ______________________________

Company ______________________________

☐ **Company**
☐ **Home** Address ______________________________

____

City/State/Zip ______________________________

Phone # ____________________ P.O. # ____________________

Mail To: **R.S. Means Company, Inc.,** P.O. Box 800, Kingston, MA 02364-0800 **(Must accompany all orders being billed)**